Primate and Human Evolution

Primate and Human Evolution provides a synthesis of the evolution and adaptive significance of human anatomical, physiological, and behavioral traits. Using paleontology and modern human variation and biology, it compares hominid traits to those of other catarrhine primates both living and extinct, presenting a new hominization model that does not depend solely on global climate change, but on predictable trends observed in catarrhines. Dealing with the origins of hominid tool use and tool manufacture, it compares tool behavior in other animals and incorporates information from the earliest archeological record. Examining the use of non-human primates and other mammals in modeling the origins of early human social behavior, Susan Cachel argues that human intelligence does not arise from complex social interactions, but from attentiveness to the natural world. This book will be a rich source of inspiration for all those interested in the evolution of all primates, including ourselves.

SUSAN CACHEL is Associate Professor of Physical Anthropology at Rutgers University. She is a member of the Rutgers Center for Human Evolutionary Studies, and is an instructor and researcher at the Koobi Fora Field School in Kenya.

Cambridge Studies in Biological and Evolutionary Anthropology

Series editors

HUMAN ECOLOGY
C. G. Nicholas Mascie-Taylor, University of Cambridge
Michael A. Little, State University of New York, Binghamton
GENETICS
Kenneth M. Weiss, Pennsylvania State University
HUMAN EVOLUTION
Robert A. Foley, University of Cambridge
Nina G. Jablonski, California Academy of Science
PRIMATOLOGY
Karen B. Strier, University of Wisconsin, Madison

Primate and Human Evolution

Susan Cachel
Department of Anthropology
Rutgers University

CAMBRIDGE
UNIVERSITY PRESS

CAMBRIDGE UNIVERSITY PRESS
Cambridge, New York, Melbourne, Madrid, Cape Town, Singapore, São Paulo

Cambridge University Press
The Edinburgh Building, Cambridge CB2 2RU, UK

Published in the United States of America by Cambridge University Press,
New York

www.cambridge.org
Information on this title: www.cambridge.org/9780521829427

First published 2006

Printed in the United Kingdom at the University Press, Cambridge

A catalog record for this publication is available from the British Library

ISBN-13 978-0-521-82942-7 hardback
ISBN-10 0-521-82942-9 hardback

I dedicate this book to my parents, Henry Cachel and
Leokadia Piotrowska Cachel.

Contents

Preface

This book is not intended to be an introductory textbook in physical anthropology, although it addresses most of the topics found in such texts. Many of the ideas developed here were originally presented to Rutgers University students in advanced undergraduate or graduate courses or in colloquia in the Rutgers Department of Anthropology or in the Rutgers graduate interdepartmental Quaternary Studies Seminar.

The focus of this book is the fundamental relationship between humans and other Old World higher primates. Many books have been written about primate behavioral ecology, and a mountain of books have been written about human evolution. However, fewer volumes deal with both human and non-human primates, and those that do so tend to emphasize the behavioral continuity between human and non-human. I will take a different approach here, because I will emphasize profound discontinuity between human and non-human primate cognition and sociality. I will also introduce evidence from Plio-Pleistocene archeology. Archeology is the description and interpretation of human behavior gleaned from the material residues of that behavior, and the spatial and temporal context of these residues. Thus, archeology contributes a line of evidence about the behavioral component of the human phenotype that is independent from inferences of behavior based on human paleontology and functional anatomy.

The strong evolutionary relationship that unites all Old World higher primates is reflected in the existence of the taxonomic category Catarrhini, which includes humans, Old World monkeys, lesser apes, and great apes. In this book I emphasize that an understanding of the strong evolutionary coherence of catarrhine primates can illuminate a number of problems in human evolutionary history, such as the advent of bipedalism, factors affected by body size or sexual dimorphism, speciation, species richness, and extinction. However, while emphasizing the anatomical and physiological coherence of catarrhine primates, I also emphasize the behavioral distinctiveness of living and fossil humans. In particular, I will argue that the behavioral ecology of living non-human primates yields no special insight into the origins of human intelligence, tool behavior, or sociality. In this sense, I am an apostate from primatology.

Yet, how can one study the origins of intelligence, tool behavior, or sociality without invoking the evidence of the behavior and ecology of living nonhuman primates? The earliest archeological record reveals important clues about human attentiveness to the natural world and human ability to manipulate the natural world. I introduce a new model of hominization, with a distinctive type of attentiveness to the natural world being a major trigger for hominization. Climatic change is usually invoked as an important or crucial factor in human evolution, but here I downplay environmental change as a major factor in hominization. Attentiveness to the natural world influences higher cognitive functions. Rudiments of this change in cognition already appear at the beginning of the hominization process, rather than being a late arrival that culminates with the appearance of modern humans. The origins of human sociality can be inferred from a broad comparative base of mammalian social organization, creating a "composite mammal" model, rather than one relying solely on the behavioral ecology of the living chimpanzee species. Studying the forces of natural selection that mold differences in sociality among mammals allows one to speculate about selection pressures that molded early hominid sociality.

Acknowledgments

Several colleagues read drafts of this book, and commented on portions of it. These are Drs. John W. K. Harris (Rutgers University), Ryne Palombit (Rutgers University), Carmel Schrire (Rutgers University), and Matt Sponheimer (University of Colorado at Boulder). Any errors that remain are my own. Dr. Robert J. Blumenschine, Director of the Center for Human Evolutionary Studies in the Department of Anthropology, Rutgers University, provided funds for manuscript preparation and wrangled up new computer hardware when technical difficulties arose. Dr. Emma Mbua, Head of the Division of Paleontology, National Museums of Kenya, Nairobi, granted me access to fossil human and non-human primate material. Drs. Phillip V. Tobias and Ronald J. Clarke, University of the Witwatersrand Medical School, Johannesburg, and Dr. Francis Thackeray, Transvaal Museum, Pretoria, granted me access to fossil human material, and Ron Clarke and Dr. Katherine Kuman invited me to explore several of the major South African sites. Ms. Purity Kiura (Rutgers University) provided me with photos taken during her thesis research on living humans in northern Kenya. During the course of our routine work together teaching in the Koobi Fora Field School, John (Jack) Harris also took me to all of the major and many of the minor paleoanthropological sites in the Koobi Fora region, east of Lake Turkana in northern Kenya. Because Jack was involved with many of the original excavations, and because his students continue to locate and excavate sites in this area, he is a fount of information about the discovery, analysis, and interpretation of Plio-Pleistocene paleoanthropological material in the Turkana Basin.

1 Introduction

The primate order

Primates, members of the Order Primates, are one of many living orders of mammals. A perusal of any major reference book on living mammals (e.g., Nowak 1999) easily demonstrates that the number of primate species is not impressive, when compared to the whole array of living mammals. This is instructive, because, as primates ourselves, we have a natural tendency to over-estimate the place of primates in the general scheme of animal life. Humility is sometimes edifying. I will use it to press the case throughout this book that the study of other mammals can illuminate major questions in primate and human evolution.

A simplified classification of living primates is given in Table 1.1. Many alternative classifications exist. Conroy alone (1990: Appendix D) lists four versions. One feature of Table 1.1 is the separation of living primates into the suborders Prosimii (prosimians) and Anthropoidea (anthropoids). The word "anthropoid" was also freely used until the middle of the twentieth century as a synonym for "ape." Here the term "anthropoid" is used to refer to all higher primates. Some other classifications place the tarsiers with higher primates in a suborder Haplorhini. Because I believe that tarsiers have had a separate evo-lutionary history for a long time, and have no special connection with higher primates (Chapter 3), I sort the tarsiers with other prosimians. Note that the Family Hominidae in Table 1.1 contains only humans and their fossil ances-tors (hominids). Some researchers now use the term "Hominini" (hominins) for this group, because they place both humans and the great apes (pongids) in the Family Hominidae (Chapter 3). I will not use this taxonomy, because I consider that the adaptations of humans and their fossil ancestors are significant enough to warrant a family-level distinction. The African great apes and the orangutan have had separate evolutionary histories, and the orangutan should perhaps be classified in a separate family. However, I retain the traditional tax-onomy, because altering it creates a barrier to accessing and understanding over 100 years of technical literature.

At one time, animals now universally placed in other orders, such as bats (Order Chiroptera), colugos (Order Dermoptera), tree shrews (Order

1

Table 1.1. *A simplified classification of living primates*

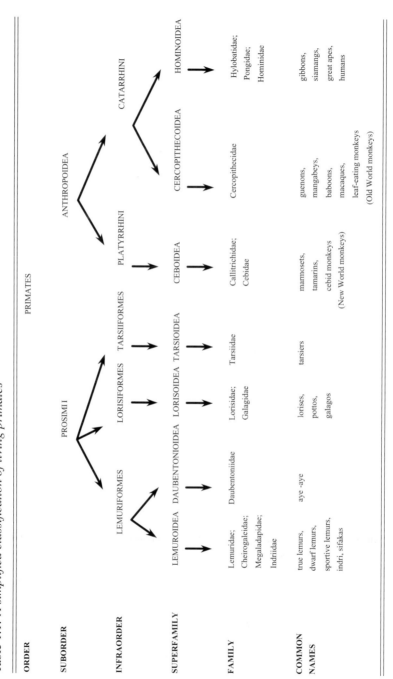

ORDER	PRIMATES						
SUBORDER	PROSIMII				ANTHROPOIDEA		
INFRAORDER	LEMURIFORMES		LORISIFORMES	TARSIIFORMES	PLATYRRHINI	CATARRHINI	
SUPERFAMILY	LEMUROIDEA	DAUBENTONIOIDEA	LORISOIDEA	TARSIOIDEA	CEBOIDEA	CERCOPITHECOIDEA	HOMINOIDEA
FAMILY	Lemuridae; Cheirogaleidae; Megaladapidae; Indriidae	Daubentoniidae	Lorisidae; Galagidae	Tarsiidae	Callitrichidae; Cebidae	Cercopithecidae	Hylobatidae; Pongidae; Hominidae
COMMON NAMES	true lemurs, dwarf lemurs, sportive lemurs, indri, sifakas	aye-aye	lorises, pottos, galagos	tarsiers	marmosets, tamarins, cebid monkeys (New World monkeys)	guenons, mangabeys, baboons, macaques, leaf-eating monkeys (Old World monkeys)	gibbons, siamangs, great apes, humans

Primates and Related Animals

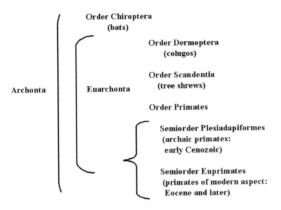

Figure 1.1. Primates are one of several closely related orders of placental mammals placed in the Grandorder Archonta. The order is composed of two subgroups: the extinct plesiadapiforms and the euprimates, primates of anatomically modern aspect that first appear in the Eocene.

Scandentia), and elephant shrews (Order Macroscelidea), were often placed in the Order Primates. Supra-ordinal designations such as Archonta and Primatomorpha are still used to link bats, colugos, tree shrews, and primates (Figure 1.1). One factor contributing to the controversy is the taxonomic position of archaic primates: the plesiadapiform primates, extant at the beginning of the Age of Mammals, the Cenozoic. In a number of features, these extinct animals do not resemble modern primates – they have long muzzles, no convergent orbits, no post-orbital bars, small brains, and many species have greatly enlarged incisor teeth, as well as other specializations of the anterior dentition. Cartmill (1972, 1974) first suggested that the plesiadapiforms should be removed from the Order Primates, restricting the order to animals that resemble living primates. Furthermore, when postcranial bones of paromomyid plesiadapiforms appeared to show the existence of a gliding membrane in the hand (Beard 1990), a link seemed to be established between plesiadapiforms and colugos (Order Dermoptera). Colugos are arboreal animals with gliding membranes connecting forelimbs and hindlimbs. Gliding membranes, however, evolve independently in many arboreal mammal groups, and their occurrence in some plesiadapiforms would not be enough to exclude them from the primate order. A recent discovery of a remarkably complete skeleton of *Carpolestes simpsoni*, a 56 my old plesiadapiform primate from Wyoming unmistakably demonstrates that this species had a grasping foot with an opposable, nailed hallux (Bloch & Boyer 2002). The species shares derived anatomical features of the foot with modern primates.[1]

However aberrant the plesiadapiforms were from modern primates, it seems clear that they should be included within the order. Being a primate at the dawn of the Cenozoic was different from being a primate today. The primate adaptive zone was different. The earliest primates were small, arboreal herbivores, and occupied niches later filled more successfully by the rodents.

Another major factor contributing to taxonomic confusion is that, unlike some other mammalian orders, primates possess no keystone feature that unequivocally identifies members of the order. Living species can experience this taxonomic confusion; fossil species, which lack soft tissue traits, and are usually represented only by incomplete dental or skeletal remains, may remain forever in taxonomic limbo, unplaceable even to order. Primate taxonomists experienced two brief spasms of hope when Meissner's corpuscles and the retino-tectal system were identified as keystone traits to identify primates. Meissner's corpuscles are specialized nerve endings embedded deep within the dermis layer of the skin that allow for delicate tactile sensitivity. However, the Virginia opossum, a marsupial mammal, also possesses these corpuscles. They are not therefore unique to primates, and can develop independently through convergent evolution in both marsupial and placental mammals. Similarly, the retino-tectal system of primates, with retinal rearrangement and neural projections that contribute to stereoscopic vision, was once thought to be unique. When fruit bats were discovered to possess the same neural anatomy, at least two major researchers argued that fruit bats were primates. They concomitantly argued that strong, powered flight evolved in mammals during two completely independent events – once among fruit bats (now allied with primates), and once among normal bats (Martin 1986a, Pettigrew 1986). Yet, a flight membrane supported by four fingers did not evolve separately in these two groups. Fruit bats are not primates.

In 1758, Carolus Linnaeus published the tenth edition of *Systema Naturae*, thus initiating binomial nomenclature and a universally recognized taxonomic system for living organisms. This volume becomes the first reference point for formal zoological nomenclature. Since that time, taxonomists have striven to identify traits that differentiate primates from other mammals (Gregory 1910, Simpson 1945, Napier & Napier 1967, 1985, Le Gros Clark 1971, Schwartz *et al.* 1978, Martin 1986b, 1990, MacPhee 1993, Szalay *et al.* 1993, McKenna & Bell 1997).

Accepting that no unique keystone feature exists, how does one diagnose primates? Linnaeus used four traits to define them: "palms which are hands," four upper incisors, two mammary glands on the chest, and clavicles. Yet, he also included bats in the order. Modern researchers (Napier & Napier 1967, 1985, Le Gros Clark 1971, Schwartz *et al.* 1978, Martin 1986b, 1990), list a dozen features that define living primates, although there are exceptions to many

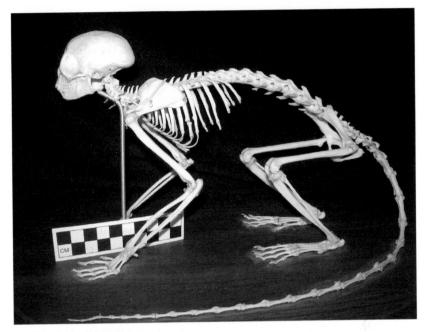

Figure 1.2. The skeleton of a New World squirrel monkey (*Saimiri oerstedii*), illustrating the generalized nature of the primate postcranium. The scale is in centimeters.

of these traits among living forms. Fossils are often only problematically identified as primates, given the fragmentary nature of fossils, the overwhelming preponderance of teeth, and the loss of most soft tissue information. Recognizing these cautions, the list of defining primate traits follows. These traits exist in euprimates, primates of anatomically modern aspect, that first appear during the Eocene.

(1) Primates possess a generalized limb structure, five digits on the extremities, and clavicles (Figure 1.2). These features are lost, reduced, or modified in many other mammal orders. For example, many mammals lose their clavicles, reduce the number of digits, or fuse the lower arm and leg bones together. (2) Primate digits are freely mobile, especially the first digit of the hand (pollex) and foot (hallux). Their extremities are thus capable of grasping. In comparison to other mammals, primate hands, even those of the plesiadapoids, have longer proximal phalanges relative to metacarpal length (Bloch & Boyer 2002: Fig. 5). (3) Flattened nails exist, rather than sharp, compressed claws. All living primates possess at least a nailed hallux. The gripping (volar) surface of each digit ends in a fleshy pad capable of fine tactile sensitivity. Primate hands and feet (and prehensile tail, if it exists) have complex, sculptured patterns on the

naked skin of the volar surface. These fingerprints (dermatoglyphics) provide friction when objects are grasped, and thus reduce the possibility of slipping. Associated with the dermatoglyphics are Meissner's corpuscles. These are organized nerve endings implanted deep within the naked skin under the dermatoglyphic ridges. Unlike nerve endings in most other mammals, Meissner's corpuscles are not protected by a surrounding membrane, and are therefore more sensitive to pressure or touch.

(4) The primate face is short, because the muzzle is reduced. Some species (e.g., the baboons, drill, and mandrill) secondarily elongate the snout. This secondary elongation is associated with lengthening of the tooth rows. (5) Unlike most mammals, primates rely less on the sense of smell. Snout reduction does not show this, because smell is affected by the size and complexity of the scrolled, paper-thin turbinal bones within the snout. Epithelial tissue that lines the turbinals is involved with both respiration and smell. A reduced emphasis on smell is apparent from the relatively small size of the olfactory bulbs in the primate brain. Furthermore, moist, hairless, glandular tissue at the end of the snout (the rhinarium) is absent in higher primates and tarsiers. Because the rhinarium is important in olfactory sensitivity, its loss highlights primate denigration of the sense of smell. When a rhinarium is present, the upper lip is immobile. A central furrow appears in the skin of the upper lip, indicating the point where the lip is anchored to underlying gum tissue. This furrow is just anterior to a foramen in the bony palate lying between the two upper central incisors. Ducts from this foramen lead back to the vomeronasal organ or Jacobson's organ. The vomeronasal organ is markedly reduced or lost in primates. Vomeronasal neurons are non-olfactory, and are linked to the accessory olfactory bulb of the brain. Vomeronasal signals circumvent higher cognitive centers, and travel to the amygdala and other neural structures that direct emotion and neuro-endocrine responses. Hence, this organ underlies a separate sensory modality that allows most mammals to sense pheromones underlying species discrimination and individual recognition, or that prime animals for appropriate sexual behavior (Leinders-Zufall *et al.* 2000, Stowers *et al.* 2002, Luo *et al.* 2003). There is no accessory olfactory bulb in the brain of catarrhine primates, and the vomeronasal organ is clearly vestigial, being represented only by a little pit in the nasal septum. Nevertheless, some researchers have argued that the vomeronasal organ is still functional in catarrhine primates, including humans. The *TRPC2* gene is essential for vomeronasal function in the mouse, and is expressed only in the vomeronasal organ. It was lost in the common ancestor of catarrhine primates at the same time that a gene duplication in the green/red opsin gene took place that allowed the development of trichromatic color vision in catarrhines (Liman & Innan 2003). The *TRPC2* gene is still experiencing selection pressure in *Lemur catta*, a prosimian, but selection pressure on this gene has

been relaxed in several species of New World monkey, including the common marmoset.

Olfactory receptor genes constitute the largest gene family in the human genome, but over 60% of these genes have been silenced, in contrast to the mouse genome, where only 20% of a similarly sized olfactory component have been functionally disrupted. When humans are compared to four other catarrhine species (common chimpanzee, gorilla, orangutan, and rhesus macaque), the human rate of inactivation of olfactory receptor genes is twice as high (Gilad *et al.* 2003). This result indicates a species-specific silencing of olfactory receptor genes in humans that may still be continuing, given the existence of many olfactory receptor genes that are polymorphic for an intact/disrupted coding region. The diminishment of olfaction, and loss or diminishment of vomeronasal signals in primates indicates their reliance on other sensory modalities.

(6) Primates emphasize vision. The eyeballs or orbits are increasingly more convergent, or rotated towards the same plane. In addition, nearly all primates exhibit frontality, because the plane of both orbits is oriented towards the front of the head, rather than facing downward or upward. Orbital convergence and frontality allow stereoscopic vision to occur. Yet, stereoscopic vision is finally achieved only when the brain integrates information from the overlapping visual fields of both eyes. Allman (1982) discovered what at first appeared to be a unique primate neural specialization: besides the systematic representation of the visual field of the opposite eye in each optic tectum, found in other mammals, one-half of each visual field also projects to the opposite optic tectum. This necessitates a re-organization of both the retina and the visual system projecting from the retina back to the brain (the retino-tectal system). The system was later found to exist in fruit bats. This neural anatomy underlies specialization for stereoscopic vision. Areas of the brain devoted to processing of visual information are relatively enlarged in primates, in comparison to other mammals. As an additional visual specialization, diurnal primates (i.e., some diurnal prosimians and all anthropoids except the secondarily nocturnal owl monkey genus *Aotus*) develop color vision. All normal Old World higher primates have trichromatic color vision. Diurnal New World higher primates have an autosomal color gene and one X-linked color gene. Only the New World howler monkeys (genus *Alouatta*) have two X-linked color genes. However, several New World monkey genera have allele variations at the single X-linked color gene locus, which would enable some females to possess trichromatic color vision (Boissinot *et al.* 1998). The relatively small size of the cornea in higher primates also underlies greater visual acuity in these animals (Kirk 2004). (7) There is some bony separation between the primate orbit and the temporalis muscle immediately posterior to the orbit. This separation may be either a bar or a septum of bone (Cachel 1979a).

(8) The auditory region (the bulla) of the primate basicranium is derived from the petrosal bone (Van Valen 1965, MacPhee *et al.* 1983). Evidence from the developing fetus is generally needed to perceive details of bulla formation. Plesiadapiform evidence can be equivocal. CT scans of the paromomyid plesiadapiform primate *Ignacius graybullianus* shows that the bulla is formed from the entotympanic bone, yet it also has a bony tube for branches of the internal carotid artery and nerves, a trait found in tree-shrews and euprimates. The internal carotid nerves in *Ignacius* are laterally positioned, which is a euprimate trait (Silcox 2003). (9) Unlike most mammals, the molar teeth of primates have a relatively simple occlusal surface pattern. There are no complicated crests or enamel folds. However, even the first primate genus, *Purgatorius*, from the early Paleocene (65.5–61.7 mya), has quadrangular molars with low rounded cusps that resemble the teeth of condylarths, very generalized contemporary herbivores. Some researchers therefore suggest that the first primates were herbivores/frugivores (Van Valen & Sloan 1965, Szalay 1968). This would signify a basic shift away from the insectivory found in ancestral mammals.

(10) Placental tissues are increasingly elaborated in primates, developing a more intimate contact between maternal and fetal blood supply. (11) When primates are compared to other mammals of the same body size, fetal and postnatal life periods are relatively extended. (12) The primate brain, especially the neocortex, is large relative to body size. Note, however, the contrary evidence discussed in Chapter 8.

The most essential primate criteria appear to involve locomotion. This has resulted in a long series of publications about the fundamental nature of primate locomotion. Gregory (1920) argued that a deep-seated functional dichotomy exists between the primate forelimb and hindlimb – the hands explore and manipulate, while the feet grasp arboreal supports, powerfully anchor animals in an arboreal setting, and drive animals forward (Figure 1.3). Napier and Walker (1967) argued that the first primates engaged in vertical clinging and leaping, a form of arboreal locomotion in which the legs thrust animals forward in great leaps from one vertical support to another, while the trunk is orthograde (held upright). During rest, the trunk is orthograde, and the abdomen is pressed against a vertical arboreal support. Cartmill (1972, 1974) argued that the first primates were arboreal animals specialized for slow, cautious movement along small-diameter supports in the terminal branches of trees or in dense brush or secondary vegetation. Both hands and feet exerted a powerful grasp, but the hands could also thrust out quickly to capture agile insect prey.

A major functional distinction between hindlimbs and forelimbs was established at the beginnings of the primate order (Gregory 1920). The hindlimbs are used principally for weight support and propulsion, and the forelimbs are used principally for reaching and grasping. It seems clear that even the most ancient

Figure 1.3. A squirrel monkey (*Saimiri oerstedii*) actively searching for insect prey in Corcovado National Park, Costa Rica. The hands are used to unroll furled leaves where insects are hiding. The feet strongly grasp arboreal supports.

members of the primate order have specializations for arboreal climbing, especially for powerful grasping with the foot (Gregory 1920, Szalay & Dagosto 1988, Bloch & Boyer 2002).[2] The early appearance of a nailed hallux signals climbing in small branches. The origins of primate locomotor traits for specialized movement in small branches is confirmed by the independent, convergent evolution of some of these specializations in a living New World arboreal marsupial, the woolly opossum (*Caluromys philander*) of South America. This species moves agilely through fine terminal branches using quadrupedal grasping, short leaps, bridging, and suspensory behaviors. And, under experimental laboratory conditions, the woolly opossum demonstrates three primate-like locomotor features. It uses the diagonal couplets gait; its arm position at hand touchdown is greater than 90 degrees relative to the horizontal axis of the body; and its forelimbs receive less peak substrate reaction forces than its hindlimbs (Schmitt & Lemelin 2002).

Allometry needs to be taken into account before assessing function (Alexander 1985). Allometric analysis of mammals across a size range from shrews to elephants reveals that primates have relatively long limb bones, and an especially long femur (Alexander *et al.* 1979). The muscles of the thigh and foot are also relatively large in primates (Alexander *et al.* 1981). The relatively large size of the hindlimb prefigures its dominance in primate locomotion. In addition, the primate center of gravity is more caudal than in other mammals; the hindlimbs therefore carry more body weight (Reynolds 1981, 1985a, 1985b). In addition, primates have a much longer stride length than other mammalian quadrupeds. This is principally caused by large angular excursions (Reynolds 1983).[3] Unlike the vast majority of mammals, primates utilize a sequence of diagonal couplets during quadrupedal locomotion – the foot contacts the substrate before the hand on the opposite side of the body (Hildebrand 1985). This strange primate gait is caused by two factors: the more caudal position of the center of gravity, and the use of the hindlimb to supply not only the principal accelerating force during each stride, but also the principal braking force (Hildebrand 1985). In other mammals, the center of gravity is cranially located, and the forelimb provides the major braking force. Thus, a suite of features characterizes primate locomotion: grasping foot, early appearance of a nailed hallux, caudal center of gravity, long femur, large hindlimb muscles, hindlimb dominance in carrying body weight, diagonal couplets gait, and hindlimb dominance during both maximal braking and acceleration. Taken all together, these features imply a fundamental primate adaptation to arboreal travel in areas where supports are small, discontinuous, flexible, and mobile. The feet grasp and hold strongly, while the hands reach out for different supports, and the center of gravity of the moving body is transferred to the opposite side.

What else can be inferred about the fundamental primate adaptive zone? The occlusal surface of the molars implies a dietary shift toward herbivory/frugivory in even the earliest primates (Van Valen & Sloan 1965, Szalay 1968). Neuro-anatomical specializations for vision and orbital convergence and frontality signal an early reliance on vision. At the same time, olfaction is downplayed, as confirmed by the reduction of the olfactory bulbs, eventual loss of the rhinarium, and reduction or loss of the vomeronasal organ. Cartmill (1972, 1974) attributed visual specializations to the fact that the first primates were visually oriented predators focusing upon solitary, agile insects. He removed the plesiadapiform primates from the order when making this argument for origins, because they lack cranial specializations for visual predation. Including plesiadapiform primates in the order signifies a dietary shift away from insectivory towards herbivory/frugivory in the earliest primates. And, of course, the earliest primates were arboreal animals adapted for moving among small branches, which is associated with a grasping foot and the early appearance of a nailed hallux. Study of locomotion in living primates documents a more caudal center of gravity, hindlimb dominance in carrying body weight, hindlimb ascendancy in braking as well as accelerating movements, and the existence of the rare diagonal couplets quadrupedal gait. Functional studies of living animals substantiates climbing in an arboreal small-branch setting as a hallmark of the primate order.

Ape and monkey bias

General awareness about the primate order is often based on only a handful of living species. The African apes and a scatter of monkey species overwhelmingly color human perceptions of the primates. If people today know nothing else about primates, they know that monkeys and apes belong with humans in a group of genetically related animals. Everyone appears to know an exact figure of genetic relatedness – over 98% shared DNA between common chimpanzees and humans – even though this is probably incorrect. In fact, the figure is more likely to be 95% (Britten 2002). The degree of genetic similarity between humans and other mammals is also remarkable. Humans and laboratory mice are approximately 90% genetically similar, and there are about 30 000 genes in each species (Chinwalla *et al.* 2002). Furthermore, studies of evolutionary development in humans and other vertebrates demonstrate the existence of highly conservative *Hox* genes that are responsible for laying-out the embryonic blueprint.

Unlike other mammals, differences between primate individuals may sometimes be granted a subspecies or species level distinction. In the middle of the twentieth century, Simpson (1945: 181) waspishly commented on the tendency of primatologists to multiply the number of primate taxa: "The importance of distinctions within the group has also been so exaggerated that almost every color-phase, aberrant individual, or scrap of fossil bone or tooth has been given a separate name, almost every really distinct species has been called a genus, and a large proportion of the genera have been called families . . . Moreover, even mammalogists who might be entirely conservative in dealing, say, with rats are likely to lose a sense of perspective when they come to the primates, and many studies of this order are covertly or overtly emotional." This remark is not overstated, because Simpson (1945: 186) later documents the occurrence of 20 genus names (26 if all variant spellings are included) for macaques alone – all now collapsed into genus *Macaca*. I have personally observed one major primate taxonomist, the late Dr. Philip Hershkovitz, arguing how minor coat color variations represented legitimate primate subspecies that I was dubious about, because they seemed to be individual variations. Chromosomal distinctions between primates are different – they signal that speciation has begun or is complete, in spite of morphological similarity. Chromosomal evidence can also be used to establish hybridization between primate species (Pieczarka *et al.* 1993).

As is the case for other mammals, new species of living primate continue to be discovered both in the field and in the lab, when karyotype differences are detected between animals that are morphologically indistinguishable – sibling or cryptic species. The New World monkey genus *Aotus* is one of the best primate examples of species multiplication occurring through karyotype differences. Once thought to be a monospecific genus (*Aotus trivirgatus*), *Aotus* may now contain as many as ten species. Relatively small Malagasy prosimians and New World monkeys represent the majority of the newly recognized primate species. However, even large primate species can suddenly emerge in countries whose natural history was thought to be well known. For example, a large-bodied macaque species (*Macaca munzala*) was discovered in 2003 in the mountains of northeastern India. Because of the vagaries of field collection, species that were long considered extinct can be rediscovered. This was the case for the hairy-eared dwarf lemur (*Allocebus trichotis*). It was first described in 1875 from a single pelt at the British Museum (Natural History); two other pelts surfaced in Paris several years later. The first living specimen was collected in 1966. In 1989, members of this species were discovered alive and thriving in lowland Madagascar rainforest. At least one major researcher (Groves 2001b) argues that many species of primate should theoretically exist, on the grounds of their ecological similarity to squirrels and fruit bats, which are speciose groups. Using the criterion of diagnosability, which tends to inflate species numbers, this researcher recognizes 356 living primate species (Groves 2001a). The number falls to 263, using a more conservative taxonomy (Nowak 1999).

If one uses the Nowak (1999) taxonomy and omits the extinct subfossil forms, the 55 living prosimian species (including the tarsiers) comprise 21% of the order. This alone should make one wary of any ape/monkey bias. Furthermore, the diversity of morphology is such that eight families are represented among these 55 species (Table 1.1). New World monkeys (platyrrhines), which remain largely unknown to the public, total 96 species, and comprise 36.5% of the order. The 112 species of Old World monkeys, apes, and humans (catarrhines) complete the analysis with 42.5%. The four African ape and orangutan species (pongids), represent 2% of the order. Thus, any ape/monkey bias is not justified by numerical prevalence.

It is also instructive to see how higher primate (anthropoid) numbers are skewed by certain genera. The living anthropoids total 208 species. Among the platyrrhines, the genera *Callicebus*, *Saguinus*, and *Callithrix* account for 6, 6, and 8% of the anthropoids, respectively. Among the catarrhines, the genera *Cercopithecus* and *Macaca* each account for 10% of the anthropoids. These five genera contain 82 species, and thus account for nearly 40% of the living anthropoids. Genus-level distinctions tend to be less controversial than species-level ones. Hence, despite taxonomic revision, there is consensus that these genera are species-rich. They thus represent evolutionary success and adaptive radiation within the New and Old World higher primates. The great apes or pongids, on the other hand, have four species teetering on the edge of extinction in the wild. Between 1983 and 2000, gorilla and common chimpanzee populations declined by more than 50%, even within their last bastion, the forests of western equatorial Africa (Walsh *et al.* 2003). Both species may be moved from endangered to critically endangered status, given that their numbers can decline by 80% in the next 10 years. Mechanized logging, bushmeat hunting, and the spread of the Ebola virus caused the precipitous drop in numbers. The lesser apes or hylobatids – largely unknown to the public and often identified as monkeys by casual zoo visitors – are much more numerous in the wild. However, the 11 hylobatid species are also listed as endangered in appendix 1 of the CITES[4] conservation document (Nowak 1999).

Evolution before natural selection

The Linnaean hierarchy was originally conceived of as a static and unchanging arrangement. It epitomizes the concept of a Great Chain of Being: primates are literally the First Ones, because they are members of the First Order (in Latin, the Order Primates). They are located at the top of the Great Chain of Being, and humans crown the primate order. There are deep cultural and theological roots to this conception, because here humans are "only a little lower than the angels," they have dominion over the entire earth, and all things are truly beneath their feet (Psalm 8: 6–9). Humans are thus the end-point of Nature, and, in fact, the very raison d'être for Nature's existence. For devout Christians of the eighteenth and early nineteenth centuries, the incarnation of God as Jesus Christ further underscored the separate and unique character of humanity.

The late eighteenth and early nineteenth centuries abounded in evolutionary speculation. Hence, the independent discovery of natural selection is not so surprising. However, Darwin, at least, was always reluctant to admit how popular and prevalent evolutionary theorizing was, since it seemed to detract from his own intellectual innovation (Browne 2002). Erasmus Darwin, Charles Darwin's own grandfather, even contributed to the general evolutionary milieu by publishing *Zoonomia*, a poetical work about the transformation of species. Furthermore, Robert Chambers' *Vestiges of the Natural History of Creation* was enormously popular in the early nineteenth century, and went through many editions. It discussed the origin of life from inorganic material, and did not

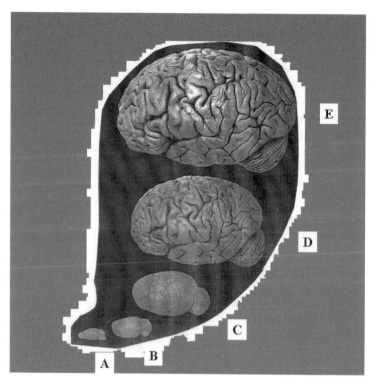

Figure 1.4. A modern Scala Naturae. The brains of some living mammals are arranged in an ascending ladder-like array. A. rat; B. cat; C. macaque monkey; D. chimpanzee; E. human. Modified from the 16 June 1995 cover of *Science*.

evade the issue of human origins. The tenth edition, published in 1858, linked humans with apes in a developmental sequence. The concept of apes, especially orangutans, as human ancestors had actually been covertly circulating in learned circles since the eighteenth century (Stoczkowski 1995).

Yet, most evolutionary speculation of the late eighteenth and early nineteenth centuries generally lacked any explanation for evolutionary change. The concept of a static Great Chain of Being or Scale of Nature (*Scala Naturae*) had been present in Western thought since Greek antiquity and the work of Plato (Figure 1.4). The Great Chain of Being became prominent during Renaissance times, when it influenced science, philosophy, politics, and religion (Lovejoy 1936). By the eighteenth century, it was deeply embedded in biology. This conception places all living things in an ascending scale with fine gradations. Organisms become increasingly more complicated along the ladder-like array of forms. This immutable array was established by God at the moment of creation, and proves the existence of the Creator. Anatomy and behavior are intricate and complicated – this demonstrates artifice. In addition, the perfect fit of anatomy and behavior to lifestyle further confirms intelligent design. Nature mirrors the mind of God. For these reasons, by the early nineteenth century, the study of natural history was often considered to complement theology. The idea of "natural theology" began in this fashion. Natural theology eased the transition to understanding how anatomical or behavioral traits could be fit or adapted to a given environment. Yet, natural theology ultimately hindered the acceptance of natural selection because it created the expectation that evolution should have both a grand design and a final promise of progress.

Most evolutionary speculation of the late eighteenth and early nineteenth centuries simply assumed without question the existence of a great Chain of Being and fastened on to it a romantic vitalism that perceived life as having an inherent tendency to perfect itself (Bowler 1986). The struggle for organic perfection and progress seemed embedded within protoplasm itself – it was a fundamental property of life – and so no additional explanation for change appeared necessary. Jean-Baptiste de Lamarck was an exception among these evolutionists. In 1809, he suggested a mechanism for evolutionary change: the inheritance of acquired characteristics. Organisms possess an innate ability to strive for perfection. Animals adjust their behaviors to particular environmental circumstances, and either use or neglect characters. Disuse eliminates characters. Characters acquired through use are passed on to the next generation. Difficulties with this mechanism became apparent almost immediately; Lamarck's idea was criticized by the vertebrate paleontologist Georges Cuvier and other contemporaries. A series of experiments by August Weismann at the end of the nineteenth century finally removed the inheritance of acquired characteristics from consideration as a viable evolutionary process. However, it re-emerged for political reasons in the Soviet Union as official state policy during the middle of the twentieth century, when it was known as Lysenkoism.

1858–1859: The advent of natural selection theory

In 1858, Charles Darwin and Alfred Russel Wallace jointly announced their independent discovery of natural selection, a workable mechanism for evolutionary change. In June of 1858, Charles Darwin received a letter from Alfred Russel Wallace, an impoverished collector of exotic plant and animal specimens, who sold his material to museums or wealthy collectors. Writing from the Moluccas, Wallace announced his discovery of natural selection. This appalled Darwin, who had delayed publication on the idea for years, amassing a mountain of data, notes, and drafts while fearing to submit a manuscript. Wallace had already published two important papers on speciation, although Darwin considered him merely a lower class technician (Browne 2002). Using two of Darwin's major sources (Charles Lyell's *Principles of Geology* and Thomas Malthus' *Essay on the Principle of Population*) and extensive field observations in the Malaysian tropics, Wallace had quickly arrived at an understanding of natural selection over a two-hour period while he was laid-up combating a fever. He sketched a draft of the idea that same evening, wrote it out fully over the next two evenings, and then sent the paper out to Darwin. To prevent Darwin from being forestalled, his friends arranged for both authors to present and publish short papers on natural selection in 1858.

The concept of natural selection is simple. No specialized knowledge is required to understand it. For this reason, discussions about natural selection and its implications swept quickly through the English-speaking world when Darwin (1859) published *On the Origin of Species.*

Four components underlie natural selection. First, organisms exist in groups of interbreeding individuals of the same species. These groups are termed populations. Individuals die, but populations persist through time and space. Organisms in natural populations are variable. Variation is the natural condition of living things. Variation is the only reality. Second, competition for resources occurs because populations increase faster than the resources needed to support them do. Organisms multiply, but their numbers are checked by famine, predation, and disease. Third, differential mortality and reproduction take place, because organisms are not equal in the competitive arena. At any point in time, certain traits of anatomy, physiology, or behavior allow some individuals to survive and reproduce better than their contemporaries. Hence, mortality and reproduction are not random. Traits that confer enhanced survivorship and reproduction are called adaptations. All fitness or adaptation is relative, because it hinges completely upon the local environment at any instant in time, and the environment is ever changing. Fourth, the fit or adapted traits are inherited. The offspring of adapted individuals resemble their parents more than they do unrelated individuals. The outcome of these processes is that traits in populations shift through time or evolve. Natural selection causes adaptation through evolutionary time, and new species arise when a successor population differs enough from the original founding population. Besides changing ecological circumstances, breeding isolation can hasten the advent of new species.

Besides natural selection, several other processes are vital for the understanding of evolutionary change. One of these is sexual selection. Darwin (1871) highlighted a special aspect of competition: competition between members of the same sex for access to mates, and choice exercised by members of the opposite sex between the competitors. This is sexual selection. Since Darwin's time, sexual selection has been investigated at both increasingly finer and increasingly grosser levels. For example, it is now being studied in terms of physical and chemical interactions occurring within the female reproductive tract. At the same time, the reproductive strategies of males and females are now revealed as often being so conflicting as to trigger the development of counterstrategies in a true arms race.

Heredity was a black box to the Victorians, although the practical experience of farmers and pastoralists through the centuries demonstrated the importance of carefully choosing parents or parent stocks in animal and plant domestication. Darwin relied heavily on the evidence of domesticated animals and plants for two reasons: domesticates epitomized the seemingly endless variability within species, and they also illustrated the heritability of traits from one generation to the next. Although the mechanism of heredity was a mystery, its existence was inarguable. Furthermore, traits were deliberately molded through human intervention. The manipulation of plant and animal reproduction by humans was artificial selection, and it mirrored the operation of natural selection. To understand artificial selection was to understand natural selection.

Because natural selection acts on normally observable traits (the phenotype), it could be discovered and fruitfully studied without anything being known about inheritance or about the genetic basis of traits (the genotype). Ironically, although the mechanism of heredity was long sought by nineteenth century workers in natural selection, the laws of heredity were actually discovered by a contemporary of Darwin and Wallace, the Czech monk Gregor Mendel. In 1866, Mendel published *Experiments in Plant Hybridization*, which expounded these laws. This publication was apparently widely disseminated – Darwin owned an uncut and unread copy (Henig 2000). Darwin himself published the results of his own hybridization experiments on snapdragons in 1868, but, like other hybridization experimenters, failed to achieve any insights into heredity. Mendel's work made no impact on nineteenth century thought, and remained unnoticed until 1900, when its import was first recognized. Failure of the Victorians to recognize the significance of Mendel's work might lie in its quantitative presentation, which was not expected by contemporary readers of natural history. Yet, even after the resurrection of Mendel's work, it ignited a firestorm of controversy. Its insistence on the particulate nature of the hereditary material (what were later termed genes with alternative forms or alleles) appeared to contradict wide-ranging biometrical work that demonstrated continuous variation in traits within populations. The ultimate acceptance of Mendel's work was delayed until 1918, when R. A. Fisher resolved the apparent contradiction with mathematics. He demonstrated that, if most phenotypic traits were caused by multiple genes with multiple alleles, these genes would have an additive effect. Traits would then show continuous variation within a population.

Additional evolutionary processes were discovered during the twentieth century. Sewall Wright (1921, 1931) examined the effect of mating systems and population size on genetic variability. He discovered that small populations are differentially affected by random changes in allele frequencies (genetic drift). Population bottlenecks also occur during colonization, and random changes in allele frequencies similarly affect colonizers (the founder's effect). Genetic drift introduces sampling error or randomness into the array of evolutionary processes.

Three major events occurred in the science of genetics during the course of the twentieth century. Early in the century, T. H. Morgan and his colleagues discovered that genetic material could be altered by mutations. In 1953, J. Watson and F. Crick elucidated the physical structure of DNA and presented a mechanism for its replication. By the end of the century, the genomes of a number of different organisms had been published. Because of the constantly repeated and now fabled genetic similarity between humans and chimpanzees, it is instructive to note the high degree of genetic similarity between humans and mice. A comparison of human and mouse genomes reveals about 30 000 genes in both organisms; the genetic similarity between the two species is about 90%, although genetic counterparts may be rearranged on different chromosomes (Venter *et al.* 2001, Chinwalla *et al.* 2002). The mouse genome has evolved 2–5 times more rapidly than the human genome, probably because of the shorter generation length of mice, and mouse genes appear more subject to physical reshuffling. Mouse genes in different locations on the same chromosome evolve at different rates. Genes supposedly code for proteins, but this definition of a gene is becoming problematical – about one-third of the shared human and mouse genes do not encode proteins. In fact, about 98% of the genome may consist of non-coding sequences. Some of these sequences may encode RNA; some may serve regulatory functions. Non-coding sequences may not only function, but they may be subject to natural selection. A recent estimate suggests that twice as many non-coding DNA sequences experience selection pressure as DNA sequences that code for proteins do (Chinwalla *et al.* 2002). Conserved, non-gene sequences in human chromosome 21 were compared across 14 mammalian species, and were found to be very highly conserved among mammals (Dermitzakis *et al.* 2003). This suggests that there exists among mammals a hitherto unsuspected class of highly conserved nucleotides outside of gene regions. They may be too conservative to function as protein binding sites. Hence, they are unlikely to regulate gene expression. The function of these sequences remains unknown. They highlight current problems in defining the role of genetic material in cell activities.

It is clear that the twenty-first century will witness another revolution in genetics. Two topics will contribute to this revolution. First, researchers will unravel the mechanisms that allow the same genetic material to produce both different products within cells and also distinctly different organisms. Second, genetics is becoming fused with embryology. Development is used to illuminate evolution. This will contribute to evolutionary theory, because it will unveil the origins of new morphology, and allow researchers to estimate the likelihood of developing new structures from old.

What of speciation – the origin of new species that entranced both Darwin and Wallace? The topic has become increasingly complex since Victorian times, and it is now intertwined with species definitions. The biological species concept promulgated by Ernst Mayr (1963) is probably now the most widely recognized definition. This defines a species as a group of actually or potentially interbreeding populations that are reproductively isolated from other such groups, and the species also possesses a distinct ecological niche. Several modes of speciation are known to exist. Traditionally, geographic barriers to gene flow were considered the most important factor in species formation, and allopatric speciation was widely recognized as perhaps the most important mode of speciation. Ecological factors operating along a spectrum of environmental differences are now receiving increasing attention. These ecological factors can operate to cause speciation among organisms even when gene flow takes place (Doebeli & Dieckmann 2003). That is, speciation does not necessarily require geographic barriers to gene flow, and can occur even in the absence of reproductive isolation. Given adaptation to highly local ecological circumstances, speciation can occur among sympatric organisms. This is sympatric speciation.

Routine karyotyping can reveal the presence of sibling or cryptic species. Alternatively, it can reveal the existence of hybrids. Modern understanding of gene flow and hybridization has led some researchers to argue about the importance of hybridization as an evolutionary process. Surveys of bird species, which are relatively well studied, reveal that hybrids regularly occur in the wild. Nine percent of all non-marine bird species hybridize – 2% hybridize regularly, and 3% hybridize occasionally (Grant & Grant 1992). Some hybridization between species may be natural. If hybrids are not adaptively inferior to either parent species, natural selection will not operate to form reproductive barriers. Hybridization may sometimes occur in primates: baboon hybridization has been recognized for a long time. Jolly (1993), however, argues that, because baboon species form fertile hybrid offspring, only one species of baboon truly exists, in spite of marked phenotypic differences. Geographic areas where gene flow is persistent indicate that parapatric speciation can occur in baboons.

The last quarter of the twentieth century saw the irrefutable documentation of natural selection under field biology conditions and in laboratory experimentation. Endler (1986) reviews studies of natural selection in the field. A major long-term study of natural selection in the lab was initiated by Richard

Lenski at the Microbial Evolution Laboratory of Michigan State University. Selection pressures are manipulated here to alter the evolution and adaptations of populations of the bacterium *Escherichia coli*. Lenski and his colleagues are also collaborating with researchers at the Digital Life Laboratory at the California Institute of Technology.[5] Populations of artificial digital organisms ("avidians") that mutate, replicate, evolve, and adapt exhibit evolutionary dynamics that strikingly resemble those of bacteria. Specialized software allows evolutionary change to be tracked in these populations of artificial organisms. Three important results of this research are that selection pressure creates quick adaptive responses, that complex functions favored by natural selection evolve from simpler functions, and that independent (convergent) evolution of complex functions can occur (Lenski *et al.* 2003).

Yet, doubts persist about the efficacy of natural selection. Alternatives to natural selection have been suggested from the beginning. Despite abundant proof of its operation both in the field and in the lab, natural selection continued to be characterized throughout the twentieth century by some researchers as weak, and incapable of effecting evolutionary change (Williams 1966, 1992). Osborn (1934) invented the concept of aristogenesis to explain the success of certain mammalian lineages, and the origins of superior adaptations within the genetic material. Species could arise independently of natural selection. Even romantic evolutionism persisted. Vitalism, for example, continued into the twentieth century with the introduction of innate forces like a vital spirit (*élan vital*) or creative intelligence to guide evolution (Bergson 1907, de Chardin 1955). Vitalism and other alternatives to natural selection were sometimes strongly invoked when the question of human evolution arose. Researchers were reluctant to face the idea that human progress is not inevitable. This is apparent in many publications in the early twentieth century (e.g., Osborn 1924, 1927). By the late twentieth century, the focus of romantic evolutionism had shifted away from the earliest hominids. Now it centered either around the origins of genus *Homo* – viewed as highly distinct from its ape-like ancestors – or around anatomically modern humans – viewed as highly distinct from the members of *Homo* that preceded modern humans, especially the aberrant Neanderthals. Unknown or unknowable mental or linguistic factors were often invoked at the point of origin for modern humans (Cachel 1997).

Essentialism versus population-thinking

Arthur Lovejoy (1936) was the first to trace two contrasting modes of thought back to Greek antiquity, and investigate their importance to biology, philosophy, and religion. Mayr (1982) fleshes out the biological impact of these intellectual differences, and portrays the history of biology as a never-ending war between those who see variation as inherent in nature (population-thinkers) and those who see variation as an inconvenience that hinders an understanding of the discontinuous and real differences between organisms (typologists or essentialists).

Natural selection could only be discovered by scholars who were population-thinkers, because they acknowledge variation within populations, which is fundamental to the process of natural selection. However, as documented by Lovejoy, treating variation as imperfection is an ancient concept that can be traced back to Plato. Certainly, variation has always been problematic for taxonomists – how different can individual specimens be without recognizing the existence of another species? Taxonomists can be baffled by animal and plant individuality. Confronting variation stored within the gigantic massed ranks of trays within museum cabinets or living outside under tropical field biology conditions can actually be a frightening experience. Everything looks different from everything else – does the variation represent individual variation, sub-specific variation, or evidence of a new species? The problem is even more confounding for fossil organisms, because the material is incomplete and subject to geological distortion, and because some parts of the phenotype, like soft tissues and behavior, are forever lost. Also, given the taphonomic winnowing that takes place during fossilization, there is no certainty that additional fossils will be found to confirm morphology or species assignment.

Before natural selection, biologists often studied anatomical differences in terms of the archetype concept. Individual specimens were collected, studied, and sorted into species. Some species resembled each other, and the resemblance was explained by their departure from an ideal type – the archetype, a concept that harkens back to Plato. Like Plato's Ideas, only archetypes were real. Specimens collected under field biology conditions were like smudged photocopies of the real archetype. The mindset of these researchers viewed life as static and unchanging. After 1859, biologists realized that the theoretical archetype could really be an ancestral species.

The contrast between essentialism and population-thinking focuses around boundaries: the boundaries between individuals, subspecies, and species in a world where variation exists. Essentialism mandates rigid boundaries. The boundaries are more real than the living organisms sorted inside them. If organisms do not sort perfectly into existing categories, new categories are created. This presages endless multiplication of categories. Population-thinkers, however, deal with variable individuals within natural populations. They do not yearn for discrete boundaries. They are content with probabilistic statements about species identification, and indeterminate species boundaries. The contrast between essentialism and population-thinking resembles the contrast between classical mechanics and quantum mechanics. Indeterminacy is a problem for both essentialism and classical mechanics.

Variation can be alarming because it is overwhelming. For some researchers, the lack of certainty about species boundaries can also be upsetting. One response has been to utilize a method of classification based on phylogenetic systematics. This taxonomic approach was invented by W. Hennig in the 1940s, and was based on certain assumptions about speciation (Cachel 1992). Hennig presented a rigid taxonomic methodology, called cladistics, which promised to generate objective phylogenies. In theory, any taxonomist examining the same material and employing the cladistic methodology should arrive at the same phylogeny. Software packages using cladistics also allow biologists to create phylogenies with relative ease. If objective phylogenies existed, they could be used to test the verity of ideas about the evolution of morphology or behavior, or test reconstructions of function and behavior in fossils (Chapter 5). In practice, however, cladistic phylogenies do not seem reproducible. Another troubling point is that many paleoanthropologists now appear content simply to create these phylogenies, and are abandoning reconstructions of fossil human lifeways (Chapter 14).

1863: Thomas Henry Huxley and the place of humans in nature

In a broad sense, it was the publication of Darwin's monograph *On the Origin of Species* in 1859 that ignited public debate and controversy about the relationship of humans to other mammals. In a narrower sense, it was the continuing altercation between Thomas Henry Huxley and Sir Richard Owen on human anatomical uniqueness that set the stage during the 1850s for the later Darwinian debates on human evolution. Owen, as anatomist and paleontologist, argued that humans were very different from other mammals. By virtue of their brain and intelligence, Owen placed humans in a separate mammalian sub-class. Huxley argued otherwise, and it was these debates, which received vast public attention, that foreshadowed furious debates about human evolution after 1859.

Addressing lay audiences, Huxley delivered many public lectures on natural selection, the kinship between humans and non-human animals, and human evolution. He then published a series of these lectures in *Evidence as to Man's Place in Nature* (1863). Huxley based the argument for primate evolution on comparative anatomy, and not fossil evidence. Although fossil primates were already known, and fossil apes had been discovered even in Europe (beginning with *Dryopithecus fontani* in 1854), there was certainly no dense sequence of fossil primate species that could persuade doubters about the reality of biological

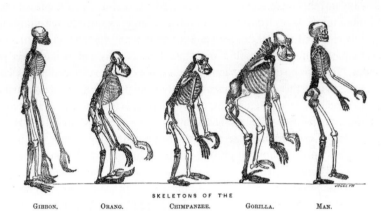

SKELETONS OF THE

GIBBON.   ORANG.   CHIMPANZEE.   GORILLA.   MAN.

Figure 1.5. The Frontispiece of T. H. Huxley's *Evidence as to Man's Place in Nature*. Skeletons of a gibbon, orangutan, chimpanzee, gorilla, and human are presented at the same scale, so that their natural body proportions are preserved. The original drawing is by Waterhouse Hawkins from material in the Museum of the Royal College of Surgeons.

evolution. This explains Huxley's reliance on comparative anatomy. With respect to anatomy, Huxley was a master of both the particular and the general, and was pedagogically gifted. In fact, he established the hands-on dissection labs from lancelet to mammal that most students use today to examine the basics of chordate biology.[6]

Rhetorical flourishes also helped to make the case for human evolution. Even the frontispiece of *Man's Place in Nature* acts subconsciously to persuade the reader (Figure 1.5). It shows a sequence of five articulated and animated primate skeletons walking across the page, in the left to right progression that a reader of English would naturally expect. The sequence begins at the left with a gibbon and ends with a human. Although only the human is an obligate biped, presentation of the other species in bipedal posture heightens their similarity to the human – a similarity that is not immediately apparent, because their proportions are very different. Huxley thus began a long tradition of paleoanthropological illustration. Gould (1989) presents a modern cross-cultural compendium of similar illustrations, as he tries to argue that a bias toward progress in human evolution taints most researchers, and thus influences their study of evolutionary processes in general. It was Huxley's goal to persuade his readers that human evolution had in fact occurred, because he lacked any fossil human evidence except for the original Neanderthal fossil finds.

In *Evidence as to Man's Place in Nature*, Huxley initiated a theoretical approach to the primate order that shapes analyses even today. He argued that possibly no other mammalian order preserved so many living examples of major morphological stages from the inception of the order through increasingly more complex forms. In fact, the order was a veritable *Scala Naturae*, a scale of nature or chain of being that in itself demonstrated the existence of evolutionary processes. "Perhaps no order of mammals presents us with so extraordinary a series of gradations as this – leading us insensibly from the crown and summit of the animal creation down to creatures, from which there is but a step, as it seems, to the lowest, smallest, and least intelligent of the placental Mammalia" (Huxley 1863: 124–125). Similar statements might be made about other mammalian orders, but Huxley chose to present the primates in this fashion because the order then becomes an exemplar of evolution. To study the living primates was to study the paleontological history of the order. The subtlety of the gradations enhanced the likelihood of evolution. For example, if it could be shown how marmosets could have diverged from other New World monkeys, then a similar process could account for the divergence of humans from the great ape stock (Huxley 1863: 125). Firmly locating humans within the primate order, Huxley argued that the order should be divided up into seven families. The family divisions were as follows: humans, the other Old World higher primates, the majority of the New World monkeys, the New World marmosets and tamarins, the lemurs, the aye-aye, and the colugos. Even Huxley had trouble defining the primates, because he included colugos in the order. Note, however, his perception of how cohesive the Old World higher primates (catarrhines) are: humans are separated from the other catarrhines, but the great apes, the lesser apes, and the Old World monkeys are all subsumed in a single family. The unity of the living catarrhines will be a theme reiterated throughout this book.

In addition to Huxley's arguments about evolution from anatomical evidence, *Evidence as to Man's Place in Nature* incorporated some anecdotal evidence about non-human primate behavior and ecology, emphasizing reports of intelligence, tool behavior, and nest building behavior. These topics enhanced the similarity of non-human primates to humans, and suggested how human behavior might have evolved. Huxley also used natural selection to explain how climatic adaptations could explain some phenotypic differences in modern humans. Lastly, he addressed the original Neanderthal fossils, using this material to make the case that a human fossil record truly existed. He presented the Neanderthal material as being more ape-like than it actually is. Again, this was a rhetorical device to establish the existence of evolution, but this presentation also began another long tradition in paleoanthropology – the stark contrast between anatomically modern humans and their predecessors.

In short, Thomas Henry Huxley's *Evidence as to Man's Place in Nature* emphasized several themes that will be presented in this book: the unity of living catarrhine primates, the evolution of tool behavior and intelligence, natural selection acting on modern humans, human evolution, and the human fossil record. The fossil record of both human and non-human primates has been greatly expanded since the mid-nineteenth century, and biologists no longer debate the existence of evolution. Yet, Huxley's approach remains a model of inspired argument.

Endnotes

1. Kirk *et al.* (2003) deny this. Yet, Bloch and Boyer (2003) strongly support the shared, derived nature of the foot in carpolestids and euprimates, and also argue that cranial and dental data confirm similarity to euprimates. The debate is reported fully online at www.sciencemag.org/cgi/contents/full/300/5620/741b and www.sciencemag.org/cgi/contents/full/300/5620/741c.
2. Hominids have obviously specialized the foot for terrestrial locomotion, and have lost this powerful grasping ability. Yet, at least one fossil hominid foot (Stw 573) retains an abducted hallux.
3. Humans are obligate bipeds, and therefore have much smaller angular excursions than other primates. Yet, their stride length is also long (Reynolds 1983).
4. Convention on International Trade in Endangered Species – a treaty signed by over 100 nations that regulates the international trade in endangered species, and imposes trade sanctions on countries that do not adhere to treaty guidelines.
5. The URL for this lab is http://dllab.caltech.edu/avida/.
6. Some of Huxley's textbook material remains in print. For example, *The Crayfish. An Introduction to the Study of Zoology*, which first appeared in 1880, was reprinted through the twentieth century, and now appears in an online version: http://www.biology.ualberta.ca/old_site/palmer.hp//thh/crayfish.htm. A glossary has been added to accommodate modern needs.

2   A brief history of primatology and human evolution

Introduction

Non-human primates receive different treatment in different cultures. On the one hand, primates may be revered. Among the ancient Egyptians, the hamadryas baboon (*Papio hamadryas*) was sacred, being one of the avatars of the god Thoth. Thoth was the god of writing and the scribe of the gods, and played a crucial role in funerary rites, because he knew the secret spells and magical formulas that allowed the newly dead to pass safely through the dangers of the afterlife. Representations of hamadryas baboons therefore frequently adorn writings from the Book of the Dead and many art objects. Mummified remains of hamadryas baboons abound at certain archeological sites. Tomb paintings also reveal that the species was tamed, and baboons were taught to climb trees to collect palm fruit for their human handlers on the ground. Egyptian paintings and other art objects also show that guenons (*Cercopithecus* spp.) were imported into the country as tribute from lands to the south, and were treated as pets (Figure 2.1).

Hanuman, one of the most beloved Hindu gods, and a protector of both village and town, is a langur monkey (*Presbytis entellus*). Hanuman is one of the major figures of the Ramayana, which details the life of Rama, an incarnation of Vishnu who became a god himself. The heroic and faithful Hanuman and his monkey followers aid Rama in his battle against the king of demons, who has abducted Rama's wife, Sita. Hanuman's efforts finally re-unite Rama and Sita, but Hanuman refuses any reward, having performed his services purely from righteousness and from devotion to Rama. The Hindu reverence for Hanuman langurs extends to other non-human primates, as well. Rhesus macaques (*Macaca mulatta*) intermingle freely with Indian villagers, and can be found even in major cites like New Delhi, where they forage without hindrance for food at markets and middens.

Religions like Hinduism and Buddhism that believe in reincarnation of the human soul after death also endorse the possible transmigration of the soul between species. As a penalty for serious sins or misdeeds, a human might be reincarnated in animal form; in contrast, a noble and heroic animal might be reborn as a human. This worldview encourages an approach to the

26

Figure 2.1. Tomb carving from the late eighteenth dynasty, Egypt. A high-ranking military officer and his wife receive funerary offerings. A pet guenon (*Cercopithecus* sp.) stands bipedally under the wife's chair. The animal is tethered by the waist to one of the chair legs. Pet monkeys appear with some frequency in Egyptian art, usually gamboling amid the furniture. From Martin (1991: Figure 122).

human/animal interface that is very labile (Figure 2.2). Humans and animals are not viewed as a dichotomy, and the question of the place of humans in the natural world never arises, because humans are self-evidently part of nature.

Japanese culture traditionally assigns the endemic Japanese macaque (*Macaca fuscata*) a special relationship with humans: it is seen as a mirror for humans, or as a special messenger between humans and the spirit world in the native Shinto religion. A monkey is a *"human minus three pieces of hair"* (Ohnuki-Tierney 1995: 302). Interestingly, treatments of the macaque reveal that the Japanese perceive emotion and the capacity to feel, rather than rationality and the power to reason, as the principal distinguishing features of humankind (Ohnuki-Tierney 1995). In Japan, humans are not abruptly separated from other animals, and the Cartesian mind/body dichotomy that prevails in the West is absent in Japanese culture (Asquith 1995). In addition, Japanese macaques are sometimes trained from an early age to act as bipedal performers in traditional traveling plays, called *sarumawashi*, monkey performance plays.

Figure 2.2. A Balinese street play. An actor portraying a monkey interacts with the Barong, a mythical entity that protects human society. Balinese temple monkeys (*Macaca fascicularis*) are sacred. They are provisioned, and habitually mingle with temple visitors.

These performances are the Japanese equivalent of Commedia dell' Arte, where traveling companies of players would traverse the European countryside giving extempore renditions of the stock comic adventures of Harlequin, Pierrot, Scaramouche, and Pantaloon. The Japanese macaque performers once had ritual status, but their principal status now is that of entertainers. Even so, the performing monkeys are not mocked, but enjoyed, and their presence at certain endeavors confers good luck. The macaques are viewed as guardian entities for horses, and scheduling a monkey performance at a stable can ensure health and protection for the horses inside. Furthermore, the Japanese performing macaques demonstrate skeletal and morphological changes that accompany bipedal training, and thus illustrate how malleable a catarrhine phenotype can be without a special genetic substrate for bipedality (see Chapter 13).

Non-human animals never receive much reverence among Western cultures. This can be related to the deep divide that the Judeo-Christian tradition places between humans and all other animals. Yet, non-human primates were especially denigrated, because they seemed to mock humanity's form and aspirations. "To ape" in the Germanic language English is to mimic or imitate, as it is in other European languages (e.g., the Romance language French: *singer*; the Slavic language Polish: *malpować*). Non-human primates were considered minions of the devil by the early Church Fathers. Saint Augustine, for example, called the devil *Simia Dei*, or "the ape of God" (Warner 1998). The cleverness of primates, and their ability to "ape" or imitate human behavior, was considered satanic. The image of primates as demonic creatures or accomplices of the Devil gradually transmogrified during the seventeenth to nineteenth centuries. Nevertheless, European folklore continued to view them as foolish, silly, vain tricksters (Rooijakkers 1995). The complex interrelationship of monkeys, apes, and humans creates a rich mélange of myth, legend, and fantasy that continues unabated to the present (Morris & Morris 1968).

The image of non-human primates as ugly, evil, and lascivious is deeply embedded within European folklore. This may be seen even in language. *Marmouset*, a French word for an ugly and bizarre image, was transferred to an entire group of New World monkeys (marmosets), which were kept as pets by Court ladies during a time when noble households routinely maintained retinues of grotesque and deformed humans for purposes of entertainment. One might think that the image of primates has been redeemed in the West since the chimpanzees at the Gombe Reserve were first presented in the 1960s as peaceful inhabitants of an unspoiled Eden. Yet, infanticide and aggression have intruded even at Gombe, and an analysis of human male behavior based on that of common chimpanzees has been described as "demonic" (Wrangham & Peterson 1996). The image of non-human primates as quintessential beasts is still embedded in Western culture, just as it was in medieval times. Non-human primates are humans devoid of any humanity. In a fundamental sense, they are lower than other animals, because they lack any redeeming noble features. They are not courageous or wise, merely clever, and they perform no useful domestic functions. Even among anthropologists, reconstructions of fossil humans are presented as ape-like and hairy, up until the advent of anatomically modern humans, when the wild hair suddenly disappears and gorilloid features are lost (Berman 1999). Popular culture is still replete with images of bestial primates such as Edgar Allen Poe's murderous razor-wielding orangutan of "The Murders in the Rue Morgue," given new life when the 1841 short story was transformed into a twentieth century film. The orangutan was the largest known ape in 1841. The larger the ape, the more bestial its behavior. The Edward Hyde transformation of Dr. Henry Jekyll has become a mega-ape in "The League of Extraordinary Gentlemen," thanks to computer-generated movie images that improve on reality. Perhaps the most famous current embodiment of bestial primate is the eponymous movie character King Kong. The original 1933 film was re-made, and a second re-make is in the planning stages. In the wild, the 50-foot Kong demands human sacrifices, and devours his screaming victims alive. When transported to the West, his barbaric behavior threatens civilization – he is the antithesis of civilized humans (Warner 1995b, 1998). Technology finally subdues him, but he scales the summit of the world's tallest building, and swats at airplanes like buzzing mosquitoes. One of the film's two directors had conceived the idea of Kong after observing a troop of baboons in Africa while on the set of another feature film (Warner 1995a). Field experience was thus translated into a classic film vision of bestial primate behavior.

At the opposite extreme, Western popular culture now often currently attempts to view non-human primates, especially the great apes, as human-like, at least in their capacity for physical and mental suffering. This viewpoint is typified by supporters of the Great Ape Project (Cavalieri & Singer 1993), who campaign for the release of captive animals, and oppose any type of captive study, even non-invasive studies of ape intelligence or symbolic capacity. Presumably, the informed consent required for human research by institutional review boards should apply in the ape case, as well. Some animal activists have argued that the U.S. constitution grants non-human primates some natural rights, including the right of self-interest, because they can be regarded as "quasi-persons," a class that once encompassed slaves denied full personhood by American law (Lebovitz 2002). Yet, at the same time, removing the boundary between human and non-human is profoundly disturbing to some individuals. Donald Griffin, who has long studied animal awareness and animal cognition, discusses (1992: 245–252) but does not endorse the opinion of ethicists who argue that dissolving the distinction between human and animal has deleterious moral consequences for humans. This is not a trivial question. In particular, classifying severely disabled humans in the same category as animals might potentially resurrect bitter debates about the relative value of human lives, and the idea that some disabled humans experience lives of such low quality that they may automatically be treated as dispensible (Groce & Marks 2001). If disabled humans are considered to eliminate the boundaries between human and non-human, might a normal ape then be ranked as of greater value than a disabled human? What rights should properly be accorded to humans and non-humans? The philosophical debates of the mid-nineteenth century about evolution and morality are continuing. It is no accident that T. H. Huxley's *Evolution and Ethics* (1894) is still in print – the debates it presents are current even now, more than 100 years later.

Antiquity and the Middle Ages

Much of our understanding about ancient knowledge of non-human primates in the Western world was unearthed by Robert and Ada Yerkes. The Yerkes compiled a monograph whose first three chapters cover knowledge of non-human primates in antiquity, the Middle Ages to the end of the seventeenth century, and the eighteenth century, respectively (Yerkes & Yerkes 1929). These two scholars were among the first twentieth century investigators to be consumed with interest about the great and lesser apes. The word "ape" was freely used by Westerners, although it is clear that no knowledge of true apes (hylobatids and pongids) existed in the West until the late seventeenth century. Furthermore, even when general knowledge of these animals began to circulate, it tended to be conflated with folklore about Wild Men of the Woods that was widespread throughout Europe (Figure 2.3). Modern tales of the yeti and Sasquatch continue the old traditions.

One of the earliest and most intriguing reports from antiquity dates from about 500 B.C., when the Phoenician city of Carthage on the North African coast sent two exploratory naval expeditions out through the Straits of Gibraltar into the Atlantic Ocean. One expedition, under the leadership of the Phoenician Admiral Hanno, sailed southward down the West African coast to set up a trading post at Kernë, possibly at Cape Arguin. During their explorations of the West African coast, Hanno and his men reported seeing strange, large, hairy, quasi-human creatures that hurled rocks at the sailors in self-defense. Interpreters called these creatures Gorillae. Three of the females were captured and killed, and their skins were brought back to Carthage. Two of the skins were displayed at the temple of Juno until the fall of Carthage (Yerkes & Yerkes 1929). Whatever these creatures were, they were not baboons, which were well known to the ancients, and were aptly called Cynocephali (the dog-headed ones) by the Greeks.

Many of these early reports fade into the realm of myth and legend. Systematic study of animal life begins later. As is true of much of biology, everything begins with Aristotle. During the Fourth Century B.C., Aristotle first defined some of the distinctive traits that separate humans from other animals. In *Partibus Animalium*, Aristotle argued that humans are bipedal, use tools, and have large brains; in *Historia Animalium*, he argued that human females lack estrous (Stoczkowski 2002). In *Historia Animalium*, Aristotle linked humans with monkeys, baboons, and apes (Barbary macaques), and claimed that dissection revealed their internal organs to be the same. This claim was not refuted until Renaissance times, and it became the source of many errors in human anatomy. Aristotle was the tutor of Alexander the Great, who founded the city of Alexandria, in Egypt, and sent an endless stream of specimens back to his old teacher. After the death of Alexander, his half-brother Ptolemy, also a student of Aristotle, seized control of Egypt and founded the Ptolemaic dynasty. Ptolemy established Alexandria as the preeminent center of ancient learning in the Western world. Scholarly research was supported and promoted by the Ptolemaic rulers for nearly 300 years. This research included study of biology, medicine, and anatomy. Just after 275 B.C., Herophilus of Chalcedon founded a school of anatomy in Alexandria, and public dissections of human bodies took place.

At about 150 A.D., Galen, a physician associated with a famous gladiator school in Pergamum in western Turkey, treated the wounded gladiators and wrote treatises on human anatomy. Because dissections of human bodies could be carried out only in Alexandria, Galen dissected monkeys, cows, pigs, bears, and dogs. He performed experiments on living animals, and discovered that the brain controls the muscles of the body via nerves. His dissections and writings about "apes" refer to the Barbary macaque (*Macaca sylvanus*), native to northwestern Africa, and well known to the ancients. Galen remained the ultimate authority on human anatomy and physiology for 1500 years, because his work was preserved by Arab scholars. His writings were first translated into Latin from Arabic during the twelfth century, and then translated from the original Greek during the sixteenth century.

Monkeys of various sorts appear in Medieval bestiaries. They have an evil reputation, and are castigated as being wicked, unruly, and lacking reason. Often they are presented in demonic form, because they were considered to be spoiled human replicas. Monstrous human races also appear in the bestiaries, and it is sometimes impossible to establish whether non-human primates or monstrous, hairy humans are being discussed. One of the most famous bestiaries was *De Animalibus*, written by the thirteenth century naturalist Albertus Magnus, who also translated the works of Aristotle into Latin. Albertus assigned humans into a category separate from all other animals. The other animals were divided into two groups: non-human primates, which were human-like, and all remaining animals. However, this should not be construed as a precursor to evolutionary thinking, because the non-human primates were ruined copies of humanity, set permanently below humans in the static Great Chain of Being. In contrast to humans, they had imperfect souls and lacked reason.

The Renaissance to the late eighteenth century

Beginning in 1500, the earliest illustrated anatomical texts appeared in print. In 1543, the Flemish anatomist Andreas Vesalius published the first profusely illustrated text on human anatomy, *De Fabrica Humani Corporis*. Vesalius worked in Italy, and local artists from Titian's workshop rendered his dissections into elaborate woodcuts. These illustrations are not only important in the history of medicine, but remain a blueprint for modern research on human gross anatomy.[1] Vesalius was convinced that accurate knowledge of human anatomy could be obtained only through dissection of fresh human corpses. He detected serious discrepancies between his human cadavers and descriptions of human anatomy published by Galen. The human sternum, for example, is not divided into seven segments. Vesalius therefore initiated a series of dissections to compare the anatomy of humans and monkeys. To the scandal of his contemporaries, many of whom were unwilling to admit that Galen could be wrong in any degree, Vesalius eventually established that Galen's descriptions of human anatomy were based on dissections of the Barbary macaque. The fascination that Vesalius's work generated among his contemporaries can be gauged by the degree of public interest produced by the modern anatomist Gunther von Hagens, who has invented a method (plastination) of preserving human cadavers by replacing the fluids within biological materials with plastics. Plastinated human bodies and body parts are exhibited in lifelike poses in a traveling anatomy exhibit called Body Worlds. Vividly posed cadavers and organs in various stages of dissection, shown with a natural range of colors, recall the artistically displayed cadavers of Vesalius. As of 2004, the Body Worlds exhibit has been viewed by nearly 13 million visitors in seven European and Asian countries. Hundreds of fascinated visitors have signed up at the exhibit to donate their bodies for display in future plastination shows (Burdick 2004).

Compilations of folklore and anecdotal traveler's tales continued to be produced during the Renaissance. In 1587, Konrad von Gesner published an illustration of a female "ape" that circulated in the literature for over 200 years. One of its last appearances, in a very human-like version, can be seen in Figure 2.3, in the creature labeled "Lucifer." Gesner's "ape" was encountered by a traveler on a visit to the Holy Land. The illustration might have been based on a sighting of a male hamadryas baboon, because these animals have large, lion-like manes, and this is a prominent feature of the illustration. As for its other features, the "ape" stands bipedally (propped up by a cane), has a long tail, possibly has ischial callosities, and has breasts like a human female. In 1640, the scholar Ulysses Aldrovandi later assembled another famous natural history compendium, including Gesner's "ape" and other curiosa.

Although gibbons frequently appear in Chinese art, both gibbons and siamangs remained unknown to Western science until relatively late. The French naturalist Georges Buffon published the first accurate description and illustration of a gibbon in 1766. Observations in the wild of spectacular gibbon arboreal leaping date back to 1697 (Yerkes & Yerkes 1929). As of the late nineteenth century, it was not common knowledge that gibbons and siamangs were apes. For example, Arthur Keith, trained as a physician and not a biologist, believed that he was observing and dissecting "yellow monkeys," rather than lesser apes, in Thailand during the 1890s (Keith 1950).

Knowledge of the great apes came earlier than knowledge of the lesser apes. Definite knowledge about great apes came to Western scholars in the mid seventeenth century with the discovery of chimpanzees and orangutans. The

Figure 2.3. Illustration from Emmanuel Hoppius, 1760, showing a typical fourteenth to eighteenth century mixture of legend and reality in European primate depictions. The figures are, from left to right, Troglodyta, Lucifer, Satyrus, and Pygmaeus. Satyrus appears to be a chimpanzee; Pygmaeus may be an orangutan, but this identification is less certain. The other two figures are mythical conflations with folklore about the Wild Men of the Woods. From Yerkes & Yerkes (1929: Fig. 12). Note the canes or sticks in two of the depictions. Spencer (1995) traces the continuity of this stick motif in primate illustrations from the fifteenth until the eighteenth centuries, and argues that it indicates that erect posture and bipedal locomotion are not normal.

first ape known to Europeans was a chimpanzee imported from Portuguese Angola and presented to the Prince of Orange. This animal was described in 1641 by the Dutch anatomist and physician Nicolaas Tulp, who called it, confusingly enough, an "Orang-outang." The true orangutan appeared later, when Jacob Bontius wrote a 1658 account of a female orangutan found in Borneo. More confusion arose when Bontius described the animal as being startlingly human-like, except for the faculty of language, and capable of bipedal walking. The accompanying figure is that of a hairy human female, not even remotely like an orangutan. In 1699, the English anatomist Edward Tyson published a monograph on his comparative dissection of a juvenile chimpanzee imported to London from Angola. He confusingly called the creature a "pygmie," although he argued that it was a new type of non-human primate, and not a human.[2] Tyson published highly accurate anatomical illustrations. In quantifying similarities and differences, Tyson found that the chimpanzee resembled humans in 48 traits, and resembled Old World monkeys in 34 traits.

The work of Carolus Linnaeus and his 1758 system of biological classification was outlined in Chapter 1. It is important to note here, however, that his sorting of humans with non-human animals into an Order Primates did not meet with universal approval. Some scholars, such as Georges Buffon, argued that Linnaeus did not give proper weight to human intellect. Humans may have resembled non-human primates in anatomy, but their ability to reason was unlike that of any other creature. Other scholars noted deep and significant differences between human and non-human primate anatomy. This was particularly true of Johann Friedrich Blumenbach, who argued that, in spite of superficial similarities between humans and non-human primates, there was a profound functional difference: humans had bipedal posture and locomotion, and other primates did not. These differences in posture and locomotion were associated with skeletal differences. In 1779, Blumenbach proposed that humans be classified into a separate order of mammals, which he named Bimana a year later. The other primates were placed in the Order Quadrumana. The division of the primate order into two-handed and four-handed species was based on human specializations for bipedal posture and locomotion.

The existence of the Great Chain of Being was unquestioningly accepted throughout this period. The static nature of the Great Chain of Being actually created a problem for scholars who were acquiring information about variation between living human groups. The Wild Men of the Woods disappeared with increasing exploration, but strange and unexpected human groups were revealed for the first time, and documented by systematic survey. It seemed impossible that climate and geographic influences alone could have caused so great a degree of human variation within the 6000 year span that then encompassed the age of the world. Consequently, eighteenth century thought saw an increasing acceptance of polygenism – the idea of multiple human creations (Spencer 1995). The different human races had been separately created. In 1863, Thomas Henry Huxley finally solved the problem of striking human phenotypic variation by arguing that natural selection was responsible for human racial differentiation. The idea of the Great Chain of Being was finally extinguished.

The orangutan always received intense scrutiny by Enlightenment scholars, because it offered them a focus for speculation about the origins of tool behavior and language. Tools and language were two principal concerns of Enlightenment naturalists and encyclopedists, because technology and language were entwined in definitions of humanity and various speculations about human history and increasing social complexity. At the end of the eighteenth century, Lord Monboddo (James Burnett) asserted that orangutans should be sorted into the human species, and speculated that, because their vocal tracts were so human-like, they might be taught to speak. The concept of apes, especially orangutans, as human ancestors had actually been covertly circulating in learned circles since the eighteenth century (Stoczkowski 1995). Fascination with the orangutan can probably be simply attributed to the fact that this species was the largest known ape until the mid nineteenth century discovery of the gorilla.

The nineteenth century

The study of comparative primate anatomy that was set in motion by Edward Tyson in 1699 came into its own during the nineteenth century. In 1835, the paleontologist and anatomist Richard Owen sorted out some of the confusion about great ape taxonomy by discovering that ontogenetic changes could be profound in chimpanzees and orangutans. What had been once recognized as species-level distinctions, might be nothing more than the anatomical differences between adult, especially adult male, and infant or juvenile of the same species. Among his many other interests, Owen became a noted specialist in mammalian brain anatomy. In 1858, he advocated that humans be assigned to a separate subclass of the Mammalia, the subclass Archencephala, by virtue of their large brains and unique intellectual powers. Owen's separation of humans into a different taxonomic category from other primates led to a spectacular series of arguments with Thomas Henry Huxley that were reported in elaborate detail in the Victorian popular press (Chapters 1 and 8). Huxley (1863) supported the unity of the primate order and the similarity of humans to higher primates (Chapter 1).

The section that appears the strangest to modern readers in Huxley's *Evidence as to Man's Place in Nature* is the one on non-human primate behavior. Huxley here attempts to draw together a mélange of traveler's reports and White Hunter's tales about primate behavior in the wild. Almost the only valid piece of evidence about great ape behavior in the wild that occurs in this section is the fact that the African great apes seem to construct some type of nest or sleeping platform of leaves. The missionary/physician Thomas Savage and the anatomist/physician Jeffries Wyman had established the existence of the gorilla in 1847. Huxley also wrote of captive great ape behavior that appeared complex and rational, and Darwin (1871) was able to cite some accounts of ape tool behavior. This continued the Enlightenment tradition of focusing on tool behavior, because this was one of the chief criteria that separated humans from other animals. However, once an evolutionary relationship was discovered between humans and non-human primates, there was an almost inevitable impetus to view the non-human primates, particularly the great apes, as highly intelligent, rational, tool using and manufacturing animals, with complex sociality, and an exalted moral sense expressed in episodes of empathy and altruism.

No field observations of primate behavior and ecology were conducted during the nineteenth century, except for one strange attempt. In the 1890s, the American zoologist R. L. Garner fashioned a large strong iron cage (the land equivalent of a shark cage) and set it out on the forest floor in Gabon, West Africa. Armed with a magazine rifle, revolver, and bush knife, he sat in the cage and tried to learn something about the behavior of sympatric chimpanzees and gorillas. Because he was not mobile, he saw very little, but he attempted to reproduce phonetically the sounds that the animals made. Garner believed that the animals were capable of real speech, and he gave the phonetic equivalents of the ape words for "cold," "drink," "death," etc. The Yerkes (Yerkes & Yerkes 1929:164) have nothing but contempt for Garner's efforts, but echoes of Garner's work probably seeped into popular culture when the American writer Edgar Rice Burroughs created an apish language for his character Tarzan, supposedly fostered by wild apes in West Africa.

The early twentieth century

Anatomical primatology

During the early twentieth century, scholars of primatology begin to divide into two main camps – anatomical and behavioral. For most of the twentieth century, primatology was largely focused on anatomy (Schultz 1976). Foremost among the early primate anatomists was the fabulously learned William King Gregory, who focused his attention on interpreting the anatomy and evolutionary relationships of fossil primates. From 1903 until 1958, Gregory published extensively on vertebrate evolution, evolutionary processes, and functional morphology (Simpson 1971). An early publication on primate phylogeny (Gregory 1916) was followed by a seminal work on dentition (Gregory & Hellman 1926) that linked modern humans and other hominoids to the extinct dryopithecine apes by virtue of a shared derived pattern on the lower molar crowns, "the dryopithecine pattern." Gregory (1920) also was the first scholar carefully to reconstruct the musculoskeletal anatomy and infer the behavior of a fossil primate (the Eocene genus *Notharctus*) using a comparative set of living prosimian primates.

From 1889 until 1892, the Scottish medical doctor Arthur (later Sir Arthur) Keith was stationed in Thailand attending to the needs of gold miners. During his stay, he identified a number of anatomical similarities between lesser apes, orangutans, and humans (Keith 1940, 1950). From his medical training, he had learned that these traits were associated with bipedal posture and locomotion. Sympatric monkeys (macaques and langurs) did not possess these traits. Keith argued that the bipedal posture of humans had its origins among a non-human ancestor whose trunk was held in an upright or orthograde posture as it moved through the trees. In this form of arboreal locomotion, the upright body was suspended from a support, and was moved by the forelimbs. Keith had observed gibbons using their forelimbs to propel themselves through the trees, and coined the neologism "brachiation" to describe this form of locomotion. The word was based on the formal anatomical term for the forearm, the antebrachium. Gregory

(1927, 1928, 1930, 1934) later promoted Keith's "brachiation" model, even after Keith abandoned it. Gregory's discussions presented no detailed reconstruction of locomotion, but were arguments for hominid origins from a pongid ancestor that was arboreal and used its forelimbs for locomotion.

Besides the use of "brachiation" in the Keith/Gregory model of human evolution, the posture and locomotion of living apes influenced models of hominid evolution throughout the twentieth century (Tuttle 1974, 1977). The following is a brief synopsis of such models. The foot structure and arboreal bipedalism of lesser apes generated a hylobatid model of hominid origins, in which bipedalism arose first in the trees (Morton 1924, 1927, Tuttle 1974). When the terrestrial knuckle-walking of gorillas and chimpanzees was recognized as a distinct category of locomotion and linked to a specific anatomy (Tuttle 1967, 1969), some researchers proposed that there had been a knuckle-walking phase of hominid evolution (Washburn 1968, Richmond & Strait 2000). Bonobos were used to model the earliest hominids, because of their gracile body build and limb proportions (Zihlman & Cramer 1978, Zihlman *et al.* 1978). Furthermore, in comparison to common chimpanzees, bonobos have reduced long bone and cranial robusticity, small relative tooth size, and small brain size relative to body size. These traits, which resemble the effects of domestication, indicate that bonobos may have been influenced by heterochrony, specifically by an extended period of growth or peramorphosis (Shea 1989). Similar traits have been noted in modern humans, especially after the advent of village life and agriculture. Bipedal posture by chimpanzees foraging for fruit in savannah settings (Hunt 1994), vertical climbing by gorillas (Fleagle *et al.* 1981), and slow, quadrumanous climbing by orangutans (Oxnard 1969, Stern 1975) have also served as models for the origins of hominid bipedality. An analysis of these ape and other primate templates for bipedal origins will be presented in Chapter 13.

A central figure in anatomical primatology during this time was the Swiss researcher Adolph Schultz, who labored heroically to increase knowledge of the anatomy of non-human primates. Schultz's publications extend from 1915 to posthumous publications appearing in the mid 1970s. Variability within primate species, growth and development, and relative body proportions remained important topics throughout Schultz's career. Schultz was associated first with the Carnegie Institution in Baltimore, Maryland, where he conducted comparative research on primate fetal morphology and development, and then with Johns Hopkins University. In 1951, he became Director of the Anthropological Institute in Zürich, and acquired a vast comparative collection of primate material. The Schultz Collection is currently the center of the Museum of Anthropology at the Anthropological Institute (Soligo *et al.* 2002). Because Schultz was a gifted artist, his illustrations are an important component of his publications,

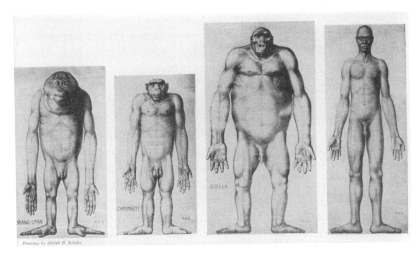

Figure 2.4. Comparative proportions of the great apes and a human (Schultz 1933). Adolph Schultz prepared these famous drawings from cadaver material. Faint outlines of the skeletons can be glimpsed within the figures. Shultz drew the originals on long window shades. The figures are reduced to the same scale to reveal differences in body proportions. These figures, along with a gibbon not shown here, are probably among the most widely recognized icons in physical anthropology and primatology. From Hooton (1942: 364).

and a number of them continue to be widely reproduced. One has even become iconic (Figure 2.4). Schultz was so impressed with the striking uniformity of catarrhine primates that he concluded that hominids had diverged from some unknown, generalized catarrhine ancestor, not too far removed from the Old World monkeys (Schultz 1936a, 1936b, 1950). He believed that hylobatids and pongids were too specialized to have given rise to hominids, but that hylobatids are more similar to hominids. The Family Hylobatidae may have originated just before the origins of Family Hominidae (Schultz 1936b: Fig. 21).

Behavioral primatology

Behavioral primatology in the early twentieth century was given a major impetus through the work of the Yale psychologist Robert M. Yerkes and his wife, Ada W. Yerkes. Robert Yerkes befriended Madame Rosalía Abreu, a fabulously wealthy and eccentric lady who maintained a large menagerie of animals, including orangutans and chimpanzees, on her Cuban estate. In 1915, the estate produced the first known record of a great ape, a chimpanzee, born in captivity. Robert Yerkes corresponded extensively with Madame Abreu about the care and behavior of her captive primates, and, after her death, Madame Abreu willed the beloved primates in her collection to Yerkes. From this nucleus, the Yerkes established the first primate captive colony at Orange Park, Florida, which became the nucleus of the current system of regional primate centers. At Orange Park, the Yerkes trained workers who later became respected pioneers in field-related primate behavior. In addition, research at the Orange Park lab resulted in the first comprehensive analysis of primate senses and intelligence (Yerkes & Yerkes 1929). Robert Yerkes had been interested in intelligence for a long time, and had been employed by the U.S. government to devise intelligence tests to screen American men drafted into the U.S. army during World War I. The comparative study of intelligence enthralled Yerkes, and he devised a series of non-verbal tests to analyze intelligence in captive primates, especially the great apes. One subject animal particularly fascinated him – a male chimpanzee named Prince Chim, who became the subject of a separate book (Yerkes 1925). Prince Chim's intelligence and sensitivity rendered him "almost human" in Yerkes' account. One of Yerkes' associates later performed an autopsy of Prince Chim, and revealed that the famous animal had actually been a bonobo or "pygmy" chimpanzee, distinct at the species level from common chimpanzees (Coolidge 1933). The calm temperament, cohesive sociality, and intelligence of bonobos (*Pan paniscus*) lead some researchers to posit that they represent early hominid behavior better than common chimpanzees do (Wrangham & Peterson 1996).

Clarence Ray Carpenter was the Yerkes' student who had the greatest impact on behavioral primatology. Carpenter pioneered the field study of primate behavior and ecology that is such an important focus of modern primatology. On Christmas Day, 1931, Carpenter initiated a study of two primate species that inhabited Barro Colorado Island, Panama – an island created when Lake Gatún was formed during the construction of the Panama Canal. Rising lake waters isolated native plant and animal species on a newly formed island which still hosts an internationally renowned tropical research station maintained by the Smithsonian Institution. Carpenter studied mantled howler monkeys (*Alouatta palliata*) and black-handed spider monkeys (*Ateles geoffroyi*) on Barro Colorado, and began to publish his investigations in the early 1930s (Carpenter 1934). In addition, Carpenter was a member of the 1937 Asiatic Primate Expedition to Thailand, which also included A. H. Schultz and S. L. Washburn. While Schultz shot wild primates for anatomical research, Carpenter (1940) managed simultaneously to study the behavioral ecology of white-handed gibbons (*Hylobates lar*).

Some of Carpenter's field methodology was taken directly from sociology (e.g., the use of sociograms to examine the frequency and pattern of social interaction). Much of the methodology was taken from the work of the classical

European ethologists Konrad Lorenz, Nikolaas Tinbergen, and Karl von Frisch. In the early twentieth century, these researchers were pioneers of ethology – the study of animal behavior in the wild, or under naturalistic conditions. The ethologists argued that the evolutionary processes affecting behavior could be studied only under these conditions, the authentic field conditions under which natural selection and sexual selection occur in the wild (Tinbergen 1951, 1963). The methodology that they endorsed was to establish and maintain contact with identified wild subjects, recording detailed accounts of diet, foraging and ranging behavior, predation, reproduction, rearing of young, and social interactions between conspecifics. They took copious written notes and photographs, and used primitive recording technology to document vocalizations. In addition, they conducted some experimental manipulations of behavior, in order to test hypotheses about the factors that affected certain behaviors. The triumph of ethology came in 1973, when Lorenz, Tinbergen, and von Frisch were jointly awarded the Nobel Prize for Physiology or Medicine for their research on animal behavior. However, during the early twentieth century, many students of animal behavior endorsed laboratory-based research and protocols, and eschewed field-oriented studies, which, to many, appeared unscientific and smacking of old-fashioned, Victorian natural history. Animal behaviorists with this mindset were not impressed by Carpenter's field observations of primate behavioral ecology.

With some exceptions, such as E. A. Hooton, who published the results of Carpenter's work in popular accounts (e.g., Hooton 1942), anthropologists remained disinterested in primate behavioral ecology. Although this disinterest now seems inconceivable, it is substantiated by primate behavior being restricted to a single column ("Psychobiology") in the chapter on primates written for an encyclopedic survey of anthropology published in 1953 (Straus 1953). Almost 10 years later, Straus's paper was reprinted in a volume of selected papers from the original survey (Straus 1962). Straus cites Carpenter's work on howlers and gibbons, but gives much greater attention to a monograph by Zuckerman (1932). Zuckerman's work was also prominent in histories of anthropology published during the 1960s (e.g., Penniman 1965). However, Carpenter's field-oriented work on primates remained largely forgotten.

During the 1940s and 1950s, Carpenter devoted his energies towards the pedagogical use of film, making films of non-human primates in the process. Carpenter also established an important primate research center close to the U.S.A. that mirrored the tropical island setting of Barro Colorado Island. This was Cayo Santiago Island, a tiny island just off the coast of Puerto Rico. Carpenter envisioned Cayo Santiago as a reserve where free-ranging non-human primate species could roam at will, and could be easily studied by behavioral ecologists. He seeded the island with a number of non-human primate species, including gibbons and orangutans. Unfortunately, the island possessed little in the way of natural vegetation. Carpenter attempted to establish a cover of tropical vegetation that would provide food for free-ranging primates. This attempt failed, and, right from the start, humans needed to provision the intro- duced primate species. Many animals died, but the rhesus macaques (*Macaca mulatta*) thrived under these conditions. The descendants of Carpenter's rhesus macaques still inhabit Cayo Santiago, and increase their numbers so well with provisioning that they need to be routinely culled down to about 600 animals – the largest number that can be supported by the area of the island. Besides providing primate subjects for biomedical research, the rhesus macaques of Cayo Santiago have been the focus of research on genetic variability and social structure, mating preferences, the establishment of dominance and dominance interactions, sex differences in behavior, matriline formation, and gerontology (Rawlins & Kessler 1986). Skeletal material from the Cayo Santiago animals forms an important primate research collection, because the material comes from identified animals with known age, sex, and health history.

Anthropologists failed to appreciate the importance of Carpenter's field stud- ies of non-human primates. To anthropologists, Zuckerman's monograph (1932) remained the *ne plus ultra* of primate behavioral research until the early 1960s. Why did Zuckerman's work exercise such a profound influence on anthro- pology for almost 30 years? It was not concerned with field-oriented stud- ies, and this was probably viewed as a positive quality during the hey-day of Watsonian behaviorism in the U.S.A. Unlike Carpenter, Bingham, Nissan, and other students of the Yerkes, who observed (or tried to observe!) natural behavior under field biology conditions, Zuckerman studied captive hamadryas baboons at the London Zoo under controlled or quasi-experimental conditions that superficially appeared more rigorous or "scientific," and did not smack of nineteenth century natural history studies. To this he added a few weeks of South African field observations and anecdotal reports of baboon behavior. Yet, there are additional factors that explain Zuckerman's influence. Zuckerman (1932) argued that a profound gulf separated the societies of non-human pri- mates and humans. Humans possessed culture, and the non-human primates did not. Culture allowed humans to control the aggressive and sexual impulses that Zuckerman believed pervaded non-human primate society. From his obser- vations of hamadryas baboons at the London Zoo, Zuckerman concluded that all non-human primate societies were held together by continuous sexual activ- ity, dominant males monopolized as many females as possible, and aggressive dominance interactions culminating in fights, injuries, and deaths character- ized nearly all social interactions. During a one-month period, from the end of July to the end of August 1927, 15 of 30 adult females were killed in violent melees. Calm was finally restored in October 1930 by removing the 5 surviving females, so that the captive group contained only adult and one immature males. Zuckerman's perspective agreed with Sigmund Freud's view that human culture sublimates aggressive and sexual impulses that would otherwise overwhelm human sociality. If it were not for culture, the sexual drives, aggression, and endocrine fluxes that permeated non-human primate social life would also overwhelm human sociality. Because Freud's ideas were a dominant influence in Western thought during the early twentieth century, and remained so in America until late in the twentieth century, Zuckerman's echoing of Freudian thought heightened the impact of his ideas. Finally, anthropologists had been trained to believe in the uniqueness of culture – it was a species-specific trait for humans. Hence, anthropologists writing about non-human primates (e.g., Sahlins 1959) tended to embrace Zuckerman's views about the vast gulf between human and non-human societies. Sahlins uses phrases like "the subhuman primate horde," which reverberates with Freudian impact, although Freud is not cited. Economic interactions, division of labor, and kinship are some of the cultural traits that Sahlins believes distinguish human society from non-human society.

Zuckerman did not realize that the hamadryas baboon behavior that he observed at the London Zoo was pathological, and was exacerbated by the Monkey Hill enclosure in which the animals were displayed. Zuckerman visited the primate facility at Orange Park, Florida, but he was not impressed by either the Yerkes' work or the work of Carpenter and other field-oriented students trained by the Yerkes (Zuckerman 1978). A second edition of *The Social Life of Monkeys and Apes* was issued in 1981. This edition was unchanged from the first one, except for several appendices and a long postcript that reiterated the author's belief that his original work of 50 years ago was upheld, and was far superior to efforts by subsequent researchers (Zuckerman 1981).[3]

It is noteworthy that *Anthropology Today* contains two chapters on fossil humans (de Chardin 1953, Weinert 1953) that also appear strange to early twenty-first century eyes. They are compendia of fossil species, with no discussion of evolutionary processes. Africa appears to hold some very early fossils, but these are only reluctantly accepted as pre-human. Weinert (1953) equivocates about whether the australopithecines are apes or hominids. This was soon to change.

The "new" physical anthropology

In 1951, S. L. Washburn articulated the notion that a major change would soon overtake physical anthropology – a new physical anthropology would emerge like a phoenix from the ashes of the old (Washburn 1951, 1953). He argued that evolutionary processes and population genetics would eventually

Table 2.1. *Contrasts between the Old and New physical anthropology*
(*after Washburn 1953: 716*)

The Old approach		The New approach
	Purpose	
Classification is paramount		Understanding evolution
Differences described		Classification marginal
Classification solves all problems		Cause of differences paramount aim
	Theory	
Theory unimportant		Theory crucial
"Facts speak for themselves"		Hypotheses important
		Hypotheses must be consistent and verifiable
	Technique	
Measurement 80%		Measurement 20%
Some morphology		Many techniques to solve particular problems
	Interpretation	
Speculation		Testing of hypotheses

turn physical anthropology away from description and typology and toward the study of natural selection and adaptation. Animal behavior and anatomy formed adaptive complexes, subject to evolutionary change. The clearest presentation of Washburn's advocated new policy appears in a table outlining an approach and tactics for the new physical anthropology (Washburn 1953: Table 1). This table contrasts purpose, theory, technique, and interpretation between the old and new physical anthropology (Table 2.1). In fact, the new physical anthropology foreshadows by about 10 years a comparable strategy outlined for archeology. This strategy is variously known as The New Archeology, Middle Range Theory, or Processual Archeology (Binford 1968). Washburn remained skeptical that fossils alone would reveal the particulars and minutiae of evolution. For this reason, he advocated experimental studies of functional morphology and the detailed study of animal behavior in the wild. These studies would generate predictions that could be tested against the fossil evidence, provided that this evidence were forthcoming.

1959 – *annus mirabilis*

Australopithecines recognized as the first hominids

In the first quarter of the twentieth century, the anatomist Raymond Dart announced the discovery of a unique fossil primate specimen from the site of Taung, in South Africa. Dart (1925) originally classified the specimen as a member of a separate primate family, the Homo-simiadae, intermediate between the pongids and living humans. Soon afterwards, however, Dart argued that this juvenile specimen appeared to be a very early hominid, although different enough from living humans to be placed in a separate subfamily, the Australopithecinae. Most of Dart's contemporaries regarded the fossil from Taung, and other australopithecine fossils discovered during the 1930s in South Africa by Robert Broom, to be members of an extinct pongid subfamily. The arguments about whether the australopithecines were hominids or aberrant apes gradually were resolved nearly 25 years later, when influential researchers like Le Gros Clark (1947, 1950) accepted the hominid status of the australopithecines. Some major researchers (e.g., Gregory 1930, 1949) had already done so, without affecting the consensus opinion. Gregory also argued that better candidates for early hominids were unlikely to emerge: ". . . if *Australopithecus* is not literally a missing link between an older dryopithecoid group and primitive man, what conceivable combination of ape and human characters would ever be admitted as such?" (Gregory 1930: 650).

Why did many researchers resist the hominid status of the australopithecines? The juvenile status of the type specimen, and incomplete knowledge of the later South African discoveries certainly contributed to the hesitant acceptance of the australopithecines as hominids. Study of the actual fossils was important. Le Gros Clark visited South Africa in 1947 in order to compare australopithecine fossils with measurements collected from more than 100 pongid specimens. He left South Africa convinced of the hominid status of the australopithecines (Le Gros Clark 1947). Le Gros Clark faced opposition from scholars who argued that hominids had diverged at an early date from the other catarrhines – the earlier the date of presumed divergence, the greater the opposition. Frederick Wood Jones, who had long held that hominids had separated from other placental mammals at the beginning of the Age of Mammals, was more dubious than William L. Straus, who proposed a later hominid separation. However, Le Gros Clark's principal opponent was Solly Zuckerman (1950a, 1950b), who vehemently argued that the australopithecines were pongids, not hominids. The argument focused around interpretation of the australopithecine dentition: whether length, breadth, height, or ratio measurements sort australopithecines with pongids (Zuckerman), or whether overall patterns of shape and form ("the total morphological pattern") sort australopithecines with hominids (Le Gros Clark). The argument was finally resolved when Bronowski & Long (1952) performed multivariate analyses of australopithecine teeth, and concluded that they were hominid-like. These multivariate statistical analyses, routinely accomplished today with computer software, were performed in the 1950s by heroes armed only with slide rules and adding machines. Le Gros Clark (1967: 32–36) later published full details of the widely publicized and stressful controversy. Some major researchers (Oxnard 1975, 1984, Ashton 1981) who studied under Zuckerman in the Department of Anatomy in the Medical School at Birmingham still consider the australopithecines to be a unique and highly divergent group of fossil hominids. They are pongid-like in many features, and are probably widely separated from the origins of genus *Homo* – in fact, they may have nothing to do with genus *Homo*, because they represent a failed parallel experiment in bipedal locomotion.

Probably the most important factor that made researchers reluctant to accept the australopithecines as hominids was their small relative brain size, which undeniably falls within the pongid size range. That brain size was the crucial factor is proven by the widespread and contented acceptance of the taxon *Eoanthropus dawsoni*, discovered at the English site of Piltdown in a series of finds beginning in 1912. Using fluorine dating and detailed examination of the Piltdown material, Joseph Weiner and his colleagues (Weiner *et al.* 1953, Weiner 1955) conclusively proved that this presumably ancient fossil material constituted a deliberate hoax. The large brain and hemispherical asymmetry of the taxon appeared modern precisely because modern material was used in the constructed hoax. However, the pongid-like lower jaw and canine tooth disturbed only individuals who argued that they had no association with the cranial material. Most scholars were comfortable with the juxtaposition of pongid and hominid traits, as long as the brain size was in the modern range. The reverse situation characterized the australopithecines: hominid dental traits coexisted here with a pongid-like brain size. Because the Piltdown and australopithecine finds illustrated divergent trends, it was difficult to accommodate both in analyses of hominid evolution. Researchers like Gregory, Le Gros Clark, and Washburn, who easily accepted the hominid status of the australopithecines, were also dubious about the meaning of the Piltdown material, even before the discovery of the hoax was announced.

The discovery of "Zinjanthropus"

Of the myriad of fossil human finds discovered in the twentieth century, only a handful can rival the importance of and achieve the popular impact of the original Taung australopithecine find in 1925. Included in this handful of important finds is Mary Leakey's 1959 discovery of the type specimen (OH 5) of *Australopithecus boisei*, initially called Zinjanthropus, at Olduvai Gorge, Tanzania. Mary Leakey had been conducting an archeological excavation at a very rich locality in Bed I at Olduvai when she discovered the OH 5 specimen (Figures 2.5 and 2.6). A paper diagnosing the new taxon appeared about a month

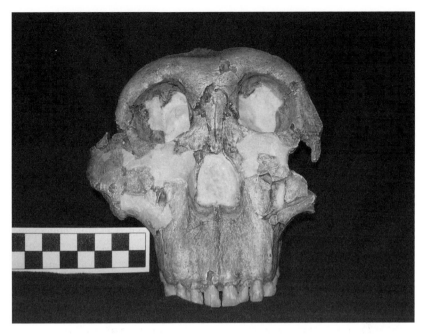

Figure 2.5. Cast of the craniofacial portion of OH 5, the type specimen of *Australopithecus boisei*. Additional cranial fragments exist. The scale is in centimeters.

Figure 2.6. Two major figures in paleoanthropology at Olduvai Gorge, Tanzania. Dr. John Desmond Clark (left) studied the Pliocene and Pleistocene archeology of Africa. Dr. Richard Hay (right) unraveled the stratigraphy of Olduvai Gorge. They stand at the edge of the gorge, by the Second Fault, North Slope.

later (Leakey 1959). Phillip Tobias later described the find in great detail, in a classic monograph (Tobias 1967). Beginning in 1960, the National Geographic Society lavishly funded research at Olduvai, and the Leakeys, who had been conducting small-scale research on the paleontology and archeology at Olduvai since 1931, never again lacked for funds. Mary Leakey carefully documented the spatial distributions of animal bones, hominid-modified bones, and artifacts on "living floors" that were thought to preserve evidence of the behavior of the earliest tool-using hominids (Leakey 1971). The OH 5 cranium was found on such a "living floor," now called the FLK Zinjanthropus site. The Oldowan Industry, representing the oldest known artifacts, takes its name from the early cultural material preserved at Olduvai Gorge.

When the OH 5 cranium was provided with a model of the absent mandible (e.g., Tobias 1967: Plates 41–42), the reconstructed skull had such strikingly large jaws and posterior teeth that newspapers of the time dubbed it "Nutcracker Man" – the posterior teeth and jaws appeared fashioned for crushing food like an old-fashioned nutcracker. The specimen appeared similar to the robust australopithecines previously unearthed by Broom and Robinson in South Africa, but it had an even more exaggerated masticatory system. As is true of other australopithecines, its brain size was pongid-like. At the same time, OH 5 was in association with Oldowan artifacts on a presumed "living floor." The long sought maker of the Oldowan artifacts had apparently been discovered. In addition, the age of the material was presumably very great, because it was found in Bed I at Olduvai in association with Pliocene–Pleistocene fossil fauna. Potassium/argon dating, a chronometric dating technique, ultimately established the age of the FLK Zinjanthropus site. Bed I was 1.8 my old (Leakey *et al.* 1961). It was previously thought that the earliest hominids appeared at 1 mya. Now the span of hominid evolution had been nearly doubled.

The idea that *Australopithecus boisei*, that most bestial australopithecine with a pongid-like brain size, was the maker of the Oldowan Industry ultimately proved troubling. Louis Leakey had long argued that tool behavior was a fundamental aspect of the hominid phenotype. He initially accepted that the association of the OH 5 fossil, modified animal bone, and Oldowan tools at Olduvai indicated that *Australopithecus boisei* was the maker of the Oldowan Industry (Leakey 1960). At the same time, the discovery that free-living common chimpanzees at Gombe, Tanzania, exhibited tool behavior when they extracted insects from termite mounds conferred an urgency to tool behavior (Goodall 1964). Nevertheless, the OH 5 specimen was undoubtedly an australopithecine, a fact that could not be erased by sorting it into the new subgenus "Zinjanthropus." When additional fossil material was discovered from Bed I and lower Bed II at Olduvai, L. S. B. Leakey, P. V. Tobias, and J. R. Napier created a new species, *Homo habilis* (Leakey *et al.* 1964). This species had a larger cranial capacity than an australopithecine, and its hand and foot bones resembled those of modern humans in several features. *Homo habilis* was the maker of the Oldowan Industry. It was the culture-bearing hominid at Olduvai Gorge, and other early archeological sites. Many researchers, such as John Robinson, who had long worked with australopithecine and other hominid fossils in South Africa, considered that the new Olduvai material was australopithecine-like, and did not represent a different taxon.

The existence of early genus *Homo*, contemporary with, and perhaps antecedent to, the australopithecines was finally established in the early 1970s. In July of 1972, a hominid cranium was discovered in the Koobi Fora region east of Lake Turkana in northern Kenya; its existence was publicly announced at a November symposium of the Zoological Society of London (Leakey 1973a, 1973b). This specimen (KNM-ER 1470) has a cranial capacity of 752 cm^3, substantially larger than that of an australopithecine. Controversy over the existence of early genus *Homo* subsided. Some researchers felt that the 1470 fossil indicated a very early divergence between *Homo* and the australopithecines, especially because first reports had the 1470 fossil dating to older than 2.6 mya. It was found below the KBS volcanic tuff, initially dated to 2.6 mya. As an example of this viewpoint, Solly Zuckerman, now Lord Zuckerman and Secretary of the Zoological Society, expressed his delight at the 1470 specimen in a discussion following the presentation of the material.[4] Re-dating of the KBS tuff to 1.8 mya and corroborating biostratigraphic dates at Koobi Fora now yield a much younger date for the 1470 specimen.

A decade after 1959, the International Omo Expedition organized by the University of Chicago paleoanthropologist F. Clark Howell discovered many hominid teeth and two lower jaws in southern Ethiopia (Howell 1969). The material was clearly older than the fossil finds from Olduvai, and may have been 3 to 4 my old. That is, the material was perhaps twice as old as the Zinjanthropus find. The span of human evolution was doubled again. Yet, the Omo fossils did not achieve the glamour of the Olduvai find. Why was this? Was the public jaded by fossil finds? This is not likely, given the continuing public attentiveness to human paleontology. Most of the finds consisted of isolated teeth. Nothing was as complete as the OH 5 cranium, and this may have been a factor in the relative public indifference. However, the major factor distinguishing the Olduvai and Omo material was the apparent absence of archeology in the Omo region. Even newspaper accounts singled this factor out at Omo (e.g., Snider 1969). No stone tools were present with the initial hominid finds, and the archeological discoveries at Olduvai, as well as the establishment of the taxon *Homo habilis* at Olduvai, led to the expectation that tool behavior was important in diagnosing fossil material as hominid-like. Hominids the Omo fossils might have been, *but they did not have tools.* Stone tools dating to 2.4–2.34 mya were eventually recovered from Members E and F of the Lower Omo Shungura Formation (Merrick & Merrick 1976), but excavated sites with dense accumulations of artifacts, animal bones, and human fossil remains, similar to the archeological sites at Olduvai Gorge, have never been located in the Lower Omo region.

The baboon renaissance

A long tradition of using living hunter-gatherers to theorize about the origins of human society began during the Enlightenment (Kuper 1988). Before the late 1950s/early 1960s, anthropological reconstructions of early hominid sociality routinely used living human hunter-gatherers to model the lifeways of fossil humans. A classic example is *Ancient Hunters and Their Modern Representatives* (Sollas 1924), in which the societies of fossil humans from Europe and Asia are reconstructed using ethnographic data from living hunter-gatherers. The ethnographic studies are used as direct analogs to reconstruct the behavior of fossil humans in vivid and fine detail. The book is better than any Neanderthal novel currently in print. And, from page to page, the discussion easily switches between living humans to long extinct fossil humans. Yet, after a general recognition of the hominid status of australopithecines was achieved in the early 1950s, it was clear that no living humans could be used to reconstruct the lifeways of these early hominids. No matter how rudimentary the technology possessed by any living human group, no one could argue that their behavior and ecology could be easily transferred to the lifeways of fossil hominids with a brain size only one-third that of living humans. Only variables that broadly determined niche structure could be examined. A very influential example of such an attempt from the early 1950s is a paper by Bartholomew & Birdsell (1953). An outline is presented here of variables determining australopithecine ecology. This outline is generated from general factors determining the distribution and abundance of other mammals. Yet, this is a very sere outline. Detailed specifics of life and behavior are missing. How could one supply these for the lifeways of fossil humans?

I use the term "baboon Renaissance" for this period of time, because it is a rebirth of the field-oriented primate behavioral studies begun by C. R. Carpenter and then neglected for a quarter century. And baboons were the major focus of research. How did this happen? It happened because S. L. Washburn, then at the University of Chicago, realized the inappropriateness of using living hunter-gatherers for reconstructing the lifeways of the australopithecines, and selected baboons as an alternative model for such reconstructions. As a student of E. A. Hooton, who was impressed by Carpenter's work, Washburn had reason to remember the primate field studies that Carpenter had conducted in the 1930s.

Furthermore, both Carpenter and Washburn had been members of the 1937 Asiatic Primate Expedition to Thailand. They had been compatriots in the field, and so Washburn had been able to observe Carpenter in action.

By the late 1950s, the disjunction between the pongid-like brain size of australopithecines – by now largely viewed as hominids – and modern human hunter-gatherers was becoming too jarring for an easy transfer of behavior from living humans back into the far distant past. Just at this time, Washburn influenced Irven DeVore, a University of Chicago graduate student, to switch his interests from human kinship to the study of wild baboon sociality. Washburn chose baboons as the analogs for early hominids for a variety of reasons. Baboons are found in Africa, and a major reason for their selection – a reason no longer accepted – was that Africa was thought at the time to have been virtually immune to Pleistocene environmental change. Hence, modern African ecosystems were thought to mirror, point-for-point, the ancient ecosystems in which the australopithecines lived. Modern African ecosystems were frozen in time, and, except for the absence of australopithecines and some other extinct species moving through the landscape, could be used to reconstruct both the Pleistocene environment and species interactions. It is now known that both Africa and other tropical areas were affected by Pleistocene climatic oscillations (Bishop & Clark 1967, Butzer 1971, de Menocal 1995). Additional points favoring the choice of baboons are that they are large primates that are mainly terrestrial. They approximate australopithecines in body size and degree of terrestriality. And baboons are omnivorous animals that sometimes engage in hunting behavior. Given their dentition, some australopithecines were apparently omnivores as well, and at least some researchers believed them capable of hunting. Five features thus highlighted the value of modeling early hominid sociality using baboons. In addition, there were practical reasons for their use: the large body size and terrestrial habits of baboons made them easy to observe, and baboons are common and widespread, not endangered species confined to a few localities.

It is abundantly clear that Washburn simply substituted non-human primates for living hunter-gatherers in a one-to-one fashion as he reconstructed early hominid sociality. A spate of papers using the baboon model began to appear in both popular (Washburn & DeVore 1961b) and scholarly formats. Scholarly papers using the baboon model include DeVore (1963), Washburn and DeVore (1961a) and DeVore and Washburn (1963). Washburn convened an international symposium on early hominid sociality that was subsidized by the Wenner-Gren Foundation for Anthropological Research, and edited the volume resulting from this symposium (Washburn 1961). DeVore (1965) later edited a volume of primate field studies that included papers by such luminaries as Carpenter, Goodall, Hall, Jay, Lancaster, Mason, Petter, Reynolds, Schaller, and

Southwick. Much of Carpenter's early work was now re-published for a more appreciative audience (Carpenter 1965). By 1965, it was taken as a matter of course that non-human primate behavioral ecology would illuminate aspects of early hominid sociality.

What about the status of current hunter-gatherer studies? Have they ceased to be applied to models of human origins? The case can be made (Roebroeks 1995, Stoczkowski 2002) that current hunter-gatherer studies and paleoanthropology are still recycling themes that were omnipresent in the nineteenth and early twentieth centuries. For example, Bowler (1995) argues that, because living hunting and gathering peoples had already been exterminated or marginalized by the late nineteenth and early twentieth centuries, this leads to paleoanthropological reconstructions in which true humans encounter and inevitably exterminate non-modern fossil humans. Stoczkowski (2002) traces some current ideas about human origins back to Greco-Roman antiquity, and analyzes a century and a half of scientific ideas about human origins. It appears to make no practical difference whether living humans or non-human primates are used in generating models of hominid origins. This may be because, when non-human primates are used in the models, they are seamlessly inserted into arguments where humans once appeared. In subsequent discussions about intelligence (Chapter 8), paleoanthropology (Chapters 15–16), and early hominid sociality (Chapter 17), I will adopt a perspective that is not exclusive to primates, but one that incorporates data from other mammal groups.

During this period, the work of Japanese primatologists became known in the West. After the end of the Second World War and the ensuing ruin of the Japanese economy, the Japanese focused their research on the endemic macaque species (*Macaca fuscata*), rather than studying species in distant lands. In 1959, they established a major primate research group at Kyoto University, and published their own international journal, *Primates*. Long-term, collective observation of single troops, analysis of troop structure across generations, the relationship between rank and kinship, and routine provisioning of troops became staples of Japanese primatology. At an early stage (e.g., Kawai 1965), the Japanese noted the appearance and transmission of novel cultural traditions, such as sweet-potato washing and stone collecting, among their study animals. The high intelligence of the study animals and their complex sociality was emphasized from the beginning, probably because the typically Western dichotomy between human and animal never existed in Japan (Asquith 1995). Unlike Westerners, who rejected empathy with their subjects as being anthropomorphic and non-objective, Japanese researchers adopted a particular "sympathetic method" (*kyokon*) in order to understand non-human primate societies. The Japanese subsequently established permanent sites in many other parts of the world, in order to conduct long-term research on other species – for example, in the Mahale Mountains of Tanzania to study chimpanzees, and in the rainforests of the Congo to study bonobos.

Sociobiology and behavioral ecology

During the last quarter of the twentieth century, behavioral primatology was extensively re-vamped by the inclusion of sociobiological theory and ecology. Sociobiology is the study of how natural selection and other evolutionary processes can create sociality. Fundamental early work in sociobiology included the ideas of inclusive fitness (Hamilton 1964) and parental investment (Trivers 1972). Early theoretical work on sociobiology was synthesized in a classic work by E. O. Wilson (1975) that also contained separate detailed sections on sociality in certain invertebrate and vertebrate species, including primates. Primate infanticide was one of the first topics explored from a sociobiological perspective. Infanticide had been observed in wild Hanuman langurs since the first field studies of the 1960s. Sarah Blaffer Hrdy (1977) argued that infanticide occurred in these langurs because adult males were pursuing a reproductive strategy in which they first kill infants sired by other males and then sire a new generation. Infant killing was thus an evolved male reproductive strategy, and not a social pathology caused by overcrowding and aberrant behavior. This perspective initially generated a great deal of controversy among behavioral primatologists, but has since resulted in a substantial shift in theory among animal behaviorists (Hausfater & Hrdy 1984). As of the beginning of the twenty-first century, many aspects of behavioral primatology routinely incorporate sociobiological theory. These include topics such as sex-ratio analysis, mating systems, parental investment, altruism, friendship, and theories of social evolution.

Ecology and sociobiological theory now affect behavioral primatology in fundamental ways. However, it is important to note that there was a significant lag in the collection of ecological data. Many of the first field studies of primate behavior published in the 1960s were oriented toward the study of social behavior. They were studies of primate "sociology," in which social behavior was the sole concern, and contained very little ecological data. Even dietary information was cursory, if present at all, and this virtual absence of ecology was a serious flaw in most of the early studies. An example was the lack of detailed nutritional composition about common primate food items, such as green leaves. It might appear that leaf-eating, folivorous animals should only rarely experience food stress, because, to a human, all the world appears green in a tropical forest. However, although mature green leaves in a tropical forest are abundant, they are low-quality food items. They are very low in protein, high in indigestible fiber, and rich in plant secondary compounds that limit the bioavailability of nutrients, or may actually be toxic, such as strychnine (Glander 1982). Leaf-eating primates therefore selectively search out and consume young leaves that are richer in protein than their mature counterparts. Without knowing about the nutritional composition of leaves, the finicky eating of folivorous primates is inexplicable. The biomass of leaf-eating primates may be determined by the protein to fiber ratio in leaves that they frequently consume. In addition, several physiological and anatomical specializations may result from consumption of mature leaves, such as a lower metabolic rate, lower activity levels, and stomach or gut specializations that allow colonization by symbiotic bacteria that digest cellulose, or break down plant secondary compounds that either inhibit nutrient bioavailability or are actually toxic.

Some publications emphasizing primate ecology began to appear by the late 1970s, generally produced by individuals who had a broad training in mammalogy or zoology (e.g., Clutton-Brock 1977). Alison Richard was one of the first researchers to note the lack of detailed ecology in most primate field studies through the 1970s. Her textbook (Richard 1985) was one of the first to rectify the lack of ecology. Since then, the welding of behavior and ecology has become so absolute that equivalent textbooks (e.g., Strier 2002) take as a given the fact that studying behavior in the wild is impossible without ecological data. The impact of ecology and sociobiological theory on behavioral primatology can be assessed by examining the range of topics covered in two multi-authored synthetic works that appeared over the course of nearly 20 years (Smuts *et al.* 1986, Kappeler & Pereira 2003). In these works, there is never any question that detailed ecological research and evolutionary theory must be used to study the behavior of non-human primates.

Endnotes

1. The U.S. National Library of Medicine, National Institutes of Health, Rockland, Maryland, has mounted an exhibition entitled Dream Anatomy, which includes reproductions of the work of Vesalius and his successors. The Website for this exhibition is planned as a permanent addition to the National Library of Medicine. URL: http://www/nlm.nih.gov/exhibition/dreamanatomy.index.html.
2. The full title of the monograph shows some of the linguistic confusion – *Orang-Outang, sive Homo Sylvestris: or, The anatomy of a pygmie compared with that of a monkey, an ape and a man. To which is added, a philological essay concerning the pygmies, the cynocephali, the satyrs, and sphinges of the ancients. Wherein it will appear that they are all either apes or monkeys, and not men, as formerly pretended.* The ape is a Barbary macaque. Tyson argued that creatures of the type he dissected were the origin of the old Greek myths about pygmies in Africa.

3. This occurs even in a 1969 review of Hans Kummer's monograph on wild hamadryas baboons. "The fact that Dr. Kummer's new observations conform in significant detail to what I published in 1932 is from my point of view welcome. My own book, which dealt with the problem of sub-human primate social life generally, and which so far as family and group behaviour, breeding activity, and territorial habits of the baboon were concerned was based on observations in the wild as well as in captivity, is at present out of print, and a certain number of reports have appeared which have questioned the basic observation that the social life of most of Old World sub-human primates is organized round a unit dominated by a male – who secures for himself as many females as his position in the social hierarchy permits." (Zuckerman 1981: 451).

4. "There have always been a few anatomists, I amongst them, who have refused to be bulldozed by a single fossil, or by attempted reconstructions. The evidence has always pointed to one conclusion that no-one could deny, the independent differentiation of a human line well into what used to be called the Pliocene/Pleistocene horizon. What has been revealed to us today is that these new creatures were there first. And as I said before, and let me repeat, had today's discovery been reported in this Society when the *Australopithecus* skull [the Taung specimen] first rested on the speaker's bench in our old meeting room, any amount of time would have been saved." (Leakey 1973a: 69).

3 *The catarrhine fossil record*

The geological time scale

Through the years, many study groups have been commissioned to create an internationally recognized reference system for geological time. It is obviously necessary for researchers to have a universally recognized time scale for events in earth history. This time scale is updated on a regular basis, in order to reflect advances in dating and stratigraphic analysis. The goal is to create a global reference system that is formally defined and universally recognized. The International Commission on Stratigraphy establishes guidelines for the creation and definition of universally recognized units of geological time; the International Union of Geological Sciences is responsible for ratifying these units. Units of the Phanerozoic (542 mya to the present) are defined by a Global Standard Section and Point (GSSP) at the base of the unit. Many units also have a "golden spike" and stage name plaque mounted at the boundary limit in the formal stratotype section. The latest geological time scale is that of Gradstein *et al.* (2004). Figure 3.1 illustrates that portion of the geological time scale that is relevant to primate evolution.

Major features of primate evolution

Figure 3.2 illustrates major events in primate and human evolution with reference to the global time scale. These events are set against a background of global climate change and major tectonic events. However, Figure 3.2 does not imply that climate and tectonic change alone can account for major features in primate evolution. Later discussions in this book query whether changes in the physical environment alone are sufficient to account for evolutionary change. Biological interactions, such as interspecific competition or sexual selection, also contribute to major events in mammal evolution. This implies that biological evolution would take place even if the physical environment was somehow held static and unchanging.

The Geological Time Scale	
Cenozoic	**Age (mya)**
Holocene	0.0115
Pleistocene	1.8
Pliocene	5.33
Miocene	23.03
Oligocene	33.9
Eocene	55.8
Paleocene	65.5
Mesozoic	**Age (mya)**
Cretaceous	145.5

Figure 3.1. Geological time and primate evolution.

The shape and pattern of primate evolution

The general shape of the fossil record of any group can be discerned from diversification curves that illustrate the number of genera or species that exist at any point. Convention orients these curves as if they were presented in a geological section: the earliest time range is at the bottom, and successively later time ranges are arrayed above it. The most recent time range is at the top. The different pace of origination and extinction through time usually creates periods of taxon abundance or rarity. Occasionally, groups show other patterns, such as persistent rarity through time, or explosive initial diversification (radiation) that continues through time.

Martin (1986b, 1990) has published two primate diversification curves. Martin and his colleagues later published a similar curve that estimated the date of the last common ancestor of living primates (Tavaré *et al.* 2002). A new statistical method presented in this paper hints that primates diverged from other placental mammal orders during the late Cretaceous at 81.5 mya, which is close to the date inferred from molecular data. The completeness of the fossil record was estimated from the proportion of species actually preserved in the record relative to a model of the primate diversification curve (Tavaré *et al.* 2002: Fig. 1). Yet, there are problems with these diversification curves.

Major Features of Primate Evolution

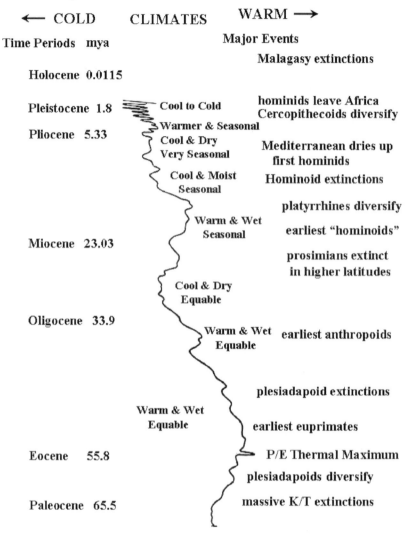

← COLD   CLIMATES   WARM →

Time Periods   mya   Major Events

                            Malagasy extinctions

Holocene 0.0115

Pleistocene 1.8   Cool to Cold   hominids leave Africa
                                          Cercopithecoids diversify
                    Warmer & Seasonal
Pliocene 5.33   Cool & Dry
                    Very Seasonal   Mediterranean dries up
                                          first hominids
                    Cool & Moist   Hominoid extinctions
                    Seasonal
                                          platyrrhines diversify

                    Warm & Wet   earliest "hominoids"
                    Seasonal
Miocene 23.03
                                          prosimians extinct
                                          in higher latitudes
                    Cool & Dry
                    Equable
Oligocene 33.9
                    Warm & Wet   earliest anthropoids
                    Equable

                                          plesiadapoid extinctions
            Warm & Wet
            Equable   earliest euprimates

Eocene 55.8   P/E Thermal Maximum

                                   plesiadapoids diversify

Paleocene 65.5   massive K/T extinctions

Cretaceous 146-65.5

Figure 3.2. Major primate radiations and extinctions and Cenozoic climatic
fluctuations. Modified from Partridge *et al.* (1995).

All of these curves show increasing diversity through time, in an expanding upside-down cone of diversity. This is achieved in two ways. First, one group of organisms, the plesiadapoids, is removed from the primate order, thus eliminating an early Eocene extinction event. Plesiadapoids do not resemble modern primates, and they are sometimes considered non-primates (Chapter 1). Nevertheless, they are always treated as close primate relatives (archontans or "primatomorphs"). Removing them entirely from consideration eliminates a large radiation and extinction event. Second, the reality of major primate extinction events that occurred at the end of the Eocene and Miocene is denied. The primate extinctions that took place at the end of the Eocene were noticed even in the early twentieth century – they formed part of the massive late Eocene extinction (*la Grande Coupure*) that the paleontologist Hans Stehlin first documented among fossil mammals of the Paris Basin. The reality and magnitude of this extinction continues to be supported by fossil evidence (Prothero & Berggren 1992). The end of the Miocene also saw the diminishment and near extinction of the hominoid primate radiation.

If one examines the problem of primate origins using molecular phylogenies, the largest known molecular data set, improved fossil dating for calibration, and a subset of small-bodied mammals resembling Cretaceous/Tertiary mammals in size, yields a date of between 69–80 mya (Springer *et al.* 2003). Certainly, it is unlikely that the earliest primates are preserved in the fossil record. However, if one is unwilling to admit plesiadapiforms into the order, the chasm of missing fossil evidence gapes even larger: it lasts until the appearance of euprimates.

There are alternative ways of viewing the fossil record. Diversity can be considered in terms of numbers of species, or as disparity in morphology over geological time. If morphological disparity or dissimilarity in anatomical design is considered, then the traditional interpretation of animal evolution is also portrayed as an upside-down cone, with disparity increasing from the past into the present (Conway Morris 1998). The expectation is the same whether species number or anatomical divergence is studied – that of increasing richness or complexity with time. Yet, when species diversity of large mammalian herbivores and carnivores is tracked through most of the Cenozoic, diversity does not seem to increase (Van Valkenburgh & Janis 1993). Stability is also discovered when rodent, artiodactyl, and carnivore genera are followed through most of the Cenozoic (Gingerich 1984). A major assessment of marine invertebrate diversity through the Phanerozoic, using the gigantic Paleobiology Database,[1] demonstrates that biodiversity has not increased through time (Schiermeier 2003).

Analyses of modern community structure and models of species diversity often explicitly assume an equilibrium condition, and the likelihood that per species extinction rates are independent of community diversity (e.g., Walker &

Valentine 1984: 890). Up until the 1980s, the governing assumption in the study of biological communities was that these communities have stable equilibria and exist close to the equilibrium point. Since then, however, much research has been oriented toward the realization that many natural populations have cyclical or chaotic dynamics (Godfray & Blythe 1990). One researcher even argues that biodiversity varies in a random fashion (Hubbell 2001). Furthermore, analysis of the mammalian fossil record suggests that evolutionary stasis or equilibrium does not exist, because selection pressures within a community are continually altering (Van Valen 1973a). Community diversity therefore does probably exercise an influence on the likelihood of species extinction, contrary to the assumption of stable equilibrium models.

The plastic species composition of modern communities, including the occurrence of empty niches, is often not recognized. These factors make the detailed reconstruction of paleocommunities difficult. A review of the literature on introduced species reveals that, in 79% of recorded cases, introduced species have no effect on resident species or community structure. Extinction followed in only 8% of the introductions, and extinction from competition could be inferred in only 0.4% of the cases (Simberloff 1981). This implies the existence of empty niches in modern communities. An analysis of fossil invertebrate taxa yields an average estimate of 54% empty niches, depending on the taxon (Walker & Valentine 1984). The widespread occurrence of empty niches in modern and fossil communities indicates that communities have a redundancy or resilience that is often not recognized. However, this implies that a community need not be an exclusive association of species that are always tightly co-evolving. When marine foraminifera are examined through a 55 my period, they sometimes show clear evidence for paleocommunities that have no modern analogs (Buzas & Culver 1994). Repeated alterations are found in these paleocommunities. Species composition appears malleable, with species responding on an individual basis to continual environmental fluctuations. This implies that there is no unified local community. Species associations are ephemeral (Buzas & Culver 1994). A corollary to these observations of plasticity in modern and fossil communities is that one should expect that paleocommunities will differ in detail from modern communities. Correspondence might be found only at the broadest level of analysis.

One cannot assume an equilibrium diversity state for fossil communities. In modern communities, the continual discovery of new species also calls into question the idea of an equilibrium state. The analysis of standing diversity at a paleocommunity level ignores the effect of time-averaging in the fossil record. Species diversity can be artificially inflated by taphonomic time-averaging, which collapses species from different environments and communities into one fossil assemblage within a geological stratum, whose period of formation can be very long. When ancient communities are known in detail, they exhibit the flux and dynamism seen in the modern world by field biologists. Two examples of these ancient communities come from North America, and typify the extremes of species origination and extinction. In the earliest Paleocene, herbivores speciate at an incredible rate (Van Valen 1978); and, in the latest Miocene (Hemphillian), an overwhelming extinction of large herbivores occurred (Webb *et al.* 1995).

Where are the first primates found? Tavaré *et al.* (2002) argue that primates originated on the southern continents. I am skeptical about this. Most of the rich land mammal localities of the Paleocene and Eocene are located in northern latitudes, and a southern origin is not implied for other orders of mammals. The proteutherian insectivores are the most probable ancestral group for primates (Van Valen 1994). The classification of plesiadapiform primates has constantly been debated. Yet, if plesiadapiform species are accepted as primates, then primates occur in the early Paleocene of North America. They are a common component of both North American and European Paleocene faunas – in fact, they occur in the same abundance within fossil localities as modern rodents do. They appear to have been occupying niches now occupied by modern rodents. That is, they were small, generalized mammalian herbivores. And plesiadapiform primates evolve rapidly in comparison to the evolutionary rates evinced in other orders. If plesiadapiform species originated in the Cretaceous, they even managed to survive the Cretaceous/Tertiary mass extinction. In any case, a new species (*Pandemonium dis*) links the early Paleocene *Purgatorius* with the basal plesiadapids (VanValen 1994).

The plesiadapiform primates and later euprimates also survived an immense global warming event at the Paleocene/Eocene boundary. A terrific upward spike in global temperature occurred at 55 mya. This is called the Paleocene/Eocene thermal maximum (PETM), and it is the most dramatic climatic fluctuation of the entire Cenozoic (Wing *et al.* 2003). Global temperature rose by 7–8 °C within several thousand years. A record of this global warming is preserved in sediments that show a worldwide geochemical anomaly – the carbon isotope excursion (CIE). The CIE was caused by the colossal release of oceanic depots of methane and CO_2 directly into the atmosphere. This resulted in intense global warming, because methane is a powerful greenhouse gas. A true greenhouse world was formed. The surface temperature of the North Sea averaged about 28 °C (Beard 2002), and the sea surface temperature rose by 5–8 °C to average 25 °C even on the Maude Rise at 65° S near Antarctica (Schmidt & Shindell 2003). Ocean bottom water temperature rose by 4–8 °C. The cause of the gigantic methane release is uncertain. Methane is formed as bacteria decompose organic material incorporated in marine sediments. One suggestion is that seafloor landslides released methane hydrates, or methane trapped in frozen water, in areas on the continental slope (Katz *et al.* 1999). Other suggestions are that a comet impact or general changes in ocean circulation were responsible for the methane release.

The most parsimonious approach to the primate fossil record is to accept the pattern of successive radiations and concomitant extinctions that paleontology documents. This pattern and shape to the primate fossil record needs no special pleading, unlike the approach advocated by some researchers, who are unwilling to admit the existence of major primate extinction events (Martin 1986b, 1990, Tavaré *et al.* 2002).

The early catarrhine primates

Catarrhine primates are generally considered to be descendants of omomyid prosimians (Family Omomyidae), an extinct family of tarsiiform prosimians (Figure 3.3). However, there are no fossils that document the transition from omomyid to anthropoid catarrhine. Omomyids appear in the early Eocene, and one species from China, *Teilhardina asiatica*, is currently the oldest known euprimate, dating to 54.97 mya (Ni *et al.* 2004). Its orbital diameter is small when regressed against skull length – in fact, *T. asiatica* falls on the extrapolated regression line for living and fossil anthropoids (Ni *et al.* 2004: Fig. 4). The implication is that this taxon was diurnal, rather than nocturnal. Because omomyids are tarsiiform primates, the fossil record of tarsiers is of interest to the study of early catarrhines. The fossil record of the tarsiers (Family Tarsiidae) is very long in Asia. The genus *Tarsius* itself has persisted for a remarkably long time. *Tarsius* may well be the longest-lived placental mammalian genus, extending from *Tarsius eocaenus* (45 mya), to *Xanthorhysis tabrumi* (40 mya), to *Tarsius thailandicus* (16 mya). This antiquity coincides with profound cranial and postcranial morphological specializations. Each hypertrophied tarsier eye is equal in size to the brain proper. These eyes are extraordinary adaptations to nocturnal visual predation, but they evolve because the tarsier has no tapetum lucidum, the shining tissue surface behind the retina that reflects the few stray photons encountered during the night back into the retina, thus multiplying dim light. Hence, tarsiers presumably evolved from diurnal ancestors that had lost the tapetum. The possible diurnal status of *Teilhardina asiatica* thus becomes even more interesting. Tarsiers are one of the few placental mammals occurring on both sides of Wallace's Line, and probably achieve this distinction by vicariance, because they are such an ancient component of Asian faunas (Beard 1998a).

Primates, including two relatively large-bodied genera and two small-bodied genera, are found in the late middle Eocene Pondaung Formation of Myanmar

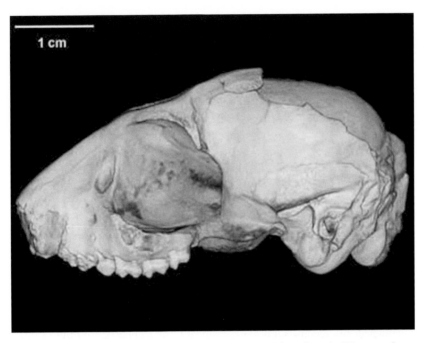

Figure 3.3. *Rooneyia viejaensis*, an extinct omomyid from the early Oligocene of West Texas, illustrating the near anthropoid status of some omomyids. Although the posterior neurocranium is missing, a natural endocast is present. *Rooneyia* was one of the largest omomyids. Image from the Digimorph NSF digital library: http://www.digimorph.org/index.phtml/.

(Burma), along with many other fossil mammals. Since the early twentieth century, the large-bodied genera *Pondaungia* (8 kg) and *Amphipithecus* (6 kg) have been discussed as possible candidates for first anthropoids, based on dental characteristics and body size. A recently discovered talus, presumably from *Amphipithecus*, based on its size, displays derived anthropoid traits (Marivaux *et al.* 2003). Beard (2004) argues that the Chinese genus *Eosimias* is the oldest anthropoid. Gebo *et al.* (2000) discuss two primate fossil calcanei from the middle Eocene of China that may represent the smallest known primates. Dating to 45 mya, these calcanei are less than half the size of the calcaneus of *Microcebus*, the smallest living prosimian genus. Given the estimated body size (within the range of shrews), I believe that this postcranial material is too small to be anthropoid. Additional small-bodied primate fossils continue to be found in China. Using this material, Gebo (2004) argues that the first primates were, in fact, the size of shrews, and he discusses the effect this would have on a wide range of variables.

The earliest catarrhine primates – undoubted anthropoids – are found in the late Eocene of North Africa and Oman. The first catarrhines are an extremely diverse group, and their phylogenetic relationships are obscure (Harrison 1987). The known taxa almost certainly have no ancestral/descendant relationships with living catarrhines. These primates are diverse, and have been sorted into four families: the Oligopithecidae, the Proteopithecidae, the Propliopithecidae, and the Parapithecidae. The parapithecids diverge from the other early catarrhines, and they are often placed in their own Superfamily, the Parapithecoidea. The parapithecids were the most abundant and diverse of these early catarrhine primates, and are distinguished from most other catarrhines by having three premolars in each half of the upper and lower dental arcades (found also in the proteopithecids), as well as unique molar crown patterns. Some species (*Parapithecus* spp.) lack permanent lower incisors, or completely lack lower incisors altogether; other species (*Apidium* spp.) have leaping specializations, indicated by a fibula that is partly fused with the tibia.

The overwhelming majority of these primates occur in a series of quarry sites within the Fayum Depression of Egypt. The Fayum lies in the Western Desert, about 100 km southwest of Cairo. During the Eocene and Oligocene, the Fayum was a prograding delta formed by a large river (the Ur-Nile) that debouched into the ancestral Mediterranean. The paleoecology of the Fayum sites has been controversial. Based largely on extensive evaporite deposition and sedimentological evidence of cyclical flooding events, I (Cachel 1975b) argued that, although gallery forest surrounded the Ur-Nile, some open-country habitats must have existed in the distal floodplain areas. Because grasslands had not yet evolved, any open-country areas would have possessed some combination of herbaceous or bushland vegetation, and could not technically be termed savannahs. The Oligocene certainly was an epoch of significant global aridity. A number of lines of evidence indicate this, including aeolian sediments in deep-sea cores and a plummeting of the Vail global sea-level curve during the Oligocene. Kortlandt (1980) argued that no tropical forest existed at the Fayum, although Bown *et al.* (1982) replied that tropical forest existed at some of the Fayum sites. Gagnon (1997) examined four successive Fayum assemblages, and concluded that they showed similarities to both modern forest and woodland-bushland habitats. Only the latest assemblage had abundant and diverse anthropoid primates, and Gagnon believes that the early Fayum assemblages had no analogs in modern communities.

Hominoid systematics

Beginning in the 1960s, immunological studies and comparative protein analysis led Goodman and colleagues to stress the close genetic relationship between humans and the great apes, especially the African great apes (Goodman 1963, Goodman & Moore 1971, Goodman 1973). This conclusion has since been confirmed by mtDNA and nuclear DNA analysis. Goodman and his colleagues believe that the close genetic and phyletic relationship between living humans and the African great apes should be reflected in taxonomy – humans, the two chimpanzee species, and perhaps the gorilla, should be classified together in a single genus, genus *Homo* (Goodman & Moore 1971, Schwartz *et al.* 2001). Recent analysis of functional or coding DNA in humans, the African great apes, orangutans, Old World monkeys, and the mouse has resulted in a further taxonomic revision. Goodman and his colleagues now recommend that all of the living great apes be sorted with humans into the Family Hominidae, with the two chimpanzee species included in genus *Homo*, separated at the subgenus level: *Homo (Pan) troglodytes* and *Homo (Pan) paniscus* (Wildman *et al.* 2003). It is ironic that the molecular anthropologists have been expanding genus *Homo* by subsuming living pongids into the genus, while, at the same time, paleoanthropologists have been diminishing genus *Homo* by sorting early members into genus *Australopithecus*. Furthermore, many paleoanthropologists believe that Neanderthals are a distinct species from modern humans (*Homo neanderthalensis*). Leaving aside the chimpanzee subgenus distinction recommended by Goodman and his colleagues (Wildman *et al.* 2003), is it truly the case that Neanderthals are no more similar to modern humans than chimpanzees are? That is, that Neanderthals and chimpanzees are equivalent in their degree of difference from modern humans?

Rates of protein evolution are also used to infer the time of divergence between hominids and the African great ape stock. Assuming a constant rate of mutation and neutral selection (i.e., no natural selection) since divergence, allows researchers to infer a date using a "molecular clock" that is calibrated using known divergence times from the paleontological record. Since the 1960s, some researchers have advocated a date of divergence as late as 5–6 mya for the last common ancestor of humans and chimpanzees (Sarich & Wilson 1967, Wildman *et al.* 2003). The discovery of hominids at 6 mya (Senut *et al.* 2001) and a possible hominid at 7 mya (Brunet *et al.* 2002) obviously pushes back the time of divergence inferred from molecular or DNA studies.

The Miocene hominoid radiation

Living hominoids occur principally as relict populations of threatened or endangered species. Ancient hominoids were once much more speciose and widely dispersed. An evolutionary radiation of hominoids occurred during the Miocene. Abundant hominoid remains are found in the Miocene of Spain,

Hungary, Paşalar in Turkey, India/Pakistan, the Lufeng Basin in China, and Kenya. The Middle Miocene Paşalar site appears to contain fossils from a very limited time range, because the sediments were deposited in an unusually rapid fashion, perhaps in a single flood. The fossils also appear to derive from a restricted geographic area of about 25 km^2, and from a limited altitudinal range. They thus approach material collected from a modern community (Martin & Andrews 1993). Two hominoid species occur at Paşalar: *Griphopithecus alpani* and another, larger species that is more rare.

Primate species from the Miocene of East Africa have been extensively studied for the last 60 years, from the 1930s until the present (e.g., Hopwood 1933, Andrews 1978, Teaford *et al.* 1993). These species are known from a series of East African sites that fall within the 20–12 mya time range. It was once thought that the African Miocene hominoids might have had some relationship to the living African pongids. Pilbeam (1969), for example, argued that *Proconsul major* was the ancestor of the gorilla, while *Proconsul africanus* was the ancestor of the common chimpanzee. If this were true, these lineages would have differentiated by 20 mya. However, it is now becoming clear that nearly all early and middle Miocene hominoids are very different postcranially from living hominoids. The sole exception is *Morotopithecus bishopi*, whose postcranial remains, older than 20.6 mya, demonstrate that this is the first taxon to possess basic aspects of the modern hominoid body plan (MacLatchy *et al.* 2000). Postcranial differences effectively remove virtually all of the Miocene hominoids from consideration as ancestors of living species. This should eliminate temptation to search for ancestor-descendant relationships within this suite of taxa.

In fact, the Family Proconsulidae and other early and middle Miocene hominoids are generalized catarrhines in terms of their postcranial anatomy. The only postcranial feature of these early hominoids that appears typically hominoid-like is the absence of a tail in some well-known taxa, such as members of genus *Proconsul*. What is the explanation for tail loss in these and later hominoids? A reasonable explanation is the advent of forms of arboreal locomotion (leaping, suspensory arm-swinging locomotion) that mandate agility, although they are appearing in medium to large-bodied primates. A long tail can account for a good part of a primate's body weight. In a howler monkey species (*Alouatta caraya*), for example, a prehensile tail accounts for over 6% of total body weight (Grand 1978). Because this tail is prehensile and therefore heavily muscled, it is equivalent to the weight of an entire limb. A normal tail would not bulk so large, but it could still represent a substantial weight. One might expect tail reduction to occur if there is selection for a reduction in body mass in arboreal animals, and if the tail performs no counterbalancing activities during arboreal locomotion. Furthermore, when typical hominoid forelimb elongation and thorax modification begin to take place, indicating use of the forelimb in locomotion, a long tail may hinder locomotion by shifting the center of gravity in a posterior direction.

Researchers investigating the postcranial remains of *Dendropithecus, Kalepithecus, Limnopithecus, Micropithecus*, and *Simiolus* conclude that the morphology and inferred locomotor behavior of these small-bodied Miocene taxa have no match among living hominoids, cercopithecoids, or fossil propliopithecids or pliopithecids. These taxa are highly arboreal "scramblers." They are, in fact, best considered "primitive catarrhines," rather than hominoids (Rose *et al.* 1992). Harrison (1987, 1993) contends that all of the East African Miocene taxa (including the larger-bodied genera *Proconsul, Rangwapithecus, Afropithecus*, and *Turkanapithecus*) have no modern analogs in terms of niche, and they have no descendant species (phyletic derivatives) in the modern world. He believes that these Miocene species represent a primitive catarrhine grade from which the ancestors of modern hominoids and cercopithecoids were later derived, and he advocates the use of a Superfamily Proconsuloidea as a wastebasket category for Miocene taxa with no affinities to modern species. Some researchers believe that some derived postcranial features may link genus *Proconsul* with living hominoids (Rose *et al.* 1992). If this is the case, then niche differentiation and reduction in competition among early and middle Miocene hominoid African taxa may partly be driven by differences in locomotion.

Andrews and his colleagues argue (Andrews *et al.* 1979, Andrews 1992) that the floras and mammal faunas from Miocene hominoid sites (with some exceptions at Rusinga Island) are similar to modern rainforest biotas. Rainforest communities are inferred from the mammal faunas. Using terrestrial gastropod assemblages, Pickford (1983: Tables 5–6) concludes that some of the East African Miocene taxa preferred three different forested environments. Pickford distinguishes dry floodplain forest (*D. macinnesi, P. africanus, P. nyanzae*), drier upland forest (*R. gordoni*), and upland rainforest (*L. legetet, M. clarki*). One taxon occurs in both wet and dry upland forest (*P. major*). Yet, Miocene paleosols from these sites indicate no rainforest, along with complex vegetational mosaics of dry forest, grassy woodlands, wooded grasslands, and drier climatic conditions (Retallack 1991: Table 4.4). Paleosol evidence appears to support the more widespread occurrence of hominoid taxa, with less distinct habitat preferences than indicated by the gastropod evidence (Retallack 1991: 234). Palaeosols also demonstrate that a major shift toward drier vegetation and climate probably occurred between the early and middle Miocene.

Given the lack of agreement about even the existence of rainforest habitats and biotas, it is clear that Miocene community analysis has so far tended to occur at the broadest level. An additional complicating factor is that East African communities of the present reflect trends toward habitat fragmentation driven by major tectonic events through the Miocene, and the development of regional faunas may also have been accentuated by the occurrence of Pleistocene refugia.

Proconsul and *Kenyapithecus*, genera once routinely discussed in terms of hominid evolution, actually possessed a postcranial anatomy very different from the postcranial anatomy linking all of the extant hominoids. In contrast, European *Dryopithecus*, as demonstrated from postcranial remains of an individual specimen of *D. laietanus* (CL1-18800) at the late Miocene Spanish site of Can Llobateres, did have a typically modern hominoid postcranium. The species had a body size of 15–45 kg. It had short lumbar vertebrae positioned within the thorax. Its trunk was therefore held in an orthograde position. Arboreal suspension and locomotion in *Dryopithecus laietanus* is confirmed by the discovery of a hand with long metacarpals and phalanges, and a long forelimb with a straight humeral shaft (Moyà-Solà & Köhler 1996). The remains of this skeleton were widely dispersed, and Begun (2002) has some doubts about the association of the cranium with the postcranial material. Regardless of which taxon this postcranial material is eventually attributed to, the anatomy implies that the terrestrial knuckle-walking locomotion of living chimpanzees and gorillas is a specialization that evolved from an ancestral anatomy like that of *Dryopithecus*. The anatomy of *Dryopithecus* also implies that hominid bipedalism evolves directly either from arboreal suspension and arm-swinging locomotion or from vertical climbing, rather than evolving from terrestrial knuckle-walking (Chapter 13). Given the *Dryopithecus laietanus* postcranium, Begun (2003) argues that European members of this taxon must have dispersed back into Africa after evolving modern hominoid anatomy. These migrant European hominoids then become the ancestral stock of both hominids and the African great apes.

However, this complicated scenario of dispersion (out of Africa into Europe and back again) may not be necessary. Another Spanish species, *Pierolapithecus catalaunicus*, has been discovered still earlier in time (Moyà-Solà *et al.* 2004). An associated cranium, dentition, and postcranial material date to 13–12.5 mya, and demonstrate that *Dryopithecus laietanus* was not unique in evolving more modern hominoid postcranial anatomy. This might imply that contemporary African hominoids were also independently evolving such anatomy, although the African fossil hominoid record is a mystery at this period. The thorax of *Pierolapithecus* has a derived hominoid anatomy, but its hand has a mixture of primitive and derived traits. The hand is short, as are the fingers, and the fingers are relatively straight. However, the ulna has retreated from contact with the carpus, which is a derived hominoid feature (Moyà-Solà *et al.* 2004). These traits may indicate the relative sequence of locomotor behaviors that emerge during hominoid evolution. Furthermore, these traits imply that specializations for an orthograde trunk develop early in hominoid evolution, and suspensory arm posture and locomotion develop late, and probably independently in several lineages. The sequence of locomotor behaviors has implications for the origins of bipedal locomotion (Chapter 13).

Searching for the ancestors of living hominoids

The search for ancestors of living hominoids has occurred throughout the twentieth century, and continues today. The postcranial remains of *Morotopithecus bishopi*, older than 20.6 mya, demonstrate that this is the first taxon to possess some basic aspects of the modern hominoid body plan, particularly thorax shape (MacLatchy *et al.* 2000).

It is clear that the hylobatids diverge first from the rest of the hominoids. No fossil species has yet been identified as a hylobatid ancestor. The extinct Family Pliopithecidae, often the focus of paleontological attention, is currently eliminated from consideration as a potential ancestral hylobatid group (Tyler 1991). Moyà-Solà *et al.* (2004) believe that the anatomy of the Spanish genus *Pierolapithecus*, as well as molecular evidence, indicate that hylobatids diverged from the other hominoids before 14.5 mya. However, a likely ancestor for the orangutan has been identified in the late Miocene of Thailand, where the orangutan lived during the Pleistocene. This 9–7 my old ancestor is a new genus and species, *Khoratpithecus piriyai*, related to the Miocene Asian genus *Lufengpithecus* (Chaimanee *et al.* 2004). There are strong dental similarities to the living orangutan, including relative enamel thickness, enamel crenulation, and degree of sexual dimorphism. The absence of an anterior digastric muscle is a particularly telling derived characteristic shared with the modern orangutan. Fossil tree trunks of palms and dipterocarps indicate the presence of tropical rainforest. A related middle Miocene species, *Khoratpithecus chiangmuanensis*, dating to 13.5–10 mya, also occurs in Thailand, and has extensive associated paleobotanical evidence (Chaimanee *et al.* 2003). Middle Miocene pollen spectra show high biodiversity. There was an environmental mosaic of tropical freshwater swamps, and a lowland forest dominated by *Syzygium* in the level with the most fossil hominoid remains. C4 plants are absent, and there is some seasonality in rainfall. The vegetation most closely resembled that found today in the Sudanese headwaters of the White Nile – the African tropical freshwater swamps of the Sudd. The vegetation from the middle Miocene Thai localities is therefore completely different from the temperate Late Miocene Lufeng flora from South China. From this evidence, the authors infer the existence of a Middle Miocene dispersal corridor between Africa and Southeast Asia (Chaimanee *et al.* 2003: 62). If this inference were confirmed, it would have profound implications for mammalian paleobiogeography. In any case, the

Arabian Peninsula was not separated from Africa at this time. Consequently, this would facilitate the movement of land animals from East Africa into Asia through a southern dispersal route.

Community structure and competition between primate species

The question of community structure in fossil hominoids brings up the question of community structure and competition in living primate species. It would be useful for paleobiological reconstructions if niche differentiation and inter-specific competition were intensively investigated in modern catarrhine groups with large numbers of closely related species, such as the hylobatids and the cercopithecids.

Resource competition between hylobatid species of the same body size is inferred for three reasons: sympatry occurs only between the larger siamang and two small species; there is ecological similarity between three adjacent allopatric small species; and these three small species demonstrate inter-specific territoriality and narrow hybrid zones at points of contact (Raemaekers 1984). However, there apparently is no evidence that the diet of the two small species is affected by the presence of the siamang in sympatric regions, and there are no species density differences in sympatric regions (Raemaekers 1984). Chivers (1984), however, argues that home range sizes may indicate resource competition, because, in an area that contains four cercopithecid species at high densities, there are large hylobatid ranges. The smallest hylobatid ranges occur at a site with only one sympatric primate species.

Sympatric *Macaca nemestrina* and *Macaca fascicularis* use different parts of the habitat, and range and forage differently. Yet, these differences are interpreted as the result of pre-existing anatomical differences, rather than as character displacement developing under sympatry (MacKinnon & MacKinnon 1980). Sympatric *Presbytis obscura* and *Presbytis melalophos* also use different parts of the habitat and range and forage differently. Anatomical differences can be related to locomotor differences. In this case, differences in habitat utilization are thought to have resulted in the evolution of locomotor morphology (i.e., the ranging and foraging differences precede the anatomical differences), but there is no discussion of character displacement under sympatry (Fleagle 1978). Stable, polyspecific associations of *Cercopithecus* species show a reduction in dietary and habitat use differences between species, rather than the increase in differences expected with sympatry (Gautier-Hion 1988). Competition occurs over fruit resources. Yet, when the availability of fruit is low, polyspecific associations do not disperse. Similarly, New World tamarin monkeys (genus *Saguinus*) frequently form polyspecific associations, and these are presumed to affect foraging success and predator detection (Terborgh 1983, 1990). However, an experimental field manipulation of food resources in polyspecific associations of *Saguinus fuscicollis* and *Saguinus imperator* to test costs and benefits concluded that each species could mitigate foraging costs incurred by the association in a different manner (Bicca-Marques & Garber 2003). Thus, field experiments to test for resource competition between these sympatric primate species yield negative results.

A review of available evidence of interactions between primate taxa (both catarrhine and non-catarrhine) concludes that niche divergence caused by inter-specific competition cannot yet be demonstrated, in spite of the fact that the structure of primate communities is generally thought to be ordered by interspecific competition (Waser 1987: 223–225). Fleagle & Reed (1996) performed a multivariate analysis of primate communities from eight localities, using nine ecological variables. The first factor of the principal components analysis accounted for 28.46% of the total variance. Leaping and body size were the variables that were most highly correlated with the first factor (Fleagle & Reed 1996: Table 1). The two platyrrhine communities from South America demonstrated the greatest clustering of species using the first two factors of the principal components analysis (Fleagle & Reed 1996: Fig. 5). The platyrrhine primate communities should therefore display the greatest degree of inter-specific competition. Yet, as discussed above, polyspecific associations often occur among small-bodied platyrrhine species. Ganzhorn (1999) assumed that competition was actively affecting the structure of primate communities. He compared communities using number of species, population densities, body weight, and diet. Ganzhorn did not actually examine these four variables in real communities, but in continents (Africa, Madagascar, Asia, South America). Comparing the ratio of body mass in adjacent species-pairs matched for weight with similar diet, Ganzhorn concluded that inter-specific competition was an important factor in primate communities, except for Asian primates. However, the body mass ratios in adjacent species-pairs with similar diet do not separate out into disparate clusters as species numbers increase (Ganzhorn 1999: Fig. 8.3). One would expect such separation – character displacement – if competition were occurring. Schluter (2000) discovered that size ratios between sympatric species with phenotypic divergence caused by resource competition were significantly larger (1.41) than when ecologically similar species were distributed evenly along a size axis (1.27). Furthermore, increasing the number of sympatric species in platyrrhine primates, for which the best real database exists, does not affect population densities in three well studied species (Ganzhorn 1999: Table 8.2).

Schluter (2000) reviews the field biology and experimental evidence for ecological character displacement. The field biology evidence consists of sixty-one published cases of closely related species within the same genus. Independent confirmation of resource competition was present in only 23% of these cases. However, Schluter (2000) also presents his own experimental evidence on stickleback fish. He demonstrates that resource competition declines as phenotypic characters diverge, that natural selection generates divergence, and that the distinct morphology and behavior of certain competitors generates the divergence. A major implication of Schluter's (2000) review is that resource competition is stronger between sympatric species whose morphological and ecological differences are slight. This confirms the existence of Hutchinson's minimal niche separation, and a threshold of divergence above which competition is reduced. However, this contradicts the assumptions of Fleagle & Reed (1996) and Ganzhorn (1999), who assume that maximal phenotypic divergence between primate species indicates resource competition. If primate species are separated in multivariate ecological space, these authors assume that competitive interaction is occurring. Yet, such divergence reflects ancient, not modern resource competition. Schluter (2000: S10) colorfully labels modern phenotypic divergence "the ghost of competition past."

In summary, if modern primate species tell one anything about primate niche differentiation, it is that the interaction between closely related sympatric species does not yield striking examples of competition with associated character displacement. This may either be the result of resource abundance or diffuse resource competition within the modern communities. Seasonal periods of severe resource stress or high primate population densities might show a different picture.

However, the likeliest explanation for the apparent absence of inter-specific competition among living primates is that the existing primate world is not a normal one. The modern primate world has witnessed two major extinction events – the extinction of the Miocene hominoid radiation and the extinction of the Malagasy prosimians. Furthermore, a great number of living primate species are in imminent danger of extinction in the wild (Chapter 4). The diminution of living species results in greater resource availability for surviving primate species, and lessened niche overlap. As a result, inter-specific competition cannot be demonstrated. However, the situation was probably very different during the glory days of the Paleocene, Eocene, and Miocene, when great primate radiations were taking place. These periods represent the normal primate world, rather than the pallid and diminished present.

Factors affecting species diversity

What factors caused species diversity in Miocene hominoids? Species diversity in African Miocene hominoid taxa is probably affected by local tectonic events in East Africa. Body size may affect competition, with competition tending to occur between animals in the same size range. The existence and degree of seasonality is probably an important variable affecting species diversity and sympatry. Paleosol evidence from the Miocene of Kenya appears to indicate an increase in aridity and more open country vegetation in the middle Miocene (Retallack 1991). If this caused an increase in interspecific competition among primates, it may not be surprising that the dentally aberrant genus *Kenyapithecus* (McCrossin & Benefit 1993) appears in the middle Miocene. Postural, positional, and ranging differences may result from foraging differences evolving among competing, sympatric primate species. Because differences in the locomotor anatomy of African Miocene hominoid taxa are minimal (Rose *et al.* 1992) or absent (Harrison 1993), this implies that inter-specific competition may have been more intense in these fossil taxa than among many recent sympatric primate species.

The end of the hominoid radiation and the rise of the cercopithecoids

Although the living Old World monkeys are far more speciose than living hominoids, the cercopithecoid radiation occurs very late in time, during the Plio-Pleistocene (Delson 1975). The first cercopithecoids appear in the East African Miocene at 19 mya, but remain rare until far later in time. They are represented only by relatively meager remains in the Miocene fossil record, where they are eclipsed by contemporary hominoids.

The cercopithecoid monkeys are derived from some unknown early catarrhine, possibly located within the generalized descendants of the early Fayum primates (Cachel 1975b). These are animals that would now be classified as members of the families Propliopithecidae or Pliopithecidae. The earliest cercopithecoids are members of the genera *Prohylobates* and *Victoriapithecus*. They are usually classified in the Family Victoriapithecidae, because they do not show the strongly developed bilophodont molars that are characteristic of all living cercopithecoids (Benefit & McCrossin 2002). Bilophodont molars confer an adaptive advantage to cercopithecoid primates, because they are able to shear tough, mature leaves (Walker & Murray 1975). Thus, cercopithecoids are able to exploit a dietary resource otherwise unavailable to primates that lack such specialized molar lophs, or shearing crests. Bilophodont molars may explain the rise of the cercopithecoid monkeys. The dentition of most hominoid species typically lacks shearing crests, which limits the efficient processing of mature leaves. This may have been a critical deficiency during the late Miocene, under conditions of increasing seasonality.

The European fossil record clearly shows that the diminishment and extinction of the Miocene hominoid radiation is related to increasing seasonality during the late Miocene. Hominoids were present in Western Europe at 14 mya, but disappear during the Vallesian Crisis at 9.6 mya, along with many taxa comprising the early Vallesian fauna. Increasing seasonality, affecting both temperature and precipitation, caused broadleaf evergreen forest trees to be replaced by deciduous tree species. Macrobotanical fossils demonstrate that humid subtropical forests are followed by deciduous temperate forests. By the late Vallesian, 45% of the flora consists of deciduous tree species (Agustí *et al.* 2003). It is possible that the appearance of a distinctive seasonal fruiting cycle caused the extinction of frugivorous hominoid primates (Agustí *et al.* 2003). This extinction occurred more than one million years before the global spread of C4 grasses between 7 and 8 mya. Forest was still present, but it was a different type of forest, in which frugivorous hominoid genera, such as *Dryopithecus*, could not survive. However, the folivorous pliopithecid hominoid genus *Egarapithecus* – not dependent on the availability of fruit throughout the annual cycle – did survive.

The extinction of Asian hominoids also appears related to environmental change. Major tectonic changes occur in Asia during the late Tertiary. Uplift of the Himalayan Mountains occurs throughout the Miocene, and is caused by the subduction of the Indian plate under southern Asia. However, a pronounced increase in uplift occurs during the middle Miocene, and affects climate on a global scale through the rise of the Tibetan Plateau. The hominoid fossil record is dense and well-studied in this region. The genus *Sivapithecus* goes extinct in the Siwalik Hills of Pakistan between 9.2 and 8.1 mya. Nelson (2003) infers that this is caused by an increasingly more intense monsoonal climate that begins at 9.5 mya. Moist, widespread forests are replaced by more open habitats until C4 grasses prevail. *Sivapithecus* survived seasonal drought that may have resembled conditions in modern Canton in southeast Asia, where the dry season can last for five or six months. However, increasingly more fragmented forest habitat heralded the extinction of this genus (Nelson 2003).

Cercopithecoid monkeys often exhibit some degree of terrestriality, and both living and fossil species can develop specializations for terrestrial locomotion. Postcranial remains of *Victoriapithecus*, the oldest cercopithecoid genus, show a degree of terrestriality similar to that of the living vervet monkey (*Cercopithecus aethiops*), and this genus was equivalent to the vervet in body size (Benefit & McCrossin 2002). The patas monkey (*Erythrocebus patas*) is the most morphologically and behaviorally derived living example of a terrestrial species. Living and fossil baboons and their ecological vicars in Asia, the macaques, typify the ability of Old World monkeys to invade a variety of terrestrial habitats and exploit their resources. Colobines also develop terrestrial representatives,

Figure 3.4. Black-and-white colobus monkeys (*Colobus guereza*) coming down to the ground to forage for food by the shores of Lake Naivasha, Kenya.

such as the modern langurs, although even largely arboreal modern species can forage on the ground (Figure 3.4). Eurasian fossil colobines such as *Mesopithecus* appear to have a number of derived terrestrial traits (Delson 1975).

During the Plio-Pleistocene, cercopithecoids become not only speciose, but abundant. At South African sites, cercopithecoid monkeys may comprise 7–74% of the vertebrate fauna. The gelada baboon genus *Theropithecus* represents about 50% of all vertebrate fossils collected from excavations at Olorgesaile, in central Kenya (Figure 3.5). It is noteworthy that sites yielding cercopithecoids also yield hominids. Hominids are radiating at the same time as the cercopithecoids (Chapter 14), and habitats that drive cercopithecoid abundance and diversity also appear to be centers of hominid abundance. Furthermore, taphonomic conditions that promote the preservation of fossil cercopithecoids similarly promote hominid preservation. African pigs or suids are also radiating during this period. There is a major turnover in suid species between 2.0 and 1.6 mya (Turner & Antón 2004), and the living African suids are reduced to three species.

Figure 3.5. Fossil gelada baboon (*Theropithecus* sp.) frontal *in situ*, Ileret, Kenya. The eye sockets face directly forward. A horizon of fish bone lies below the primate fossil, which dates to about 1.5 mya. The scale is in centimeters.

Cercopithecoids may disperse widely through a broad array of habitats. The macaques exemplify this dispersal potential. The genus *Macaca* has been extant since 5 mya (Delson & Rosenberger 1984), and still remains one of the most speciose and widespread of living primate genera. Its geographic range today extends from Northwest Africa to the Phillippines, and it occurs in a wide array of habitats, including highly seasonal temperate habitats in Japan. The limiting factor for macaque dispersion in higher latitudes appears to be not temperature or seasonality, but access to plant foods. Deep snow cover hinders access to plant foods, so that Japanese macaques (*Macaca fuscata*) must strip bark from trees during the height of the winter season, when they cannot reach plant foods through deep snow.

Climate change in the late Miocene and the first hominids

Traditionally, in an explanation that can be traced to the nineteenth century, hominid origins are explained by triggering climatic events (Feibel 1997, Stoczkowski 2002). These climatic events are thought to involve a major period of desiccation that removes tree cover, and thus forces arboreal proto-hominids to the ground. Once these creatures are forced to the ground, they achieve bipedality, and thus make the transition to hominid status. This explanation is still current today, and so it is important to survey the paleoclimatic and paleogeographic records to discern the degree of Neogene climatic oscillation, especially in Africa.

The major global climatic oscillations associated with the Pleistocene actually first appear during the Tertiary, or even during the late Cretaceous. Fluctuating sea-level estimates gleaned from both the New Jersey coastal plain and the Russian platform appear to be caused by the growth and decline of late Cretaceous glaciers (Miller *et al.* 2004). There is no tectonic explanation for these sea-level changes, and there are several sequences that link oxygen isotope 18 increases in ocean sediment cores to lowered sea-level. The only reasonable explanation for this data pattern is that growing and decaying ephemeral Antarctic ice-sheets 5–10 million km^3 in extent caused rapid global sea-level variation during the late Cretaceous (Miller *et al.* 2004). Thus, climatic events traditionally offered as an explanation for hominid origins actually begin during the greenhouse world of the late Cretaceous. Climatic fluctuation and environmental disruption are not unique to the late Neogene, but actually pre-date the Tertiary.

East African and Red Sea rifting began at 25 mya, and created extensive tectonic changes in East Africa (Harland *et al.* 1989: Fig. 7.3). Loess deposits in China document aridity and the beginning of desertification in Asia at 22 mya (Guo *et al.* 2002). The East Antarctic ice sheet appears at 14 mya, and remains stable since that time (Burckle 1995, Kennett 1995). During the latest Miocene, global temperature fluctuations and cooling became pronounced. A fall in global sea-level resulted from cooling and encapsulating of sea water when the Antarctic ice sheet expanded between 6.2 and 5.7 mya. This was the most critical period of unusual Cenozoic climatic and oceanic events (Frakes *et al.* 1992). The greatest known extinction of large herbivorous land mammals occurs during this time in North America (Webb *et al.* 1995). At about 6 mya, lowered sea-level and tectonic movements between the African and Eurasian lithospheric plates closed off the connection between the Atlantic and Mediterranean basins. This occurred between 5.96 and 5.33 mya (Krijgsman *et al.* 1999). The missing water inflow from the Atlantic could not be compensated for by the rivers that debouched into the Mediterranean. High temperatures and dry climates in the Mediterranean basin caused an evaporation rate that exceeded any inflow of water. The Mediterranean began to dry up, becoming first a series of shallow evaporitic lakes, and then a gigantic salt pan (Hsü 1983). One million cubic km of evaporites were deposited in the Mediterranean Basin. This Mediterranean desertification, known as the Messinian salinity crisis, reached its apogee at 5.5 mya. A pronounced uplift of about 1 km between the African and Iberian lithospheric plate may have been a major precipitating factor (Duggen *et al.* 2003). Events may have been hastened by local pre-Messinian and Messinian tectonism in the Apennines and elsewhere.

The Messinian salinity crisis certainly affected local climate. Deep sea cores from the Red Sea and the Persian Gulf also demonstrate extreme aridity during this time period. It is tempting to view both the Plio-Pleistocene cercopithecoid and hominid radiations as the result of the Messinian salinity crisis. Hominid origins have also been linked to this episode. Yet, it now appears that the first hominids may antedate the salinity crisis. Within the span of 2 years, three new hominid taxa that are older than 5.5 mya have been announced. These are *Sahelanthropus tchadensis* at 6–7 mya (Brunet *et al.* 2002), *Ardipithecus ramidus* at 5.2–5.8 mya (Haile-Selassie 2001), and *Ororrin tugenensis* at 6 mya (Senut *et al.* 2001). Only *Orrorin* has persuasive evidence of bipedality in the morphology of three femora (Pickford *et al.* 2002). *Sahelanthropus* has no postcranial remains, and the argument for bipedality in *Ardipithecus ramidus kadabba* rests on a single toe bone (Haile-Selassie 2001).

Global climate becomes warmer and sea levels rise during the early Pliocene (Frakes *et al.* 1992). This is sometimes termed the "Pliocene Golden Age," dated to 3.6–3.3 mya, although climatic variability still exists (Shackleton 1995). Global cooling takes place in the late Pliocene. Ice sheets are found at both poles at 3 mya, and widespread aridity occurs in low to middle latitudes. After 3.0–2.6 mya, seafloor records demonstrate an astronomical forcing of climate. Extreme climatic oscillations begin with the advent of the Pleistocene at 1.8 mya, and the rhythm of the continental ice sheets starts 900 000 years ago. However, marked climate fluctuations had been occurring globally ever since the late Miocene, and so Pleistocene climatic fluctuations represent an intensification of events taking place during the late Cenozoic. The stratigraphic context of dated hominid fossil finds has been linked to the global record of climate change and reversals in the earth's magnetic field (Feibel *et al.* 1989). Paleosols from the Gona area of the Awash River, Ethiopia, indicate an increase in grassland from 3.4 to 1.6 mya (Quade *et al.* 2004). An initially predominantly forested habitat develops an increasing amount of grass cover through time. Grass was always locally abundant on the Awash floodplain, but increases to 50% or more of the groundcover beginning at 2.7 mya. Paleosols from the Turkana Basin, Kenya, record three episodes of aridity and concomitant increases in the abundance of C4 tropical grasses; these episodes occur at 3.58–3.35, 2.52–2, and 1.81–1.58 mya (Wynn 2004). In general, it appears that the savannah habitat that now characterizes the Serengeti ecosystem did not become widespread in Africa until 1.8–1.6 mya (Figure 3.6).

As noted above, the traditional explanation for hominid origins is a climatic one. The idea that changes in the physical environment drive evolution

Figure 3.6. Giraffes traversing savannah in the Serengeti ecosystem, Masai Mara Reserve, Kenya.

is common, although simplistic. "... [T]here is not a single biological event from the past two million years that could not be attributed to 'climate change'" (Penny & Phillips 2004: 520). Although vertebrate paleontologists since the time of W. D. Matthew (1914) have identified climate change as the powerhouse of evolution, this idea is associated in modern times with the vertebrate paleontologist Elisabeth Vrba (1992). Vrba (1995) presents a detailed, expanded, and revised account of her climate-driven model of evolution, called the "turnover pulse" hypothesis. "Is physical change the necessary pacemaker of speciations and extinctions, or do living entities drive themselves to evolve and disappear even in its absence? ... According to some models ... this biotic force at the level of organisms constantly works toward speciation and extinction, irrespective of population structure and even in the absence of physical change. That is, the biota is akin to a perpetual motion machine that inexorably drives itself to evolution" (Vrba 1995: 24–25). However, this model simplistically denies the existence of competition and other biotic generators of evolution (Cachel 1998). Far from being some unidentified, almost vitalistic force, competition between species fighting for resources creates an arms race within an ever-changing biological realm. This is The Red Queen's hypothesis. Van Valen (1973a) arrived at it by asking himself whether evolution would stop if all change in the physical environment stopped. Unlike Vrba, he answered "No," and the answer has been supported by the fossil record and by field biology observations. In fact, ecological interactions occurring under field biology conditions are responsible for some of the instances of "contemporary evolution." Vrba's theory also incorrectly assumes that natural selection is weak, and allopatry is not sufficient per se for generating new species without the added factor of climatic forcing (Vrba 1995: 27). With regard to the matter at hand – hominid evolution – no clear turnover events reflected in species origin or extinction appear to follow climate changes.[2] The climatic forcing model has been tested. Vrba had predicted that a turnover pulse should occur in the African mammal fauna between 2.8 and 2.5 mya. A test of the model in three regions of the Lake Turkana Basin over the last 4 my, with well-dated, abundant mammal fossils, shows that 58–77% of the mammal taxa are replaced between 3 and 2 mya. Diversity increases from 3 to 2 mya, and then declines. However, the expected turnover pulse between 2.8 and 2.5 mya does not occur (Behrensmeyer *et al.* 1997).

Endnotes

1. About 8000 museum collections are currently subsumed within this growing data base, which is expected to yield major insights into the nature of evolution throughout Phanerozoic time. The URL for the Paleobiology Data Base is www.paleodb.org.
2. However, aridification episodes may have an effect on the archeological record (Chapter 16).

4 Primate speciation and extinction

Primate speciation and extinction in the geological past

The topic of this chapter is the speciation and extinction of living primates. However, two important points must be discussed now. The first is simple to state, but subject to intense philosophical and taxonomic debate: in my opinion, fossil species do not differ in kind from living species. Modern processes that cause or contribute to speciation or extinction also operated in the geological past. These processes include ecological differences or karyotype novelties that are difficult or currently impossible to discern from the fossil record. Processes observable today only under field biology conditions or in the lab were responsible for speciation and extinction in the past. Paleontology extends current processes back into geological time, and thus there should be no division between neontology and paleontology, or a distinction between definitions of living species (the Biological Species Concept) and fossil species (the Phylogenetic Species Concept). Processes responsible for speciation and extinction include natural selection, which some researchers label as always weak or ineffective (e.g., Eldredge & Gould 1972, Gould & Eldredge 1977, Gould 1980). Selection rates documented in the wild can be so strong that they can cause phenotypic change within several generations, leading to the idea of "contemporary evolution" (see Chapter 17). Given "contemporary evolution," distinctions between living and fossil species become artificial.

The second important point is whether primate extinctions and speciations in the geological past conform to known patterns of mammalian evolution. Do primates differ from other mammal groups in their evolutionary patterns and processes? Some researchers consider that the primate fossil record shows constantly increasing diversification, with no major radiations or extinctions (see Chapter 3). However, the consensus opinion is that major radiations and extinctions occur in the primate fossil record, just as they do in most other mammalian groups. In comparison to later parts of the Cenozoic, the primate fossil record is very dense and well dated during the early Tertiary. The pattern of primate evolution during this time appears no different from that of other mammals (Gingerich 1976).

Placental mammals in the early Tertiary exhibited explosive evolutionary radiations as ecosystems reorganized after the catastrophic mass extinction at the Cretaceous/Tertiary (K/T) boundary. The earth's collision with a ten-kilometer wide asteroid is almost certainly the principal cause of this extinction, although other factors contributed to the collapse of ecosystems. Two of these factors were massive flood basalts on an unprecedented scale and global cooling before and across the K/T boundary, as indicated by macrobotanical remains and foraminifera (Wilf *et al.* 2003). Yet, evolutionary rebound rapidly occurred even in North America, whose landscapes and ecosystems had been devastated by the asteroid impact. Ground zero for the impact was at Chicxulub, just off the Yucatan Peninsula, where a buried impact crater 180–200 km in diameter has been detected by gravity anomalies. The asteroid entered the earth's atmosphere at an angle, and destroyed the interior of the North American continent. Tsunami deposits in Texas, and shocked quartz and spherule layers in sediments from Colorado to New Jersey at the K/T boundary are evidence of the devastation. Placental mammals had been rare in Cretaceous North America, whose mammal communities were dominated by marsupials and multituberculates. Yet, placental mammals underwent a phenomenal evolutionary radiation in the earliest Paleocene of North America, where the fossil record is extensive and well-dated. If the North American genus *Purgatorius* is accepted as a primate, then the primate order, in the form of plesiadapiform primates, is one of the first placental mammalian orders to appear after the K/T mass extinction (Van Valen & Sloan 1965, Van Valen 1994). As explained in Chapter 1, I am identifying the plesiadapiform species as primates, although they are not primates of modern aspect (euprimates). It is possible that primates survived the K/T mass extinction. A specimen of *Purgatorius* was dated to the late Cretaceous (Van Valen & Sloan 1965). Although this specimen may postdate the K/T boundary, another line of evidence indicates that primates existed in the Cretaceous. The latest attempt to create a grand phylogeny for all of life (a "Tree of Life") puts primate origins at 70–80 mya (Benton & Ayala 2003). The plesiadapiforms may not have resembled modern primates (euprimates), but they appear to have been an evolutionarily successful group.

Early Tertiary fossil primates are found in temperate areas of North America, Europe, and Asia that would be inhospitable to most non-human primates today. During the early Eocene, fossil primates even occur on Ellesmere Island, Canada. This is the highest known latitude of non-human primate occupation (McKenna 1980). However, primates occupy these areas at times when the paleoclimate and paleoenvironment were tropical or subtropical. That is, like the majority of modern primates, early Tertiary primates appear to have inhabited tropical habitats, and appear to have required forest cover. In areas where the paleoenvironment can be reconstructed in detail (e.g., the Bighorn

Basin of Wyoming, the Paris Basin, the Messelgrube in Germany), primates appear to have inhabited lush tropical rainforests. Major forests are documented to exist even in very high latitudes, although, given the pronounced seasonal differences in sunlight, these forests were deciduous.[1] At 40 mya, during the Eocene, stands of a giant conifer, the dawn redwood (*Metasequoia*) grew on Axel Heiberg Island, in northern Hudson's Bay. The earliest rainforest also occurs in what is now a temperate region. In the early Paleocene, a strikingly diverse tropical rainforest was established in the Denver Basin of Colorado at 64.1 mya, only 1.4 my after the catastrophic mass extinctions that took place at the Cretaceous/Tertiary boundary (Johnson & Ellis 2002). The leaf litter preserved here on a buried forest floor most nearly resembles the leaf litter found today on the rainforest floor of Manu, Peru, at the foot of the Andes in the western Amazon Basin. Orogenic uplift of the Laramide Front Range in Paleocene Colorado may have duplicated modern conditions at Manu, where the Andes create local conditions of heavy rainfall. Because no known Cretaceous floras preserve the leaf morphology typical of rainforests, this early Paleocene locality represents the first documented site with a tree species diversity and leaf physiognomy that are found today only in tropical rainforests.

The Paleocene, the first division of the Tertiary, saw the diversification of plesiadapiform primates in a remarkable radiation. Global climate was warm and equable in this geological epoch, with no pronounced distinctions between the equator and higher latitudes. However, a terrific upward spike in global temperature occurred at 55 mya, at the Paleocene/Eocene boundary. A worldwide record of this is preserved in sediments that show a geochemical anomaly called the carbon isotope excursion (CIE). The CIE was caused by the colossal release of oceanic depots of methane and CO_2 into the atmosphere, which resulted in intense global warming. A greenhouse world was formed, in which the temperature of the North Sea averaged about 28 °C (Buchardt 1979). The global warming of the CIE was associated with the immigration of new land mammal groups out of Asia into North America and West Europe (Beard 2002, Bowen *et al.* 2002). Asia was a powerhouse of mammalian evolution, and witnessed the origin of many new taxa. Immigration from Asia resulted in a fundamental biotic turnover event in North America and Europe (Beard 1998b). Among the land mammals that dispersed west across Eurasia and east across the Bering land bridge were the first euprimates, which originate in Asia.

Plesiadapiform primates declined during the Eocene, and went extinct, possibly because they were competitively inferior to rodents, which first appear in the late Paleocene of Asia. When productive fossil localities in the Old and New Worlds are examined, the relative proportion of plesiadapiform primates among other mammal groups far exceeds the relative proportion of primates within modern communities. They resemble modern rodents in their abundance.

From this alone, it is possible to infer that plesiadapiform primates occupied rodent-like niches. They were small mammalian herbivores. However, they lacked the specialized traits that enable rodents to gnaw, shred, and grind vegetation. Plesiadapiform primates lacked self-sharpening incisors, complex molar crowns, and the extremely specialized masseter muscles found in rodents.

Diversification of euprimates took place during the early Eocene. Where the fossil record is dense and well dated, euprimate evolution provides evidence of anagenesis or phyletic transformation of fossil species (Bown & Rose 1987). A pronounced decline in global temperature occurred at the end of the Eocene. This drop in temperature was associated with major extinctions in many mammalian groups. This distinctive Eocene/Oligocene extinction event was called *la Grande Coupure* by early twentieth century paleontologists. The Oligocene was an epoch of extreme global aridity. Perhaps because of the nature of the geological record – e.g., a severe reduction in sediment deposition – the diversity and abundance of many mammal orders declines at this point. Primates are no exception. Hominoid primates appear at 22 mya in East Africa. As discussed in Chapter 3, hominoids undergo a remarkable radiation during the early Miocene, and disperse widely through Africa, Europe, and Asia. They are already in decline by the mid-Miocene. Most hominoid species are extinct by the early late Miocene, which is already suffering the first climatic oscillations that culminate during the Pleistocene. *Oreopithecus*, dating to 6–7 mya, is a genus from Tuscany that represents the last vestige of the hominoid radiation. Because *Oreopithecus* appears to possess some bipedal specializations, it will be discussed in more detail in Chapter 13. New World monkeys have such a sparse fossil record that it is difficult to discern periods of platyrrhine abundance and rarity, or to estimate diversity. Because the evolutionary history of the New World monkeys takes place completely within tropical rainforest, the absence of sediment traps can account for their impoverished fossil record. It is very clear, however, that Old World monkeys are rare until the late Miocene, and experience a radiation during the Plio-Pleistocene that is contemporary with the hominid radiation. Old World monkeys were the only non-human catarrhines to adapt to open country conditions. Fossil apes from the Eurasian Mio-Pliocene move to preferred forested habitats as seasonality increases, rather than adapting to changing habitats (Eronen & Rook 2004).

Prosimian primates were one of the few non-volant terrestrial mammal groups to colonize the island of Madagascar, which was an isolated mini-continent for the duration of the Tertiary. Madagascar does not have a deep mammalian fossil record, and so the evolutionary history of the Malagasy prosimians remains dim, except for the subfossil species of the Holocene. The first human colonization of Madagascar occurred at about 350 BC, and human populations were large and dispersed throughout the island by AD 1000 (Burney *et al.* 2004). Human colonization interrupted and terminated a great radiation of Malagasy prosimians. Because no mammalian ungulates or specialized folivores had ever colonized Madagascar, the limited number of colonizers evolved to fill empty niches. Endemic prosimians radiated to fill terrestrial baboon-like niches and arboreal sloth-like or koala-like niches. Their extermination occurred after human colonization, although some species may have lingered until the seventeenth century, or even possibly until the twentieth century (Burney & Ramilisonina 1999). Recently dated evidence from archeology, paleontology, and palynology indicates that prosimian extinctions occurred in several stages, principally as a result of human disturbance of the landscape through deforestation and fire. Prosimian extinction did not occur immediately after human colonization, but many species were extinct by about AD 1200 (Burney *et al.* 2004: Fig. 3A). The latest dates are for the sloth-lemur *Palaeopropithecus ingens*, which persisted until the seventeenth century. In this sense, Madagascar is no different from other islands colonized by humans at a late date. Endemic insular species rapidly become extinct after human colonization.

The magnitude of the threat posed to insular animals by human colonization can be illustrated by events on an island continent, Australia. The prehistoric human colonization of Australia about 46 000 years BP was followed by the almost immediate extinction of the Australian megafauna (Miller *et al.* 1999, Roberts *et al.* 2001). Extinctions occurred across all classes of animals – reptiles, mammals, and birds – as long as the animals were above 100 kg in weight. Six of the seven genera weighing between 45 and 100 kg also went extinct. Megafauna in West Papua, New Guinea, connected to Australia by a land bridge during the Pleistocene, were also affected. So rapid were the extinctions that they appear to be synchronous with human colonization, given the level of time discrimination available in the paleontological and archeological records. Arid climate pulses have been ruled out as a factor in these extinctions. Consequently, there is no doubt that human hunting or anthropogenic habitat destruction or alteration was responsible for the megafaunal extinctions in Australia (Miller *et al.* 1999, Roberts *et al.* 2001). Because even an island continent is not immune to major extinctions after human colonization, this illustrates the level of threat faced by organisms on far smaller islands, such as Madagascar, and the islands of the Philippines and Indonesia.

A survey of living mammals suggests that parapatric speciation occurs more frequently in small mammal species than in large ones; small mammal species may also exhibit more modes of speciation than large mammal species do (Searle 1996). A general review of the influence of body size on animal speciation confirms that ecological isolation associated with micro-habitat differences may be sufficient to disrupt gene flow and create new species in small-bodied taxa (Bush 1993). If this holds true, then one would expect that Paleocene and early Eocene primates, which were generally small in body size, would have experienced more parapatric speciation and more diverse modes of speciation than living primates. Living primates, with some exceptions, can be categorized as large mammals. The importance of parapatric speciation may therefore have declined in primate evolution after the acquisition of large body size. The primate shift away from parapatry and multiple modes of speciation may have occurred during the middle to late Eocene, when primate body size tends to increase.

Speciation in modern primates

The systematist Ernst Mayr (1963) wrote a classic monograph on animal speciation. Since the publication of Mayr's monograph, speciation has been the subject of two other important treatments (White 1978, Otte & Endler 1989). Mayr argued that the principal mode of animal speciation is allopatric speciation, in which geographic barriers or simple physical distance interrupt gene flow within a population, and generate new species when reproductive isolation is complete. Sympatric, parapatric, and peripatric speciation were additional modes of speciation recognized by Mayr. Because reproductive boundaries are crucial to the establishment and maintenance of species, the appearance of new mate recognition systems has also been emphasized as a mode of speciation (Paterson 1993). The co-evolution of pathogens and hosts may also drive speciation. Nunn *et al.* (2004) detect a positive correlation between primate diversification rates and the richness of parasite species in primates. However, it is not clear how this correlation is generated. Do the parasites drive primate speciation, or do the primates drive parasite speciation?

What degree of morphological variation is expected with the appearance of a new species? The existence of cryptic species shows that separate species need have no detectable morphological differences. The relationship between morphological variation and speciation has been investigated in platyrrhine primates. Among New World monkeys, skull differences are associated with the level of the genus or higher taxonomic categories, but genetic drift may account for some species differences (Marroig & Cheverud 2004). Hence, speciation in platyrrhine primates need not necessarily be associated with adaptation, but can also be caused by random events.

Allopatric speciation is documented to occur in primates. Careful examination of the provenance of primate pelt and skeletal material in museum collections shows that neotropical river systems have been important in creating platyrrhine subspecies, which are incipient species (Hershkovitz 1977). A relationship between river boundaries and the range size of platyrrhine species

Figure 4.1. A tributary of the Río Sarapiquí bisects lowland tropical rainforest at La Selva Biological Reserve, Costa Rica. Even small rivers may be important factors in platyrrhine primate speciation, if they form geographic barriers to gene flow.

has also been confirmed by later researchers (Ayres & Clutton-Brock 1992). Neotropical river barriers that affect platyrrhine species are not necessarily only major rivers like the Amazon and Rio Negro; even relatively minor tributaries can serve as geographic barriers (Figure 4.1). This seems strange, given that riparian forest often forms a continuous, tunnel-like cave over small rivers (Figure 4.2). One would expect agile, arboreal platyrrhines easily to ford these small rivers by traveling over them in the canopy. It is possible that areas for safe canopy travel over these rivers are limited – that is, canopy with sturdy, rugged trackways over the river is relatively scarce. In addition, the prevalence of palm trees, which have no branches or terminal twigs, only giant, slippery leaves, and relatively weak, non-woody lianas in neotropical forests assures that canopy

Figure 4.2. Riparian forest along the Ishasha River, Kivu Province, Eastern Congo.

Figure 4.3. Two black-handed spider monkeys (*Ateles geoffroyi*) in Palo Verde National Park, Costa Rica. An adult female and a juvenile traverse the rainforest canopy. Note the use of the prehensile tail to bridge canopy gaps. Photo courtesy of Dr. Kellen Gilbert.

travel is more hazardous for platyrrhines than for primates in Africa and Asia (Figure 4.3). The presence of prehensile tails in several platyrrhine primate genera has been explained by the structure of the Neotropical rainforest canopy.[2] Rivers must create an effective geographic barrier, if gene flow limitations have already begun. As is true for most primates, platyrrhine primates do not swim. Hence, even a shallow, slow-moving stream can be a formidable barrier. Furthermore, many seasonally flooded neotropical forests exist in South America. Two or three meters of water can cover the forest floor for months, and enormous, column-like tree trunks rise up from the calm, quiet surface waters like pillars within a high, dim cathedral. These shallow surface floods create additional hazards and barriers for platyrrhine primates. Animals encounter water barriers in a contingent, accidental fashion. This may explain why platyrrhine primate species do not necessarily show a relationship with differences in cranial morphology, although adaptive cranial differences occur above the species level (Marroig & Cheverud 2004).

The dispersion of non-human primates across water barriers is problematic, given their general inability to swim. Because primates cannot swim across major water barriers, rafting on floating mats of vegetation remains the only option for dispersal. It has long been considered that the prosimian colonization of Madagascar occurred through rafting. Madagascar has been isolated from continental land masses at least since the mid-Cretaceous, and certainly for the duration of the Tertiary. Although the Mozambique Channel that separates Madagascar from the southeast African coast is not very wide, the channel is deep, with powerful contrary currents. Even humans did not reach Madagascar until late in time, and then they arrived from Indonesia. Using only traditional boating technology, it is easier to cross the width of the Indian Ocean than it is to cross the Mozambique Channel. The non-human colonization of Madagascar is therefore difficult and unlikely, and depends on what G. G. Simpson termed "sweepstakes dispersal." Mammal evolution on Madagascar has thus produced strange suites of endemic, non-volant species, including prosimian primates.[3] In spite of their morphological diversity, Malagasy prosimians are closely related. From this it is inferred that only a single prosimian colonization event occurred. Perhaps only a solitary pregnant female that resembled the living mouse lemur (*Microcebus*) accidentally rafted from the African coast on a mat of vegetation to become the mother of all Malagasy prosimians (Charles-Dominique & Martin 1970). If this appears impossible, analysis of macaques on the islands of Southeast Asia demonstrates that at least one macaque species managed to cross an ocean gap by rafting over deep water on floating mats of vegetation, and macaques are far larger than mouse lemurs (Abegg & Thierry 2002). Platyrrhine primates appear on islands of the Lesser Antilles as far back as the early Miocene (MacPhee & Iturralde-Vinent 1995). Because the geologic history of the Caribbean is extremely complex, it is not clear whether these primates dispersed over an ocean gap from mainland South America, or were part of a terrestrial fauna isolated on blocks of lithosphere separated by plate tectonic movements. The Greater Antilles and the Aves Ridge – a now submerged area of land west of the Lesser Antilles – formed a great peninsula linked to South America between 50 and 30 mya (Iturralde-Vinent & MacPhee 1999). The history of higher primate evolution is also affected by sweepstakes dispersal. South America was an isolated continent through most of the Tertiary. The first primates to appear in South America dispersed from North America across a chain of volcanic arc islands that existed in the Central Atlantic between North and South America. I consider this colonization route much more probable than a South Atlantic route that seeds primates into South America from Africa (Cachel 1981). The Madagascar and South America examples reveal that primates have a potential for explosive speciation in novel environments that lack competitors.

Eldredge and Gould highjacked Mayr's idea of peripatric speciation to ground their evolutionary process of punctuated equilibria (Mayr 1963, Eldredge & Gould 1972, Gould & Eldredge 1977, Gould 1980). Punctuated equilibria was purported to be an alternative to natural selection. In peripatric speciation, a very small, isolated peripheral population became genetically and phenotypically distinct from its founding population, sometimes simply through chance or accident (genetic drift). As sponsored by Eldredge and Gould, peripatric speciation was driven by chance. They argued that peripatric speciation was divorced from natural selection or adaptation. Natural selection was not needed to create a new species. As environmental changes occur, the new species might spread through the geographic range of the original ancestral species. It was once thought that speciation was promoted in small populations. However, it is now known that speciation can take place even among organisms with a broad geographical range. In fact, a concept called centrifugal speciation makes exactly this point, and argues that these are the conditions likely to generate new species. Given the differences in niche a species exhibits across its range in different habitats, the appearance of novel species is more probable.

On theoretical grounds, one might expect to see more elaborate mate recognition systems and sympatric speciation in primates than in other animals. This is because primates in general are tropical, K-selected animals. Kingdon (1988, 1989, 1997) emphasizes that mate recognition systems must exist among African guenons (genus *Cercopithecus*). Although the pelage on most of the guenon body is uniformly dull, the head, face, and rump are characterized by brilliant coat colors and patterns that are distinctive to each species. In the dim, twilight rain forest where these species live, a unique flash of patterned color and a typical vocalization may be the only sensory signals differentiating one

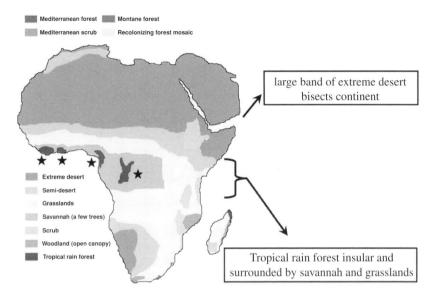

Figure 4.4. African habitats during the Last Glacial Maximum (18–21 000 years BP). Four small islands of lowland tropical rain forest (marked by star symbols) are surrounded by savannah woodlands and Serengeti-like grasslands with no trees. The northern third of the continent is completely bisected by a large band of extreme desert. Modified from a Quaternary Environments Network map (http://members.cox.net/quaternary/).

guenon species from another. This allows animals quickly to recognize members of the opposite sex as being the appropriate species before courtship displays or mating behavior takes place. Such subtle choice creates reproductive isolation. Hence, the guenon head, face, and rump colors, as well as vocalizations, function as mate recognition systems, and underlie a recognition concept of species in these primates. Guenons are a speciose group of primates (Gautier-Hion *et al.* 1988). Their evolutionary radiation has been very recent, and may well have been initiated after Pleistocene climatic fluctuations began to disrupt rainforest habitats. Through the Pleistocene, small, undisturbed forest nuclei remained intact as the African equatorial rainforest contracted with climatic oscillations (Figures 4.4 and 4.5). These forest nuclei served as refuge areas, and centers of origin for new, endemic species. As the rainforest expanded with ameliorating climate, new, closely related species would encounter each other. Hence, the importance of establishing mate recognition systems to maintain reproductive boundaries between species.

Sympatric speciation has been the focus of increasing theoretical and empirical work (Dieckmann & Doebeli 1999, Treganza & Butlin 1999). Sympatric speciation created by ecological barriers to gene flow has also been documented

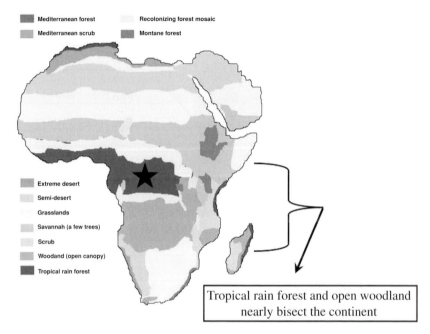

Mediterranean forest | **Recolonizing forest mosaic**
Mediterranean scrub | **Montane forest**

Extreme desert
Semi-desert
Grasslands
Savannah (a few trees)
Scrub
Woodland (open canopy)
Tropical rain forest

Tropical rain forest and open woodland
nearly bisect the continent

Figure 4.5. African habitats during the early Holocene (9000 years BP). A broad band of tropical rain forest (marked by a star symbol) and open woodland nearly bisects the continent. Expansive grasslands are present in the modern Sahara, and extreme desert occurs only in a narrow band on the southwestern coast. Modified from a Quaternary Environments Network map (http://members.cox.net/quaternary/).

(Doebeli & Dieckmann 2003). Groups within a population diverge because of ecological differences, which are established initially because resource competition causes segregation between competing individuals. When karyotyping of organisms in the wild during the 1960s first revealed unexpected variability, sympatric speciation was assumed to occur. It was argued that sympatric speciation was facilitated by the spontaneous production of novel karyotypes and chromosomal rearrangements (White 1978). Such chromosomal novelties are now known to occur in humans and common chimpanzees. Chromosomal inversions selectively prevent recombination. These incompatible chromosomal segments may act as reproductive barriers. Genetic differences would accumulate in the rearranged chromosomes, and thus facilitate speciation. This is parapatric speciation, or speciation with gene flow, in which geographic isolation does not occur. This type of karyotype difference – chromosomal rearrangements or "chromosomal speciation" – may lie behind hominid divergence from sympatric chimpanzees (Navarro & Barton 2003, Rieseberg & Livingstone 2003).

Species that are parapatric have ranges that are in contact, but that do not overlap. Parapatric speciation occurs when gene flow or hybridization between two species that are in contact creates a stable hybrid zone. This hybrid zone may ultimately generate a new species if reproductive barriers are established between the new species and the original founder species. Clifford Jolly has long argued that parapatric speciation is an important factor in baboon evolution, and he has also suggested that parapatric speciation may occur in hominid evolution (Jolly 1993, 2001). He draws a distinction between morphological form and behavior (phenostructure) and genetic material (zygostructure), in order to make the point that baboons show extensive phenotypic variability without developing enough genetic differences to cause reproductive barriers between the different baboon morphs. Fertile hybrids can occur between different species of the genus *Papio*. Phenotypic variability is a common occurrence in species with a broad geographic distribution, because the niche of such species can differ across the geographic range, as members of the species encounter significantly different habitats. The modern baboon evidence has general implications for interpretation of the fossil record. The different living baboon morphs would certainly be identified as different fossil species, but the genetic similarities between members of genus *Papio* would not be discerned.

Comparison with other mammals, however, suggests that parapatric speciation is unlikely to occur in large-bodied mammals like baboons or hominids. The "chromosomal speciation" outlined above may be an exception (Navarro & Barton 2003, Rieseberg & Livingstone 2003). Small mammal species often are distributed as separate entities in proximity to each other through a particular geographic region. They are parapatric, and parapatric speciation is facilitated under these conditions. In general, small mammal species are expected to demonstrate more parapatric speciation than large mammal species, and it is possible that they commonly exhibit more modes of speciation than large mammals (Searle 1996).

The evolutionary ecology of birds has been studied even more than that of primates. In addition, genetic testing of wild birds has confirmed the relatively high frequency of hybrid individuals. Among birds, seasonal environments contribute to species richness, because of the temporary presence of migrant species in breeding communities (Hulbert & Haskell 2003). The proximity of closely related species increases the chances of hybridization, and it therefore appears reasonable to assume that sympatric speciation would be an important mode of speciation under these circumstances. Because non-human primates do not exhibit seasonal migrations or engage in communal breeding associations, the degree of hybridization and resultant opportunity for sympatric speciation would be relatively limited in primates, as opposed to birds.

Competition in a seasonal environment with seasonal variation in resource abundance may be one of the factors that contributes to species richness. A theoretical model for the coexistence of competing species under these circumstances demonstrates how species diversity is generated by trade-offs between foraging efficiency and maintenance (Brown 1989). However, the firm association between primates and non-seasonal, tropical environments assures that primate species richness would not be caused by this mechanism.

Extinction in modern primates

Among mammalian orders, the highest proportion of endangered species occurs in the Order Primates (Jernvall & Wright 1998). Mammalian and primate extinctions do not occur randomly, with species being mowed down as if they were accidentally killed by a fortuitous spray of bullets. Instead, mammalian and primate extinctions preferentially affect higher taxa – an entire genus becomes extinct (Purvis *et al.* 2000). Because higher taxa are removed, biodisparity or phylogenetic history is reduced. Thus, let me confess, to begin with, that I have no confidence that any conservation efforts now being mounted – no matter how heroic – will stave off the imminent extinction of most non-human primate species in the wild. There is no "wild" left anywhere in the world. In many areas, it is even difficult to reconstruct what the original ecosystems were like – that is, what was "natural" about a particular area. Managed reserve lands, if surrounded by a humanly altered landscape, will not be able to compensate for the loss of original habitat. As explained below, species diversity is directly proportional to geographic area. Thus, the encapsulation of animals within tiny, island-like reserves surrounded by a sea of modified landscape inhabited by humans, their domesticates, and non-native, invasive species will not duplicate the natural engines driving speciation, and will not save threatened or endangered animals. Some non-human primates (e.g., some baboon and macaque species) will survive in a human-dominated environment as vermin or pests raiding human crops and garbage dumps. But, in the future, the majority of non-human primate species will be found only in zoos or colonies, where their reproduction will be managed. The only glimmer of optimism that I can offer is that at least the declining status of these primate species has already been recognized. At least they will be preserved, unlike the tragic case of another large mammal species, the thylacine, or Tasmanian wolf. Originally found throughout Australia and New Guinea, by historic times, the thylacine was restricted only to Tasmania. Deliberately exterminated by sheep ranchers in Tasmania, thylacines finally appeared only in zoos. The last living thylacine was an adult male called Benjamin. He was the object of intensive scrutiny by photographers and film-makers, and finally expired in the Hobart zoo in 1936. At least a similar fate – dying last survivor ignominiously ringed by a myriad of recording devices – will not occur for endangered primate species. Many efforts are now afoot to create thriving captive primate populations. In addition to these efforts, advances in reproductive technology can also assure primate species survivorship.

Primate extinctions, especially the extinction of hominid species, have been the focus of much research (e.g., Walker 1984). In general, anthropologists and primatologists, unlike mammalogists and vertebrate paleontologists, seem relatively unprepared to face the specter of extinction. The vertebrate paleontologist Alfred Sherwood Romer, for example, famously characterized the earth as "a charnel house of species." But anthropologists and primatologists appear to be dubious about extinct species. The perception is something along the lines of "Obviously, X must have been adaptively or competitively inferior, because it is extinct. What was wrong with species X that caused it to go extinct?" Given this mindset, it is ironic that a major anthropogenic extinction event will occur within the next few decades that will rival the "Big Five" mass extinctions of the geological past. So severe are the prospects of extinction that the course of evolutionary processes may be affected – in fact, a colloquium on the future of evolution was convened in 2000 by the U.S. National Academy of Sciences.

The prospect of the looming elimination of tropical habitats is critical for primate extinctions. The elimination of tropical habitats, however, also has an impact on evolution itself. Tropical regions have been the center of animal and plant diversity for the last 250 my. Current field biology research establishes that the heart of biodiversity is centered in tropical areas (Allen 1956, Leigh *et al.* 1982, Janzen 1983, Whitmore 1984, Kingdon 1989, Whitmore 1990, Richards 1996, Leigh 1999). The destruction of tropical habitats signals the end of the ancient foci of speciation or generators of evolution. Extinction will occur without communities being replenished by new species, because extinction rates will finally outpace origination rates (Myers & Knoll 2001). Biodiversity losses are the highest when extinction risk is high, and speciation rates are low. Besides the reduction in species diversity, global biomass will decline precipitously with the destruction of tropical forests. Global climate will be affected, because photosynthesis scales in a positively allometric fashion with tree size, and local rainfall will decrease when canopy cover is removed (Ozanne *et al.* 2003). It is taken as inevitable that large mammals will become extinct in the wild – there will be no wild habitats, or habitats unmodified by human presence. A series of predictions emerged from a colloquium on the current biotic crisis that was convened by the U.S. National Academy of Sciences. Among these predictions were the following: the speciation of large mammals will end; surviving species and biota will be blended together; morphological variety, as reflected in the elimination of higher taxa (genera, families, orders), will decline; and opportunistic pest or weed species will multiply (Myers & Knoll 2001).

A broad mammalian comparative perspective allows insight into general factors that affect primate extinction. Habitat fragmentation, tropical niches, diet, and body size are four major factors that affect the probability of mammal extinction.

Habitat fragmentation is the leading cause of species extinction in the modern world. Organisms on island habitats are particularly vulnerable to extinction, because they always suffer from a more restricted area and a more limited resettlement pool than continental habitats do (MacArthur & Wilson 1967, Diamond 1984). The isolation of island species renders them more prone to extinction. If island species are endemic, which is often the case, the total diversity of a group may be significantly reduced. This explains concerns about the status of the Malagasy prosimians and primates on the islands of the Philippines and Indonesia. Continental habitat fragmentation is nothing other than a type of island formation occurring on a continental scale. Even reserves and protected areas suffer, if they are surrounded by extensively altered habitats. In effect, the reserves become islands (Newmark 1996). If a population within a reserve is already small, the insularization of the reserve may cause local extinction, because new animals cannot be recruited from outside the reserve (Ginsberg *et al.* 1995, Woodroffe & Ginsberg 1998). In theory, given the known relationship between geographic area and species diversity, the future prospects of species in reserves are grim. Unless the scale of reserves mimics that of natural habitats, future speciation will be compromised by the inevitable loss of species diversity in vestigial patches of reserve (Rosenzweig 2001).

Deserts and arid environments create barriers for primate movement, unless they disperse along major rivers surrounded by riparian forest (Figure 4.6). Similarly, high altitude is a barrier to primates, not merely because of the physical obstruction created by mountainous terrain. The absence of trees and impoverished vegetation at high altitudes create impossible conditions for dispersing primates. However, in Africa north of the equator, all non-human primate species actually increase their geographic range with increasing altitude (Cowlishaw & Hacker 1997). Because geographic range expansion is a response to climatic variability in species that can acclimate to seasonal change (Stevens 1989), altitude may incorporate several dimensions of climatic variability in this region of Africa. The empirically generated expectation is that tropical species will have a more narrow latitudinal range of distribution than species in temperate latitudes (Stevens 1989). Note that this narrower latitudinal range is not necessarily equivalent to a smaller geographic range – latitudinal range directly addresses latitudinal spread, which affects the number of climatic and

Figure 4.6. The Chalbi Desert of northern Kenya. Gabra women draw water from the Casa waterhole. Among non-human primates, only olive baboons and bushbabies sometimes penetrate and occupy this area.

habitat zones that are occupied by an organism. Higher primates of the Southeast Asian Sunda Shelf islands that have narrow latitudinal ranges have a heightened risk of extinction (Harcourt & Schwartz 2001). In addition, a larger body mass increases the probability of extinction, although this probability is undoubtedly exacerbated by insular conditions, which necessarily limit the area of occupation. The narrower latitudinal range of tropical organisms is called Rapoport's rule. It implies that there is less tolerance for seasonal changes in tropical organisms.

Given the importance of forests and tree cover for primates, it is not surprising that primates are most speciose in tropical rainforests or deciduous tropical forests. Open country habitats, in contrast, are much more impoverished in primate species (Bourlière 1963, 1985). Terrestrial primates are rare. In open-country, they do not contribute much to the biomass of tropical ecosystems (Figures 4.7 and 4.8). Fragmentation of tropical rainforest or deciduous tropical forests therefore creates a major crisis for the survivorship of primate species. Rainforest cover is rapidly disappearing. Traditional slash and burn agriculture can eliminate rainforest. The degree to which traditional agricultural practices eliminate natural forest cover is dependent upon local human population pressure. A major portion of the global expanse of rainforest was

Figure 4.7. Bushland savannah near the Ishasha River, Kivu Province, Eastern Congo. Primate species are rarer here than in forested habitats. Before elephants were hunted and poached out in this region during Belgian colonial rule, grasslands were more widespread. Elephants preferentially eat leaves and bark, and destroy trees. They thus remove forest cover. Because of the decimation of elephants, forest is returning to this area.

Figure 4.8. Sympatric zebra and impala on the Laikipia Plateau, Central Kenya. The biomass of ungulates is high in open grasslands, but the biomass of primates declines precipitously here.

removed in just a few decades of the twentieth century, facilitated by the spread of industrialized logging, and other recent technological innovations (Richards 1996).

Fruit is an important food resource for many primate species. However, there is a general lack of synchronized, seasonal fruiting in the tropics. Because of the diminished or virtually absent seasonality of the tropics, tree fruiting is highly irregular. Flower opening may occur only after near drought conditions, or the return of rain after drought. Even if observed for years, a single tree may have no established fruiting schedule. Members of the same tree species in close proximity may bear fruit years apart from each other. Hence, the distribution and abundance of trees that are currently fruiting is also important in the maintenance of primate species or their extinction (Marsh & Mittermeier 1987).

It is well established that geographic area is directly proportional to the number of species found within that area. The relationship is linear or virtually linear (MacArthur & Wilson 1967, Rosenzweig 1995, 2001). Hence, the removal of, for example, an average of 5% of the rainforest in a given area per year should remove an average of 5% of the rainforest species in that area per year. Worldwide assessment of the rates of decline of tropical habitats allows one to rank the continents in terms of diminishing geographic area of tropical habitats. Given the broad expanse of the Neotropics that still remains, South America will remain a stronghold for tropical species in the near future; the prospects for Asia and Africa are considerably worse. The most critical area for the survivorship of tropical habitats and species is Madagascar, where species appear doomed. Because Malagasy prosimians comprise 44% of the terrestrial mammalian taxa, their demise will likely cause ecosystem collapse in Madagascar without prospects of a recovery (Jernvall & Wright 1998). Furthermore, the unique character of many Malagasy prosimian niches ensures that a simple invasion of vacant prosimian niches by other organisms is not possible. Some lemurs, for example, serve as important plant pollinators, and this behavior, along with the morphology that underlies it, is not easily reproducible.

If current primate faunas are treated as if they were paleontological faunas located in geological time, the impact of declining primate biodiversity can be assessed on a global basis. Jernvall & Wright (1998) assess worldwide ecosystem richness at three times (present, first period after present, second period after present) given the inevitable extinction of primates. Using taxon-free measures of ecological diversity such as body size, activity-patterns, and diet, they predict the effects of primate decline and loss on different tropical ecosystems. The probability of extinction was given by whether a species was characterized as critically endangered (ranked as first to go extinct), or vulnerable or threatened (ranked as second to go extinct). Jernvall & Wright (1998) conclude that Malagasy prosimians will become extinct first, with a catastrophic and irreversible effect on Malagasy ecosystems. Primate extinctions in Africa and Asia will be staved off a little, but they will have a major effect on tropical ecosystems in those continents. South America will be the continent least affected by primate extinctions, because of the ecological similarity of platyrrhines, and because members of other mammalian orders are ecologically equivalent to the platyrrhines that will become extinct.

It is an irrefragable fact that greater species diversity occurs in the tropics than in higher latitudes (Leigh *et al.* 1982, Janzen 1983, Whitmore 1984, Whitmore 1990, Richards 1996, Leigh 1999). And limited exploration of the rainforest canopy unleashes an avalanche of unsuspected new species (Ozanne *et al.* 2003). This latitudinal species gradient – with the highest species numbers in equatorial regions, and a steady diminishment in species with increasing latitude away from the equator – is a classic relationship in evolutionary ecology. It has been maintained through the Phanerozoic, deep into geological time. When geologists attempt to draw paleomaps of the surface of the earth, species richness is one of the lines of evidence that they use to position ancient fragments of lithosphere, and reconstruct the primeval continents and oceans. The smaller size of tropical niches also automatically assures greater alpha diversity because of their greater degree of specialization, but there are additional factors that contribute to species diversity in the tropics. Modern tropical species have smaller home ranges, measured over a latitudinal gradient, than temperate species. This correlation has been called Rapoport's Rule (Stevens 1989). Modern temperate species also have more generalized niches than tropical species, largely explained by the greater seasonality associated with higher latitudes. Rapoport's Rule suggests that temperate and circumpolar organisms have broader environmental tolerances than tropical organisms. However, tropical species are more vulnerable to the occurrence of seasonality or the degree of seasonality than are temperate or high-latitude species (Stevens 1989). Thus, global or local climatic changes causing increasing seasonality have a greater impact on tropical organisms.

Diet also affects the likelihood of extinction. Extreme dietary specialists will obviously become extinct if their specialized food resource is eliminated. The giant panda's dependence on bamboo is a classic example. However, even bamboo lemurs (genus *Hapalemur*) are not as specialized in diet as the giant panda. Folivorous primates are eclectic in terms of leaf species eaten, although they preferentially consume individual leaves or leaf parts (petioles) that are high in protein. In general, primates are omnivores. Omnivores tend to be significantly more rare than herbivores or carnivores in natural communities, and rarer than expected in computer simulations (Van Valen 1973a, Rosenzweig 1995). This rarity reflects the hazards of omnivorous diet. Contrary to received anthropological wisdom, being a generalist or jack-of-all-trades is not a common road to survivorship in the competitive arena of evolution (see Chapter 17).

Body size is also a factor in extinction. Large mammals suffer declines in population numbers that are more severe than in small mammals. This is because their initial population numbers are always lower, and their intrinsic rate of increase is lower than in smaller species (Newmark 1996, Owen-Smith 1998, Pimm *et al.* 1998). Nearly half of all living mammal species are rodents; thus, the average mammal is the size of a rodent. Humans tend to think about pachyderms as being large mammals. In actuality, primates qualify as large animals, given the size of the average mammal. The mean body size of living primate genera is 8.120 kg (Table 11.1). Prosimian primates (1.265 kg) are smaller than anthropoids (11.548 kg). However, using the figures given in Chapter 1, 79% of all living primate species are anthropoids. When this number is broken down, 42.5% are catarrhines and 36.5 are platyrrhines. Catarrhines (20.055 kg) are larger than platyrrhines (3.041 kg). Hence, by virtue of their body size, primates are more vulnerable to extinction than mammals in many other orders.

In living mammals, the average mammalian genus contains 4.2 species, the average primate genus contains 4.2 species, and there is a median of 2 species in each case (Purvis *et al.* 2000: Footnote 17). Primate speciosity thus exactly matches the mean value for mammals. Conroy (2003) agrees that primates rank as average with respect to mammalian speciosity. However, he expects a higher degree of speciosity. Using a data set of 249 living species and 62 genera for which recent body mass data are available, he discovers that 45% of the species are less than 2 kg in weight, and thus rank as small mammals. There should theoretically be more primate species. But, 37% of Conroy's genera are monospecific, and 63% contain three or less species. Two Old World monkey genera (*Cercopithecus* and *Macaca*) far exceed any other in species richness (Conroy 2003: Fig. 3A). Taken as a whole, the order appears impoverished in species – neither small nor large genera are speciose. Only one or two species occur in both six of the ten smallest genera and in eight of the ten largest genera (Conroy 2003: 789). This impoverishment must surely reflect both the extinctions of the Miocene hominoids and the very recent extinctions of the Malagasy subfossil prosimians. It is not accidental that the two most speciose genera occur among the Old World monkeys, because the cercopithecoid monkeys experienced a substantial evolutionary radiation during the Plio-Pleistocene that has not yet been diminished. When living primates are surveyed, the order has been cut down by the scythe of two extinctions – one very recent – and speciation in the remaining primates has been hindered by the pace of habitat loss in tropical ecosystems. Thus, unlike Conroy (2003), I do not think that primates are strange exceptions to the negative correlation between animal body size and species diversity. The Order Primates is not robust, with regard to its present vigor and future expectations. When one examines living primates, one finds mainly remnant groups that have managed to escape the scythe of extinction that cut through both prosimians and anthropoids within the last 15 my, and only few groups that form speciose nuclei. The order has been amputated by extinctions, and thus analysis of living primates produces strange results, when contrasted with other animals. A dispassionate observer from Mars would not gauge the future prospects of the primate order very highly.

Because of their large body size, non-human anthropoid primates contribute a substantial amount to the vertebrate biomass within tropical rainforests. Besides primates, among large, forest-dwelling vertebrates, only forest antelopes, tapirs, and suids (e.g., peccaries, forest hogs) contribute any significant biomass.[4] Human hunter-gatherers employing traditional hunting technology therefore routinely hunt non-human primates. Hunting technology has now been modernized in many areas, and extensive trade networks have formed to disseminate vertebrate meat. Consequently, non-human primates have become an important component of the current bushmeat trade. Skinned primate carcasses now often appear for sale at butcher shops in small tropical market towns, and even in markets within large cites like Nairobi and Kinshasa. A recent survey of markets in Ghana documented seven primate species sold as bushmeat. The olive baboon (*Papio anubis*) was the most common primate species sold (Brashares *et al.* 2004: online supplementary Table 1). At 26 kg, it was also the largest primate species offered for sale. Hunters tend to target species above a certain size range, and primates, being large, tropical forest vertebrates, are a general focus of predation. Bushmeat hunting of mammals in West African reserves increases when ocean fish species, an alternative source of protein, decline because of commercial overfishing (Brashares *et al.* 2004). Human predation on non-human primates can therefore be an important factor contributing to their extinction, especially under conditions of limited sources of dietary protein for human populations.

Field biology research on very large mammals demonstrates that they have a profound impact on the general ecology of their community (Owen-Smith 1998). The world of arboreal animals is difficult to explore from ground-level, and life in the high forest canopy remained mysterious for a long time. A variety of technologies have been devised to explore forest canopy: ropes, harnesses, permanent walkways, canopy cranes, and helium balloons. Initial exploration of the rainforest canopy reveals the intricate loom of relationships between arboreal species (Perry 1986, Moffett 1993, Ozanne *et al.* 2003). For example, primates serve an important role as fruit dispersers. The removal of primate species from tropical forests therefore affects tree reproduction and diminishes tree diversity.

Field biology work illustrates specific factors that are important in primate extinction. One instructive case is offered by vervet monkeys in two protected areas of Kenya, the Amboseli Reserve and Segera Ranch, in the Laikipia District of central Kenya. Vervets (*Cercopithecus aethiops*) are a widely distributed primate species in sub-Saharan Africa. They are generalized in diet, and are very tolerant of human presence. In fact, vervets are often classed with vermin and other crop pests, because they raid gardens and fields. A significant portion of their diet may come from domesticated plants, and they may be encountered boldly foraging for garbage in villages and on the outskirts of large cities. Nevertheless, vervets are going extinct in these two protected areas. At Amboseli, the problem appears to be caused by climate change and water table alteration that destroys local *Acacia* trees. From the 1970s through the 1990s, open woodland at Amboseli has been transformed into grassland. Although generalized in diet, vervets depend heavily on *Acacia* tree species for food (Figure 4.9). They also need some forest cover, because they absolutely require night time shelter off the ground, which is provided by trees. Extinction of vervets at Amboseli essentially occurs because, generalized in diet though they are, vervets cannot abandon their dependence on *Acacia* trees. Twenty-five species of *Acacia* are found in Amboseli. Against a background of declining numbers of *Acacia* trees in Amboseli, vervets increase the size of their home ranges to maintain the same degree of *Acacia* dominance in their diet. They increase home range size, rather than substituting non-*Acacia* species in their diet. The overall general quality of resources within the home range consequently suffers a broad decline (Lee & Hauser 1998). At Segera Ranch, vervet population numbers have fallen, because of persistent, long-term leopard predation (Chapter 8).

Tropical habitat fragmentation and alteration – mostly anthropogenic – is a serious factor in non-human primate extinctions. At the same time, species that are tolerant of human presence may boldly raid human fields and steal domesticated crops. I have personally witnessed baboons, vervets, and black-and-white colobus monkeys engaged in African crop raiding. Adult baboons audaciously sit by the side of major roads, fearlessly eating maize filched from local gardens. They cunningly watch traffic, and seem never to be hit by passing cars. Direct human persecution of crop raiding primates may be important, as they are perceived as vermin in this context. The degree to which non-human primates depend on domesticated plants for food can be assessed by general surveys of primate diet (e.g., Tappen 1960). The proximity of fields to forest or

Figure 4.9. Open *Acacia* woodland, Laikipia Plateau, Central Kenya. The species is *Acacia drepanolobium*. The round objects on the tree are woody galls containing ants. Local patas monkeys (*Erythrocebus patas*) preferentially feed on these ants.

forest corridors is an important factor that facilitates crop raiding by primates (Naughton-Treves *et al.* 1998). Raiding of garbage dumps may also be important. In some cases (e.g., olive baboons at Gilgil, Kenya), researchers have quantified primate consumption of human food at garbage dumps, and discovered that human food scavenged from dumps can contribute significantly to primate diet. Such scavenged food may alter activity-patterns, because animals do not need to forage widely to access high nutrient foods. A relatively sedentary lifestyle may develop (Quick 1986).

The threatened or endangered status of many non-human primate species is documented by their appearance on CITES (International Convention on Trade in Endangered Species), IUCN (World Conservation Union), and USESA (United States Endangered Species Act) Red Lists produced by conservation organizations. As one might predict, pongids receive much more attention than other primate species, and special organizations exist that are devoted solely to pongid survival in the wild or the rehabilitation of captive animals. But human destruction and disturbance of tropical habitats is so severe that no "wild" or natural areas exist any longer. Therefore, how can free-living animals remain or rehabilitated animals be returned to the "wild"? In less than a 20-year span, from 1983 to 2000, gorilla and common chimpanzee populations declined by more than 50% within the forests of western equatorial Africa

(Walsh *et al.* 2003). And these forests are the last citadel of wild African pongids. Free-living pongid populations may be miniscule. The Cross River gorillas offer a good example. Although a single osteological analysis maintained that the Cross River gorillas are a separate subspecies, this has not been borne out by genetic studies. Realistically, the animals appear to be doomed in the wild. Fewer than 250 animals exist – there are as little as 20 animals in some areas – and the tiny population sizes create perfect conditions for genetic drift to occur. Bushmeat hunting is also taking place.

It may be instructive to examine species persistence, rather than species extinction. What are the conditions that allow species to persist, rather than perish? An ability to coexist with humans and to survive in highly altered environments is obviously important. Many natural habitats have been seriously degraded, particularly in tropical islands like Madagascar and Zanzibar. Yet, Kirk's red colobus monkey (*Colobus kirkii*) has survived in Zanzibar in spite of extensive habitat degradation and a burgeoning human population (Struhsaker & Siex 1996). Kirk's red colobus monkeys exhibit some behavioral responses to habitat degradation that are strange for colobines. They exploit the resources of mangrove swamps, and are terrestrial to a far greater degree than other African leaf-eating monkeys. They also consume charcoal filched from traditional iron-working forges. The eating of charcoal may help to detoxify consumed plant foods that are alien to the colobine digestive system, but that are being ingested out of necessity as starvation fall-back foods (Struhsaker *et al.* 1997). Powdered charcoal is used as a first aid device by humans who have taken poison, because the charcoal adsorbs or neutralizes some poisons. In any case, the Zanzibar red colobus monkeys demonstrate how behavioral plasticity and adaptability can offset changing environmental conditions that would otherwise be too extreme for survival.

The assessment of extinction risk has become an important focus in conservation biology. Chance or demographic accident, such as the production of a vanishingly low number of females, obviously plays a role in extinction events (Lande 1993). By their very nature, the impact of random catastrophes that affect major evolutionary events can only be assessed after the fact (Cachel 2000a). However, it is clear that some species are more prone to extinction than others are, and sometimes this predisposition can be linked to ecological factors. One such factor is the degree of tolerance for human alteration of natural habitats. Even closely related primate species can differ in their tolerance for humans and human disruption of the environment. Richard *et al.* (1989) divide macaque species into "weed" and non-"weed" categories. The weed species persist and even thrive under conditions of extreme environmental disruption, and may eagerly consume agricultural products. As for the non-weed species, primate survivorship in disrupted environments may be highly dependent upon a combination of local conditions. For example, primate species that are declining or extinct elsewhere, such as common chimpanzees, persist under highly altered conditions on Tiwai Island, Sierra Leone (Fimbel 1992, 1994).

Endnotes

1. There is an extensively documented late Cretaceous flora on the northern Arctic Slope of Alaska, which was then located at a latitude between 75 and 85° N. It is difficult to understand how a polar forest could arise, given seasonal differences in sunlight, but deciduous angiosperm leaves are abundant (Skelton 2003).

2. This explains the relative prevalence of prehensile tailed vertebrates, including several prehensile tailed primate species, in the Neotropics. Seven vertebrate families (including Family Cebidae) have prehensile tailed species in the Neotropics (Emmons & Gentry 1983). Prehensile tails, which can support the entire weight of an animal's body, serve to anchor animals in an unsteady canopy environment. The ancient presence of elephants and elephant relatives in Africa and Asia, and their relatively recent colonization of Central and South America after the establishment of the Isthmus of Panama, may account for the difference in liana structure. Living elephants pull down and destroy trees, as they eat leaves and bark. Elephants use lianas to tug down forest trees. Old World lianas, especially in Africa, have responded to this long-term selection pressure by becoming large, thick, and woody – i.e., resistant to elephants.

3. Endemic insectivores, viverrids, and giant elephant birds represent some of the remarkable Malagasy species that evolved from a limited number of colonizers. The viverrids illustrate not only the diversity of mammal radiation on Madagascar, but also its biodisparity. The falanouc (*Eupleres goudotii*) is mongoose-like, the fanaloka (*Fossa fossa*) is fox-like, and the fossa (*Cryptoprocta ferox*) is leopard-like. Each of these genera is monospecific, and sorted into a separate subfamily: Subfamily Euplerinae, Subfamily Fossinae, and Subfamily Cryptoproctinae, respectively. An extinct endemic order (Bibymalagasia) is convergent to the African aardvarks. The only nominal ungulate present is an amphibious artiodactyl, the pygmy hippopotamus.

4. Arthropods, especially ants, which are key canopy herbivores, contribute far more to tropical forest biomass than vertebrates do.

5   *Anatomical primatology*

Introduction

Anatomy and morphology are linked. Anatomy is the description of biological structures. Morphology is the philosophical interpretation of these structures in terms of general categories such as archetypes, and, later, in terms of evolutionary processes such as adaptation. Anatomical primatology has a number of components that contribute directly to the study of primate and human evolution. These are functional morphology, ontogeny and anatomical genomics, and phenotypic variability. All three of these areas were studied during the twentieth century, and are the focus of continuing research. However, the end of the twentieth century witnessed an epidemic of skepticism about natural selection, adaptation, and genetics. The disbelief was fostered especially among paleontologists, but also seeped into other realms of biology. Disbelief in natural selection and adaptation have a direct impact on the study of morphology, because adaptation is the foundation of functional morphology. In addition, phylogeny reconstruction and cladistic systematics downplayed the utility of morphology and trivialized anatomical research.

Phylogeny and cladistic methodology

During the 1980s, it appeared that the death-knell had sounded for functional morphology. How did this happen? One trend that was largely responsible for the diminishment of functional morphology was that zoology and paleontology emphasized phylogeny and cladistic methodology. During this time, the study of anatomy was often pursued only for purposes of acquiring a shopping list of character traits for phylogeny reconstruction. Once a phylogeny was obtained, it was assumed that research was complete – the phylogeny itself was the culmination of research. And the phylogenies had to be reconstructed in a particular way. They had to be reconstructed using cladistic methodology.

Cladistic methodology was the invention of the taxonomist Willi Hennig, who argued that a taxonomic system that reflected evolution contained the most information and was of the greatest utility. When Hennig's monograph

Phylogenetic Systematics was translated into English (Hennig 1966), it exercised a profound effect on modern taxonomy. The methodology advocated by Hennig was rigorous, and was ultimately based on a peculiar idea of speciation. But cladistic methodology pledged that objective phylogenies could be created, and that they could be independently replicated. In theory, researchers anywhere in the world who used cladistic methodology on the same datasets would conclude with the same phylogeny. It is important to note that true phylogenies were supposed to be produced. That is, cladistic methodology was supposed to yield a picture of the actual course of evolutionary events – what evolution created – and was envisioned as not just another mode of classification. The phylogeny was not an hypothesis about evolutionary events, because reciprocal illumination and other devices yielded the true picture (Hennig 1966). In support of the methodology, many software packages (e.g., PHYLIP, PAUP, MacClade) have been produced since the 1980s, and are widely available. This makes it extremely simple for researchers to create cladistic phylogenies or cladograms by using the software to analyze their datasets. A discussion of tree-building methods is now a routine component of many papers. Cladistic methodology can be applied not only to phenotypic traits, but also to amino acid sequences in proteins and to DNA nucleotide analysis.

I give here a brief exegesis of cladistic methodology, although many textbooks and monographs explore it fully (e.g., Eldredge & Cracraft 1980, Wiley 1981, Forey *et al.* 1992). The establishment of clades or lineages is emphasized. Hence, the name of the system. The evolutionary sequence of divergence from a common ancestor always occurs through dichotomous branching. Phylogenies therefore consist of a series of bifurcations. The ancestral species branches into two sister species, and the ancestral species goes extinct at the moment of dichotomy. Primitive traits (plesiomorphies) must be differentiated from derived traits (apomorphies). Sister groups are determined only by the joint possession of shared derived traits (synapomorphies). The strict dichotomous branching of lineages emphasizes speciation as the very heart of evolution. The speciation event always creates two sister species, and diminishes the ancestral species so much that it goes extinct. Discontinuities are stressed. This is reflected in the binary coding of traits (i.e., 0,1) that occurs in cladistics. Continuous variation is rejected. The only valid taxa are those that contain all the descendants of a common ancestor and the ancestor itself. These are monophyletic taxa, and the invalid taxa are termed either polyphyletic or paraphyletic. When many species and traits are being examined, many possible phylogenetic trees (cladograms) can be generated. Several mathematical methods can then be used to select the best phylogenetic tree. One method that is often used is to invoke parsimony, and to choose the tree that minimizes the number of changes necessary to produce shared derived traits.

The adaptive significance of characters or traits is not important in cladistics, only the polarity of trait transformation, and the ability to differentiate primitive from derived traits (Eldredge & Cracraft 1980). Adaptive traits that evolve through parallel or convergent evolution cross the boundary lines of clades, and therefore distort cladistic analysis. In cladistic methodology, traits or characters cannot be weighted using functional morphology or adaptation. In fact, functional complexes that reflect the entrance into novel adaptive zones are rejected as taxonomically useless. Other classification systems (i.e., evolutionary systematics) use novel adaptations as the basis for establishing higher taxa such as families or orders, but this procedure is rejected by the cladists. Furthermore, because lineages or clades are the primary focus of attention, grades or levels of structural organization that are based on adaptation are anathema (Cachel 1992). Grades may not always be clades, because adaptive traits can evolve in separate lineages through parallel or convergent evolution.

The cladistic methodology promised that any researchers given the same body of data would independently arrive at identical phylogenies. If this were so, a phylogeny could become an autonomous datum against which to assess the course of behavioral or anatomical evolution. For example, a phylogeny that unequivocally showed the separate evolution of a particular behavior or morphology in separate lineages would demonstrate that the behavior or morphology had adaptive significance in particular environmental circumstances and would allow researchers to ferret out the forces of natural selection responsible for the origins of the trait. Yet, cladistic phylogenies are not necessarily replicable. Cladistic phylogenies have proven so divergent that some researchers employing the cladistic methodology now eschew any connection between cladograms (that were originally supposed to be phylogenies) and evolution. These researchers practice cladistics without implying that a phylogeny generated by evolution will be revealed (Scott-Ram 1990). The resulting school of classification is called transformed cladistics. Some paleoanthropologists who were formerly dedicated to cladistic methodology (e.g., Wood 1991, Wood 1992) have since turned to using adaptive complexes in hominid classification (Wood & Collard 1999a, 1999b). Other paleoanthropologists also employ adaptive complexes or evolutionary ecology in the analysis of hominid evolution (McCollum 1999, Wolpoff 1999, Conroy 2002, Cela-Conde & Altaba 2002). This agrees with general studies of morphological characters or traits that are correlated or integrated into a functional complex. This phenomenon is termed morphological integration (Olson & Miller 1958). Traits or characters that are morphologically integrated do not assort independently from each other – they are linked together – and therefore characters within adaptive complexes need to be weighted when conducting phylogenetic analysis (Emerson & Hastings 1998). Hence, researchers now give adaptive complexes special weight, and use new adaptive zones or strategies to signal the advent of new genera or higher taxa. This procedure follows the school of classification called evolutionary systematics (practiced by Gregory, Simpson, Mayr, Le Gros Clark, etc.) – a school of classification that was disparaged by the cladists as not being rigorous and objective, and as possessing more art than science.

Thus, contrary to the original promise, cladistic phylogenies cannot necessarily be duplicated. There are two major reasons why identical phylogenies do not necessarily appear. The first reason is that the polarity of traits, or the direction of evolutionary transformation, is not easy to determine. The second reason is that parallel and convergent evolution create homoplasy, the bête noire of cladistics. Homoplasy occurs when organisms that are not closely related possess similar traits (Sanderson & Hufford 1996). These traits appear similar, although they are not homologous, or derived from a common ancestor. However, traditional methods of analyzing morphology emphasized many methods that allowed researchers to define homologous structures, differentiate primitive from derived traits, infer closeness of descent, and to reconstruct the characteristics of the ancestral species (e.g., Gregory 1951, Maslin 1952, Simpson 1961: 87–106). In addition, new perspectives offered by evolutionary development also promise to unravel some of the problems of detecting and defining homology (Minelli 2003).

Another major problem is the choice of datasets for cladistic analysis. Analysis of the same species using different anatomical areas yields different cladograms (Cachel 1996a), presumably because of mosaic evolution – different areas of the phenotype evolve at different rates. Mammalian species diagnoses are often based on cranial and dental data. Therefore, a great dataset of craniofacial and dental traits has been amassed for catarrhine primates. This is especially true for hominoids and some Old World monkey species (macaques, baboons, mangabeys, drills, and mandrills). When cladistic phylogenies derived from craniofacial and dental traits of these catarrhines are compared with those derived from molecular evidence, there is no concordance (Collard & Wood 2000, Gibbs *et al.* 2000). If one assumes that the molecular phylogeny is the true one, and that any phylogeny generated from craniofacial and dental traits is suspect, one must then abandon virtually all hope of assessing phylogeny from the fossil record. The vertebrate fossil record is, largely, a record composed of teeth. A long trail of teeth leads back to the origin of vertebrates, and the diagnosis of vertebrate species and genera is based primarily on craniofacial and dental traits. However, a three-dimensional morphometric analysis of the hominoid temporal bone examines a complex shape, and creates a phylogeny that is in agreement with a molecular phylogeny (Lockwood *et al.* 2004). Similarly, one major researcher (Oxnard 1984) has demonstrated that, if multiple anatomical regions are used to generate a phylogeny, rather than a single region (e.g., tooth or shoulder structure), the resulting phylogeny agrees with one generated from molecular evidence. In addition, when many soft-tissue traits ($n = 197$) are used to assess phylogeny in living hominoids, the results agree with molecular evidence (Gibbs *et al.* 2000). Molecular phylogenies that incorporate multiple lines of evidence also appear to be more concordant than those based on a single genetic locus, loci from a single chromosome, or a single type of DNA (Templeton 2002). As a result, the first molecular phylogenies based on albumin or mtDNA are now being revised to incorporate much more genetic evidence. Hence, whether anatomical or molecular evidence is being examined, the more traits or genetic loci that are assessed, the less divergent the resulting phylogenies are.

Hawks (2004) performed Monte Carlo simulations on a phylogenetic model, in order to investigate the effect of the number of traits, sample size differences, trait frequency, and non-independent traits on a cladistic phylogeny using parsimony. A large number of traits were used, traits were statistically independent, and their genetic substrate was known. Yet, the true phylogeny was produced only erratically, and, if less than 10% of the traits were non-independent, a false result was produced. Small sample sizes and a small number of independent traits almost guarantee a false result – a disaster for paleoanthropology, where the database is small, and only a few independent traits can be gleaned from fragmentary fossils. Concordance between cladistic analyses can also be found in phylogenies that do not use the cladistic methodology. If hominid fossil samples were statistically adequate, new fossil specimens or species would not radically transform cladistic trees (Hawks 2004: 217). Yet, this often occurs in paleoanthropology.

Some researchers are disillusioned about the ability of morphological traits, especially craniofacial and dental traits, to reveal phylogeny (Collard & Wood 2000). Other researchers believe that an understanding of the genetic basis of morphological traits will infuse practitioners of the cladistic methodology with a new respect for morphological complexes and their function when phylogenies are generated (Lovejoy *et al.* 1999). However, linking morphology to its genetic substrate may be difficult to accomplish. For example, molar crown morphology has been the basis for many species diagnoses among mammals. Yet, study of the genetic bases of molar crown morphology has yielded no clear picture of the link between molar crown morphology and genetics, or even an understanding of the genetic basis for tooth identity. Most genetic mutations that affect the dentition cause teeth to be small, deformed, or missing. These mutations thus yield virtually no information about the genetic basis for normal tooth morphology. Furthermore, it is currently not known how the intricate development of a particular dental trait within an individual is associated with the range of dental variability documented within a population (Jernvall & Jung 2000).

Recent experiments on mouse dentition highlight the relative ease of convergent evolution, and the high probability of homoplasy (Kangas *et al.* 2004). Dental traits – sometimes the only evidence of fossil taxa – may thus confound phylogenies. Lower molar crown development was compared in normal mice and two mutant mouse strains. The three strains differ in the expression of the protein ectodysplasin, which affects growth and development in most ectodermal organs. Many dental traits are simultaneously affected by ectodysplasin activity levels. These traits include tooth number, tooth size, cusp number and shape, and transverse and longitudinal cresting (Kangas *et al.* 2004). Of the 19 dental traits examined, traits that were either non-polymorphic or had a small level of polymorphism responded most reliably to ectodysplasin signaling. Yet, these are traits that would have likely been chosen by taxonomists to classify different taxa. Individual dental traits are not determined by separate genes. Instead, a single protein signal simultaneously affects multiple dental traits. These dental traits, which are routinely analyzed by paleontologists and coded for cladistic analysis, are not independent from each other. If a cladistic analysis – which assumes the independence of traits – is performed, phylogenetic information is concealed. The true phylogeny will be buried. In fact, Kangas *et al.* (2004: 214) recommend an alternative to the normal cladistic approach of equal weighting of characters in fossil teeth, especially when transitional morphology is rare or not preserved in the fossil record. If dental traits that are most sensitive to ectodysplasin (i.e., those relating to lateral cusp position and cresting) appear to covary, then a different method of coding taxonomic characters should be employed. However, this may not be possible, given the small sample sizes routinely encountered in fossil material. The outstanding significance of this mouse experimentation is that it highlights the relative ease of convergent evolution, and the occurrence of homoplasy in mammalian dental traits.

Like anatomical, molecular, or genetic data, behavior can also be used to generate phylogenies, although behavior has been used far less frequently than other categories of evidence. The greatest utility of the cladistic methodology is its ability to screen large amounts of data from an unstudied group or from a relatively neglected area of phenotypic or genetic variability, and to reveal unsuspected patterns. A striking example of such a revelation occurs when cladistic methodology is used on elements of primate social organization (Di Fiore & Rendall 1994). The analysis reveals the peculiar and derived status of Old World monkey sociality with respect to other extant primates. Old World monkey societies exhibit far greater amounts of female driven intra-group competition. This revelation is important for models about early hominid sociality, because it implies that the ancestors of hominids need not have lived in social groups where strong female dominance hierarchies, female–female food competition, or competition between matrilines existed (Chapters 15 and 17).

Genetic data are now routinely used to reconstruct population history, through inference of gene flow, dispersal, or isolation by distance. These inferences can be tested using the technique of nested clade analysis, where genetic data, cladistics, and geography test hypotheses about past population history (Templeton 1998). Population migrations or expansions can be detected. Templeton's invention of nested clade analysis in order to test for variance versus dispersal or for isolation by distance is an example of the productive use of cladistics. Another productive use of cladistics is the examination of how morphological divergence is partitioned during adaptive radiations. Morphological disparity in fossil or living organisms can be examined through time by using a molecular phylogeny to assess disparity occurring in lineages from the root of the phylogeny along through its various speciation nodes (Harmon *et al.* 2003).

Padian (1995) argues that functional hypotheses should not be subordinated to phylogenies. ". . . [F]unctional explanations can be developed without recourse to specific phylogenies. In fact they *should*, because they depend on more than phylogenetic plausibility. Functional hypotheses, if they have any intellectual content, have to be able to stand or fall on their own merits, and must be testable on their own terms. Then they can be compared to plausible phylogenies in order to arrive at the most robust explanation of all the available evidence." (Padian 1995: 271–272). Padian then suggests a five-step protocol for the analysis of function, beginning with identifying an adaptation in a purely functional manner, and then listing the taxa that share the adaptation. Phylogenetic analysis of these taxa and appropriate outgroups can then yield functional stages that can be compared between groups that share the adaptation (Padian 1995: 272).

It has been claimed that phylogenies generated by the cladistic methodology are the touchstone against which to test all evolutionary research, because they reveal absolute knowledge of evolutionary events (Harvey & Pagel 1991). But, if the phylogenies are not replicable, or if phylogenies generated from different lines of evidence are not in concordance, how may one assess the adaptive nature of a particular behavioral or morphological complex? Is it possible that researchers have been pursuing an endless tautology? The tautology is as follows: if a complex exists, it is adaptive; it is adaptive because it exists. This tautological fallacy is exactly what was claimed by Gould & Lewontin (1979) to taint anatomical research. However, use of all available evidence, and comparative analysis of groups that share adaptations, frees researchers from the tautology (Padian 1995). In addition, extant close relatives can be used as a comparative cohort even when reconstructing soft tissue such as muscles in fossil taxa (Witmer 1995), and the mutual interactions between phylogeny and anatomical modeling can reveal function in fossil forms (Weishampel 1995).

A final problem introduced by cladistic methodology is the insistence that any morphologically or molecularly distinct organisms should be recognized as species. This is the phylogenetic species concept (Eldredge & Cracraft 1980). Species are populations that differ from all other populations by at least one taxonomic trait, and that have "a parental pattern of ancestry and descent." If species can be diagnosed, then the phylogenetic species concept can be applied. Reproductive isolation is not necessary, and geographically distinctive morphs can be recognized as separate species. Ecological information is ignored, because adaptive distinctiveness (and adaptation itself) is rejected. However, the phylogenetic species concept insists that the species is more real than the individual organisms incorporated within it. These organisms are variable, but the species boundaries established under the phylogenetic species concept are fixed and unvarying. This is a return to typology. Groves (2004) clearly articulates the impact of the phylogenetic species concept on primate taxonomy. "A species has one or more fixed differences from other species; it is 100% different; so one asks not how much difference is necessary to decide whether a population rates as a species, but what proportion of individuals differ? Any kind of character will suffice, be it color, size, vocalization, or a DNA sequence, as long as there is a reasonable supposition that the difference is heritable . . . . The difference between the PSC and most other specific concepts is that under the PSC we look for pattern; under other specific concepts we look for process." (Groves 2004: 1110–1111). Thus, the phylogenetic species concept rejects evolutionary theory, including ideas about speciation. Furthermore, Groves (2004: 1112) argues that a major advantage of the phylogenetic species concept is that "We do not need to speculate on the functional significance of specific differences." Hence, Groves identifies adaptation as a problem – researchers only speculate about the adaptiveness of traits. This clearly implies that there are no objective or systematic methods of studying adaptation.

The phylogenetic species concept and cladistic methodologies also multiply the number of species, because, if organisms are diagnosably different, they are identified as different species. Application of the phylogenetic species concept to vertebrates has resulted in 48% more species being recognized on average than use of the biological species concept has (Isaac *et al.* 2004). This is true for primates. Groves (2001a) uses the phylogenetic species concept in his monograph on primate taxonomy, and doubles the numbers of primate species. Isaac *et al.* (2004) note that primates are a classic example of "taxonomic inflation," because the taxonomic doubling effect occurred even though only about 30 actual new discoveries were made, and most of these were originally described as subspecies.

Differences established by new analytical techniques and methodologies, such as mtDNA, Y chromosome haplotypes, microsatellite DNA, and allozymes, appear to drive much of the recent species expansion (Crandall *et al.* 2000). However, paleontologists studying morphology may also generate a plethora of new fossil species. The very dense, stratigraphically controlled, and well dated Eocene primate record from the Bighorn Basin appears to demonstrate anagenesis, rather than bifurcating cladogenesis, but species boundaries may be indistinct to paleontologists (Rose & Bown 1993). However, researchers studying living organisms and interested in conservation issues confront many practical problems when species are multiplied. New species generated by the phylogenetic species concept tend to arrive on lists of threatened species. Thus, of the 24 primate taxa just added to the 2000 IUCN Red List of threatened species, 17 are new species created under the phylogenetic species concept, while only 7 of these primate species have actually changed their conservation status (Mace *et al.* 2003).

Adaptation and the "adaptationist program"

Initiating another trend that created difficulties for functional morphology, Steven J. Gould and Richard Lewontin (1979) characterized the study of adaptation as an antiquated, nineteenth century endeavor. Lewontin (1974) had long been skeptical about the ability of natural selection to cause evolutionary change, because he viewed natural selection as being uniformly weak. Lewontin also harbored doubts about whether any evidence for adaptation existed at all, including among living humans. For example, in a book devoted to human variation and variability, Lewontin (1982) denied the occurrence of any adaptations in living humans. And if adaptation could not be found in humans, the most well studied of all organisms, it could not be found anywhere. Adaptation did not exist. Gould & Lewontin (1979) scorned the study of adaptation as generating merely a collection of "just-so" stories in the style of Rudyard Kipling's famous stories for children. They disparaged the study of adaptations as "the Adaptationist Program," a mode of thought in which all structures and organisms are viewed as adaptively optimal. They claimed that the study of adaptation was tautological: whatever existed was adaptive, and it was adaptive because it existed.

The Gould and Lewontin denigration of adaptation was so successful that the study of adaptation precipitously declined during the 1980s. Major evolutionary researchers felt obliged to justify the study of adaptation (Mayr 1983). Mayr's *The Growth of Biological Thought* received first a disparaging review in *American Anthropologist* (Tattersall 1984), and then a defense in the same journal

(Cachel 1986). In his Presidential Address to the American Society of Naturalists, Janis Antonovics (1987) advocated a dismantling of the neo-Darwinian synthesis, which he tied to "the Adaptationist Program", and a simple-minded faith in the existence of natural selection. Because of this philosophical tumult, the discipline of vertebrate morphology dangerously faltered, only to be galvanized into activity again after about a decade of hesitation (Liem 1989). Rose & Lauder (1996b) entertainingly describe a lecture that took place during this period in which a beleaguered speaker was obliged to substitute the word "banana" for the word "adaptation," in order to avoid being interrupted at every turn by outbursts from the audience, protesting the use of this forbidden word – so controversial had the idea become. The Gould & Lewontin (1979) disparagement of adaptation was finally dismantled into its rhetorical constituents (Borgia 1994, Queller 1995). Yet, even now, studies of adaptation sometimes begin with an obligatory apologetic statement about the value of such studies.

However, as of 2004, a revolution has occurred in the study of morphology. Functional morphology has been newly resurrected, in large part because of genetic studies. In their attack on adaptation, Gould & Lewontin (1979) did not envision that genetics would someday be used to study morphology and adaptation. Both Gould and Lewontin were suspicious of genetics. Gould argued that the neo-Darwinian synthesis was flawed because paleontologists had yielded to geneticists in assessing evolutionary processes and rates (Gould 1983, 1984). Gould considered that genetics was acceptable only with the possibility that macromutations or systemic mutations might instantaneously produce a new species or a higher taxonomic category (Gould 1980). This idea had been bruited before by Richard Goldschmidt (1940), an early twentieth century evolutionist. Goldschmidt argued that a new species could theoretically arise from a strikingly altered individual whose radically altered phenotype was caused by a macromutation or systemic mutation. This is not population-thinking, but essentialism (Chapter 1). Essentialism accepts that species can arise with a single newly transformed individual, rather than requiring a distinctive, variable population. Consequently, essentialism is a return to the typological thinking that prevailed before the advent of natural selection. Field biology evidence that documents the reality of natural selection, or the role of ecology in incipient speciation is disparaged in favor of typological species, untrammeled by variability or gradations between species.

Lewontin (1974) provided evidence of surprising amounts of unsuspected genetic variability in natural populations. Ironically, however, from this he drew the conclusion that natural selection was always too minute a force to winnow out genetic variability. If natural selection were uniformly weak, it was not responsible for macroevolution – i.e., it could not create the major patterns of evolutionary change. As the product of natural selection, adaptation was also suspect. Gould and Lewontin were jointly mistrustful about the supposed political consequences of genetic research (Gould 1978, Lewontin *et al.* 1984, Lewontin 2000). Both were associated with the Sociobiology Study Group of Science for the People, which hurled ideological thunderbolts at sociobiology and behavioral genetics. Gould and Lewontin's comical apprehension about social policy was fueled both by a belief that genetics, especially behavioral genetics, inevitably shades into biological determinism, and by Marxist ideology that mandates the infinite malleability of human nature. The attack on adaptation (Gould & Lewontin 1979) was launched as part of the major campaign against sociobiology (Segerstråle 2000: 101–126). If one asks how two prominent twentieth century evolutionists could become deeply embroiled in attacks on natural selection, adaptation, and behavioral genetics, the answer is that they were motivated by a political agenda. They believed that the ends (eliminating sociobiology) justified the means (trivializing and discarding the neo-Darwinian synthesis).

As of the early twenty-first century, the biomechanics of vertebrate bones and joints has been galvanized as a research area through the introduction of studies examining stress and strain in living organisms. Anatomy is now the linch-pin between genetics, cell biology, and physiology. Genetics has given morphology a novel direction through the introduction of comparative evolutionary studies and anatomical genomics, and investigation of the anatomical consequences of gene expression or disruption. A whole sub-set of reports has been accruing on the phenotypic consequences of mouse gene manipulations (knock-outs or knock-ins). The pattern of expression of genes or proteins in developing tissues is of interest, and evolutionary development ("evo-devo") is revealing the secrets behind the anatomical patterning of the embryo. A gene or protein's patterns of expression in developing tissue can sometimes now be followed through time and space within an embryo.

Finally, although Gould and Lewontin decried the influence of the geneticists on the New Synthesis or the neo-Darwinian synthesis of the early twentieth century, the study of genes is now creating a major revolution in evolutionary theory. Specifically, comparative analysis of genomes is now revealing the genes underlying keystone adaptations and adaptive differences.[1]

Studying adaptation

Researchers continued to study adaptation after 1979, even during the height of the post-spandrel era (e.g., Hildebrand *et al.* 1985). There are five general approaches to the analysis of adaptation. First, an hypothesis is advanced about the function of a trait, predictions are generated, and these predictions are tested through experimentation with living organisms (e.g., through examination of physiological response during locomotion). Second, phenotypic differences occurring through a geographic range can suggest that a species is experiencing adaptation to environmental variables. A genetic basis may ultimately be identified for a certain variable phenotypic trait, and its expression may be shown to vary with both genetic and environmental differences. Third, a changing phenotypic trend through time is linked with concomitant changes in the physical or biological environment. Studies of living organisms demonstrate that the same phenotypic response occurs as an adaptation to the same environmental changes. Fourth, experimental alterations in the phenotype can be shown significantly to affect normal function of the organism in a predictable fashion. Fifth, comparison of different species may reveal how adaptation has produced similar morphologies and similar lifestyles through convergent or parallel evolution.

It is important to note that methodologies for the study of adaptation yield results that are verifiable in terms of biomechanics, mathematical modeling, and laboratory testing of physiological response. Methodologies for studying adaptation continue to be refined (Gingerich 1993a, Rose & Lauder 1996a). Multiple lines of evidence are also now commonly used in the study of adaptation. The concordance of several lines of evidence is more persuasive than results produced by a single method of study.

Optimization theory is another approach used to study adaptation. It should not be confused with the discovery that the best possible designs exist in nature – a viewpoint mocked by Gould & Lewontin (1979) as like that of *Candide's* Dr. Pangloss, who continually avowed that this was the best of all possible worlds. The world is not being viewed as the product of intelligent design. Instead, optimization theory is used by mathematicians to discover the theoretically best possible behaviors or structures. Mathematicians happily study optima in an abstract fashion, and are pleased to discover approximations to optimal design in nature. Animal behaviorists have often applied optimization theory to activities like feeding, territoriality, competition, and reproduction. Optimal skew theory has even been applied to predict optimal group size, stable group size, and the division of resources in social foraging groups (Hamilton 2000). Engineers employ optimization theory when assessing optimal designs for weight-bearing structures like bridges, roads, or buildings. In *Optima for Animals*, Alexander (1996) illustrates how optimization theory also can be applied to biological structures when studying bone strength, bone shape, and the mechanics of locomotion.

Alexander (2003b) surveys the many ways in which locomotion can be studied. These include kinematic analysis utilizing video, establishing the center of gravity, measuring forces, measuring energy consumption, recording the electrical activity of muscles, and examining the material properties of bone and cartilage. To this one can add a long list of methodologies: the measurement of wet or dry muscle weights, physiological cross-section of muscles, strain gauge analysis, finite element analysis, Fourier analysis, experimental manipulations of physiology, morphological phenotype, and behavioral phenotype, and analyzing the degree of morphological integration. Biomechanical analysis of bone has been particularly detailed. In fact, because bone is a living tissue, it is extraordinarily malleable. The internal architecture of bone can illustrate its use, given known properties of structural design that respond to weight-bearing, locomotion, and any activity imposing loading regimes. Cross-sectional properties of long bones can indicate relative bone strength and variation in activity patterns within members of the same population (Ruff & Hayes 1983). The thickness or compactness of cortical bone, the geometry of cross-sectioned or CT-scanned long bones, the distribution of cancellous bone and its biomechanical significance, and computer simulations of how bone reacts to particular stresses are all used to infer function (Thomason 1995). Statistical software is available to quantify and to compare the compactness of bone in a cross-sectioned image (Girondot & Laurin 2003).

The functional morphology of fossil species

Functional morphology is predicated upon adaptation. Any assault on adaptation vitiates the study of function. Without function, morphology becomes mere descriptive anatomy. Difficult as it is to study function in living species, it is even more difficult in fossil species – difficult, but not impossible. In the early nineteenth century, Georges Cuvier used the muscle scars on fossil bones literally to flesh out the musculature of extinct vertebrates. William King Gregory was one of the first twentieth century researchers to emphasize functional morphology (Simpson 1971). Gregory began by studying the extinct titanotheres of the American West. He had been given the assignment of studying these animals by his superior, Henry Fairfield Osborn, President of the American Museum of Natural History. Gregory therefore initiated a series of comparative dissections of living perissodactyls, members of the same order as the titanotheres, in order to infer musculo-skeletal relationships, muscle leverage, and muscle activity in titanotheres. He later studied living Malagasy prosimians, in order to reconstruct the functional anatomy of *Notharctus*, an extinct Eocene primate (Gregory 1920). Figure 5.1 illustrates some typical muscle maps that Gregory reconstructed, in this case, for the humerus of *Notharctus*. This work allowed him to infer specifics of locomotion in this long-extinct primate. Rainger (1989) summarizes William King Gregory's contributions to the study of functional

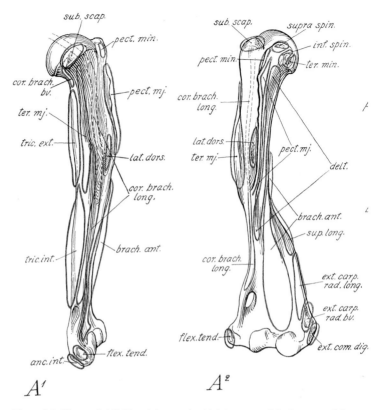

Figure 5.1. The medial (left) and the anterior (right) aspects of the humerus of the fossil Eocene primate *Notharctus*, showing muscle maps based on a comparative analysis of living lemurs. From Gregory (1920: Plate 28).

morphology, and his use of functional morphology in inferring the lifeways of fossil vertebrates. Rainger (1991) analyzes how vertebrate evolution studies at the American Museum of Natural History, under Osborn's tenure as president of the museum, generated an enduring fascination with prehistoric life throughout American culture. Osborn's conviction that human origins had occurred in the Gobi Desert led the American Museum to sponsor the first fossil-hunting expedition to the Gobi, where Cretaceous dinosaurs and mammals (but no hominids) were discovered. Osborn's conviction also focused academic and popular attention on human fossil finds from Java and China. During the early twentieth century, Asia appeared to be the fount of human origins.

There are a number of enduring problems in the study of fossil organisms. One problem is the existence of morphology that has absolutely no modern equivalent or analog. A good example of such morphology is the recent discovery of four-winged, arboreal, gliding dinosaurs that were ancestral to birds (Xu *et al.* 2003). The hind wings were lost when active flight became possible. There appears to be some consensus that fossil members of taxonomic groups with many extant species offer the greatest likelihood of inferring function, because a large array of living related species can be studied (Lauder 1995). This would appear to damn the hopes of paleoanthropologists. It also appears to explain continuing arguments about the lifeways of the earliest hominids, the australopithecines. An enduring and intractable problem is the fragmentary nature of fossil finds. This, in itself, creates problems of reconstructing morphology and inferring function. However, given the current impetus on the use of cladistic methodology to reconstruct phylogeny, another problem is that fragmentary fossils create missing data – this leads to unresolved phylogenies. There appears to be no easy solution to the problem of missing data when phylogenies are generated using cladistic methodology (Kearney & Clark 2003, Norell & Wheeler 2003, Wiens 2003).

New imaging techniques have been productively used in studying fossil morphology. Three dimensional laser camera scans allow specific areas to be easily quantified. CT scans can reveal the internal architecture of bone or allow endocranial or other volumes to be assessed without any damage to a fossil. For example, high resolution X-ray CT scans have been used to establish the fact that *Tremacebus harringtoni*, a 20 my old extinct platyrrhine primate from Argentina, had small olfactory bulbs (Kay *et al.* 2004). This indicates that *Tremacebus* was probably diurnal (Figure 5.2).

Vertebrate paleontologists have continued to refine the study of function in fossil species, even in taxa that have no living analogs. The vertebrate paleontologist John Ostrom, for example, initiated the furor over warm-blooded dinosaurs by analyzing the fossil remains of *Deinonychus*. This carnivorous bipedal dinosaur had scimitar-like claws on the second digits of its feet; these claws were drawn up off the ground when the animals were walking or running. Ostrom (1969) postulated that *Deinonychus* had held its prey fast with forelimbs and jaws, and used the massive claw on one foot to disembowel a victim while counter-balancing on the other foot. The long rod-like tail acted to maintain balance as the head and forequarters were thrown forward. This activity seemed to require a degree of quickness and agility that made endothermy likely. Vertebrate paleontology is now replete with similar functional analyses of extinct species (e.g., Witmer & Rose 1991, Gingerich 1993a).

Analyzing the locomotion of the earliest hominids, the australopithecines, has always presented a challenge to human paleontologists, because there are no living analogs to this group, which possesses a mixture of morphologies, some strongly indicating bipedality and some indicating arboreal climbing. The traditional approach to analyzing australopithecine locomotion is based on

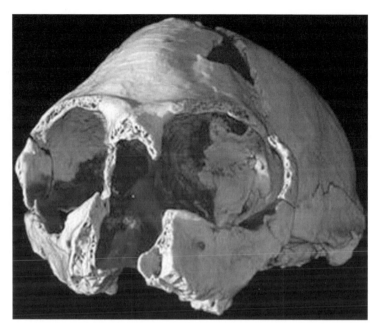

Figure 5.2. *Tremacebus harringtoni*, a fossil platyrrhine primate. Image from the Digimorph NSF digital library: http://www.digimorph.org/index.phtml/.

description and comparison of modern human, pongid, and australopithecine morphology. Yet, questions remain: to what extent were these creatures terrestrial bipeds? Were they facultative bipeds, and not habitual or obligate bipeds? Did arboreal climbing trump walking as the principal locomotor category? A novel solution for these questions was recently attempted. A very complete individual specimen of *Australopithecus afarensis* (AL 288–1, "Lucy") was used in a computer model and simulation of upright bipedal walking (Nagano *et al.* 2005). Three-dimensional laser scans of the pelvis and leg bones were fleshed out with neuromuscular modeling, and biomechanical computer simulations produced a walking gait (Nagano *et al.* 2005: Fig. 1). Muscle force and heat were used to trigger the system. Energy expenditure during locomotion was quantified. Researchers concluded that *Australopithecus afarensis* walked in a fashion comparable to modern humans – upright, with extended knees. The degree of locomotor efficiency was comparable to human youngsters of a comparable body size (30 kg).

Because the vertebrate fossil record is composed largely of teeth, analysis of dental function is a major activity of paleontologists. Laboratory experimentation is used to infer specifics of mandibular movement by examining wear patterns on tooth models (Naples 1995). Dental enamel is analyzed to determine its stress-bearing abilities when food items of different hardness and texture are eaten (Rensberger 1995). Because trophic level affects the biological segregation of stable isotopes of elements like carbon, nitrogen, and oxygen, the chemistry of prehistoric collagen and enamel yields unequivocal evidence of diet, by fitting an extinct taxon within the food web of its ancient community. This methodology has been applied to Plio-Pleistocene hominid species (Sponheimer & Lee-Thorp 1999, Lee-Thorp *et al.* 2003), as well as to Neanderthals of the late Pleistocene.

Alexander (1996, 2003a, 2003b) has productively used body size differences between closely related organisms to examine variables that affect locomotion. Not only bone dimensions, but also limb posture and stance are affected by body size. These considerations allow for a more detailed reconstruction of extinct species. Giants of a group, such as the extinct 700 kg rodent *Phoberomys*, must have stood on straight limbs, rather than crouching and scampering on bent limbs (Alexander 2003a). Contemplation of a giant taxon like *Phoberomys* leads Alexander to ask why rodents in general are so small. The answer may lie in phylogenetic constraints that shape rodents with short limbs and forelimbs suited for generalized digging. Species can escape predators by burrowing, but find it difficult to evolve large forms that would need long, straight, slender, gracile limbs in order to escape by running (Alexander 2003a). Size considerations can also illuminate the lifeways of fossil primates. Like *Phoberomys*, a primate fossil species characterized by gigantism is the extinct hominoid *Gigantopithecus blacki*, which may have been as large as 700–800 kg. There are no postcranial remains of this species, but, given its body size, the taxon must certainly have been terrestrial. In contrast, there were many very small primate species during the Paleocene and Eocene. By virtue of their body size alone, these species could not have subsisted on low quality vegetation.

One of the most interesting methods in the study of the morphology of extinct species is a taxon-free approach. This is the examination of fossil organisms in morphospace, a theoretical dimension established solely through morphological traits. This multivariate space is then analyzed to assess its degree of occupation and vacancy at different times in the evolutionary history of a group. Invertebrate paleontologists have used this technique to examine the diversity of morphology at various time levels. This allows one to assess the nature of radiations, rebounds after extinctions, potential competition between organisms with similar morphologies, and reduction of competition as morphologies diverge (Conway Morris 1998, McGhee 1999).

The modern study of morphospace is equivalent to the traditional study of grades. Thomas Henry Huxley used the concept of grades when discussing the evolution of primates and other mammals, but Julian Huxley (1958) coined the words "clade" and "grade." He was the first researcher to draw a formal distinction between organisms united by common descent (a clade) versus organisms united by a level of structural organization (a grade). The parallel evolution of separate lineages through an adaptive zone or through changing levels of structural morphology hinders cladistic methodology, because homoplasy frequently occurs under these circumstances. Furthermore, in grade analysis, lineage distinctions and ancestor-descendant relationships are of less importance than adaptation. Yet, grades are once again discretely entering the realm of paleontology. In both morphospace and grade studies, morphology is being examined without concern for the individual taxon or lineage with which it is associated. Morphological shifts through time are documented, and phylogeny is unimportant. Even though the use of grades or grade analysis is anathema to cladistic methodology, the utility of the approach is demonstrated by the emerging study of morphospace.

Ontogeny and anatomical genomics

For over 50 years, the Swiss researcher Adolph Hans Schultz described anatomical variability in non-human primates. Working first at Johns Hopkins University and the Carnegie Institution and later returning to his alma mater, the University of Zürich, Schultz collected and described a vast number of non-human primate specimens. These specimens form the core of the museum collections at the University of Zürich. The Schultz Collection and the Collection of the Anthropological Institute and Museum currently contain 6802 non-human primate skeletons, crania, postcrania, and preserved or frozen cadavers – one of the largest and most diverse holdings in the world.[2] During the early twentieth century, Schultz was one of the first researchers to emphasize the analysis of reasonable sample sizes. Hence, he was one of the first researchers to appreciate the range of variability encompassed within living primate species. He also realized at an early date that allometry could influence morphology, and began to account for size differences in proportions by scaling measurements against trunk length.

Much of Adolph Schultz's work dealt with comparative fetal anatomy and comparative growth and development. Major publications dealt with ontogeny in single species, such as orangutans, gorillas, chimpanzees, gibbons and siamangs, rhesus macaques, and proboscis monkeys. A major feature of Schultz's research was that he directed attention toward the link between ontogeny and variability in primates. Much of Schultz's work dealt with revealing the manner in which every adult is the result of ontogenetic changes that are affected by both individual life history and species-specific genetic determinants. And adults are not the end-point of change, because change related to senescence also occurs.

Comparative genome analysis has the potential for illuminating the evolution of morphology, through study of the presence and absence of genes, conserved gene order, and patterns of duplication and rearrangement of genetic material. For example, recent duplications indicate that a function has been newly emphasized, while ancient duplications indicate that functions have diverged.

Evolutionary development ("evo-devo") has become a crucial new discipline within biology. The potential impact of evolutionary development on the study of evolution can be gauged by its current effect on comparative morphology (e.g., Minelli 2003). A new protocol for research is emerging. Morphological differences are first defined. Then the developmental modules that underlie the morphology are distinguished. Lastly, the genetic differences that lie beneath the developmental modules are identified. Gene expression underlies morphology, and the boundary lines between anatomical areas express different genes. Morphology is thus now not merely an endeavor that is static in time, but an endeavor that is affected by both spatial and temporal changes. These spatiotemporal changes explain the emergence of a developmental Bauplan whose rudiments are becoming known. Patterning occurs first at the level of cells within the developing embryo. Morphology of the neonate and adult emerges from the embryonic Bauplan. For example, *Hox* genes lay out the major pattern of the mammalian axial and appendicular skeleton during embryonic life, and organize the limb buds and future limbs and extremities (Muragaki *et al.* 1996, Shubin *et al.* 1997, Chiu & Hamrick 2002, Kmita *et al.* 2002, Wellik & Capecchi 2003, Papageorgiou 2004). Comparisons of mammalian vertebrae and limb structure therefore involve study of how *Hox* gene expression has changed in different taxa. In the mammalian forelimb, for example, digit I (the thumb) occurs in a separate developmental module from the module for digits II–V and the zeugopodium (radius and ulna). *Hox* d13 and a13 underlie thumb patterning, while *Hox* d11 underlies digits II–V and the zeugopodium. The developmental association of digits II–V with the radius and ulna and the separate status of digit I can explain major variations within the forelimb of fossil and living mammals, including primates.

However, evolutionary development does not offer an easy or universal solution to problems of morphological variation. Dentition and dental variability have been major topics in mammalogy since the nineteenth century. Yet, *Hox* genes do not underlie tooth patterning, because the influence of the *Hox* genes does not extend to the ancient first branchial arch, where jaws and teeth originate. A variety of homeobox-containing genes (*Dlx, Msx, Barx*) have been implicated in the establishment of tooth identity, but the timing, shape and complexity of molar tooth crown morphology is underlain by very complex genetic substrates (Jernvall & Jung 2000: Fig. 7).

Given the arguments outlined in a previous section about punctuated equilibria and morphological stasis, or skepticism about natural selection and adaptation on the part of some researchers, it is important to note that development does not necessarily constrain or channel morphology – it can underlie a significant degree of morphological variation. For example, the expectation is that intricate color patterns on butterfly wings that identify species should be highly constrained by genetics. Pattern components are segregated within parts of a wing and bounded by veins. Yet, artificial selection experiments show that these wing patterns are highly flexible. "Standing genetic variation within a laboratory population of one species is sufficient to account for the production of all phenotypes found within the genus, and even one which was not explored in any extant species" (Beldade *et al.* 2002: 846). Natural selection was identified as the principal cause of the variation in wing patterns between species, and not developmental bias or constraint.

Leigh Van Valen once defined evolution as "the control of development by ecology" (Jablonski 1999: 2114). It appears that continuing research into evolutionary development and field or laboratory research on the relationship between morphology and habitat use will confirm this definition.

Phenotypic variability

The process of natural selection is anchored in the existence of variable individuals within breeding populations. Nevertheless, variation itself can be troublesome to researchers. Is some particular structure in an individual animal influenced by age, sex, body size, position along a cline, population difference, activity patterns, diet, or disease? How many other animals of the same species must one examine in order to determine this? If this animal is extinct, what constitutes a reasonable fossil sample? How much variability can accrue before taxonomists argue that a species boundary has been crossed? If traits appear continuous across species boundaries, how does one determine species differences? A trait that is useful for distinguishing species may be subject to homoplasy. Researchers attempting to discern true ancestor/descendant relationships may be dubious about using such a trait. Furthermore, because a species diagnosis is based on the phenotype and genotype of a type specimen, the formal boundaries of a species are defined with respect to a single individual or part of a single individual. Irrespective of whether other material (paratypes) is examined before the species is defined, the species diagnosis is inevitably linked to a single individual that may have an anomalous or exaggerated phenotype.

It is clear that variation is of critical importance in understanding natural selection and adaptation. Patterns of variability and morphological integration or disruption can yield insight into shifting adaptation, even among fossil organisms. Understanding variation and patterns of variation is also of fundamental importance in systematics. Because phenotypic plasticity is the norm, taxonomists must assess the limits of individual plasticity before assigning new species rank to taxonomic material.

Although individual responses may be strikingly malleable, development must ultimately constrain plasticity. Among modern humans, for example, high altitude environments severely affect normal physiology, especially through hypoxia, caused by a low partial pressure of oxygen. Any healthy lowlander entering a high altitude area can gradually increase the number of red blood cells and capillarization of tissues. Native highlanders, however, have a larger heart and lungs, and mature more slowly. Both New and Old World highland populations experiencing long-term adaptation to high altitude also have increased placental volume and nitric oxide in the lungs. Populations on the Tibetan Plateau, which have been in a high-altitude environment for the longest time, have a genetic basis for hemoglobin with increased ability to bind oxygen. Thus, the depth of genetic constraint on high altitude adaptation is determined by the length of time over which a population has been experiencing selection pressure through hypoxia. This example might serve as a template for the analysis of phenotypic plasticity.

Endnotes

1. In retrospect, the furor mounted against adaptation, natural selection, and genetics may have been the biological equivalent of post-modernism. Certainly, some of the necessary elements are present: revisionist history, disparagement of previous research, nihilistic attitudes towards "facts," insistence that the social milieu inevitably colors and distorts ideas and interpretations, and intense scrutiny of the political ramifications of ideas. During the same time, archeology underwent post-processualism, which had similar elements. Were both these disciplines experiencing an overflow from the literary realm? In any case, because both biology and archeology contribute to paleoanthropology, paleoanthropology certainly was shaded by these trends.
2. The Website address of the Anthropological Institute and Museum of the University of Zürich is http://www.unizh.ch/anthro/.

6   *Captive studies of non-human primates*

Introduction

Captive studies of non-human primates have been a part of primatology since the days of Robert and Ada Yerkes and their students during the early twentieth century. A good portion of the famous monograph *The Great Apes* (Yerkes & Yerkes 1929) concerns analysis of pongid senses and physiology. The equivalence of human and pongid senses and physiology was demonstrated for the first time. The study of ape intelligence and comparative learning was also emphasized. Intelligence in both humans and animals was a particular interest of Robert Yerkes. During World War I, he had created a battery of intelligence tests for assessing draftees into the U.S. Army. Yerkes corresponded with Wolfgang Köhler, who studied ape intelligence and problem-solving abilities, emphasizing their creativity and insight. Yerkes extensively examined the intelligence and problem solving abilities of a beloved chimpanzee subject, Prince Nim (Yerkes 1925). This legendary animal was later revealed to be a bonobo (*Pan paniscus*), rather than a common chimpanzee (*Pan troglodytes*). Yerkes speculated that apes might be taught to communicate with humans if they were taught a gestural or sign language – this idea was brought to fruition over 40 years later. Yerkes (1943) later presented a synthesis of the years of research on chimpanzees at the Yale Anthropoid Experimental Station in Orange Park, Florida. This facility later became the first of the American Regional Primate Centers.

The influence of captivity on behavior

Given the imminent extinction of many primate species in the wild (Chapter 4), captive studies are certain to become ever more significant. There are two general types of captivity: captivity in zoos or primate centers, where conservation or species preservation is the goal of maintenance and research; and captivity in biomedical experimentation, where experimental protocol and hygiene is the goal of maintenance and research. The difference between these two captive situations is exemplified in the contrast between two publications – *The International Zoo Yearbook* and *Laboratory Animal Care*.

The behavior of domesticated animals had long been observed, and formal studies had been done on practical aspects of their management. These studies were an important component of any farm manual. Every dedicated farmer would keep such manuals on the family bookshelf. However, detailed study of the influence of captivity on the behavior of wild species began in the 1950s. Heini Hediger (1964, 1968, 1969) conducted comparative studies of captive animals in both zoos and circuses. His observations integrated ethological data with captive studies, so that a species-specific behavior in the wild, such as flight distance, could be meaningfully interpreted in light of the captive situation. Hediger documented that circus animals, that performed and interacted with humans, appeared healthier, and often fared much better mentally than zoo animals, raised in isolated confinement. The circus animals were both mentally and physically engaged. This observation established the importance of keeping animals in such a way that an active mental and physical life was maintained. Animals had evolved, after all, to behave. A captive situation that reduced an animal to the physical and mental level of a sofa cushion was bound to create illness and aberrant behavior.

With the recognition that captivity might represent the only chance for species survival for many taxa, consideration of the influence of captivity on behavior became even more marked (Markowitz & Stevens 1978). Non-human primates are the focus of much of this attention. Particular attempts are made to ensure the mental well-being of captive primates, or to create "enriched" captive environments that mimic, as closely as possible, conditions in nature (Novak & Petto 1991, Shepherdson *et al.* 1998).

Many researchers maintain the hope of re-introducing captive-born animals into the wild. This is a forlorn hope in most circumstances, because modern habitats are often so seriously degraded that it is impossible to reconstruct what the natural environment once was. Nevertheless, there have been some successes. Perhaps the best success among non-human primates concerns the reintroduction of the highly endangered and spectacularly beautiful golden lion tamarin (*Leontopithecus rosalia*) into the Brazilian Atlantic rainforest. Intense work at the U.S. National Zoo and in Brazil preceded the reintroduction, because captive-born animals had to learn how to search actively for natural food items and then return to safe core areas within a home range. Nevertheless, Benjamin Beck discovered that captive-born, reintroduced animals have shorter day range lengths, and appear less competent in moving through their home range and utilizing resources than wild lion tamarins. Intricate analysis of the movement of lion tamarin groups in a forested experimental area at the U.S. National Zoo yields information about how animals recognize landmarks and form mental maps of environmental resources (Menzel & Beck 2000).

Harry Harlow's research

A particular subset of captive primate studies is concerned with alterations in normal mental functions under captive conditions. Some of these studies may have therapeutic value in treating humans with mental illness. Classic examples of these studies are found in the work of Harry Harlow and his colleagues, carried out at the Wisconsin Regional Primate Center. Noticing that lack of social contact appeared to generate aberrant behavior in non-human primates, Harlow initiated a series of experiments on rhesus macaques (*Macaca mulatta*) that illustrated and quantified the effects of social deprivation (Harlow & Harlow 1962, Harlow *et al.* 1971). These experiments created what were known as "motherless monkeys." The behavior of these animals sometimes resembled that of institutionalized humans, leading to the insight that some abnormal human behaviors might be traced to early rearing practices, as well as social deprivation. However, some of the stereotyped behaviors that appeared in the macaques (such as rocking and self-grasping) also appear in humans with congenital conditions, such as autism. "Motherless monkeys" preferred to cling to a cloth-covered metal frame (pseudomother) for comfort, rather than nursing at a pseudomother that lacked the soft, tactile stimulation of cloth. Rudiments of a face were provided for each type of pseudomother, but the soft cloth pseudomother also had exaggerated eyes. "Motherless monkeys" would run to the cloth-covered pseudomother when alarmed and distressed, while avoiding the pseudomother that provided sustenance. Tactile comfort appeared to be a more important aspect of motherhood to the young monkeys than food alone.

Harlow and his colleagues also demonstrated that the profound effects of social isolation on captive primates could sometimes be ameliorated or reversed by introducing juvenile animals to other conspecifics. However, female macaques who were raised as "motherless monkeys" and who had no peer associations when young exhibited abnormal maternal behavior when they themselves bore young (Harlow *et al.* 1966). In fact, newborn animals had to be removed from such females immediately after birth, or they would be injured or killed by their own mothers. In fact, these mothers treated their newborn like annoying vermin. Hence, the effects of abnormal rearing could be multigenerational.

An inventory of abnormal captive behaviors

Abnormal animal behaviors in captivity can be divided into two categories. The first category comprises behaviors that are qualitatively abnormal – nothing like these are seen in nature. The second category comprises behaviors that are quantitatively abnormal. Behaviors like these are observed in nature, but the inappropriate occurrence of these behaviors, or the degree of their expression, is abnormal.

Among the qualitatively abnormal behaviors are the following. Aberrant positions exist, such as self-grasping or embracing. There is self-biting or other kinds of self-injurious behavior. Limbs are held in bizarre floating positions. "Saluting" occurs, in which a hand is repetitively brought up to the head in a military-like salute. Eating disorders exist, in which inedible material or unnatural behaviors occur. For example, paint, plaster, wood, or plastic stripped from a captive enclosure is consumed. Food is regurgitated, and the vomitus consumed, or feces are consumed (coprophagia). Sexual disorders exist. Sexual behavior may either be quantitatively exaggerated or not occur at all. Copulation occurs in non-functional positions, or in a fashion not observed in the wild. Masturbation occurs. Stereotyped motor acts take place, and may account for a large proportion of activity time. Animals pace continually. Pacing accounts for 97% of the stereotyped behaviors in mammalian carnivores (Clubb & Mason 2003). The 35 species of surveyed carnivores were caged, and limited space may account for the high occurrence of pacing behavior. In fact, home range size in nature, median daily travel distance, and body mass are highly correlated with the frequency of carnivore pacing (Clubb & Mason 2003). Pacing or stationary animals may nod or zigzag their heads back and forth. Animals sway themselves, or rock, bounce, or somersault. These bizarre behaviors also may exist in institutionalized humans or in humans born with congenital behavioral abnormalities, such as autism. This realization has implications for the understanding and treatment of human mental health.

Among the quantitatively abnormal behaviors are the following. Normal activity patterns are disrupted. Animals are hyperactive or lethargic. Animals have an exaggerated startle response, or are phlegmatic and indifferent to environmental stimuli. Diurnal species are active at night, and nocturnal species are active during the day. Animals can overeat (hyperphagia) or suffer a loss of appetite. They can drink constantly (polydipsia). Animals can be either extremely fearful or hyper-aggressive.

Many physical features of the captive environment can adversely affect animal behavior. Space can be limited, and affect normal locomotion and activity patterns. Animals do not forage for food, and may eat *ad libitum*. Animals can quickly gorge themselves. In the wild, feeding is constrained by normal food search and discovery. Food in captivity may be artificial and highly processed. Calorie and nutrient intake may be adequate, but food in captivity may not meet the species-specific normal demands of mastication (chewing of tough plant or muscle fibers, bone-cracking) or manipulation (plucking, peeling, extraction, skinning) before ingestion. Temperature and light or light periodicity can be altered from the natural condition. Altitude or humidity can be different in captivity. The absence of normal seasonal patterns of temperature and precipitation changes can affect reproduction.

The captive situation can isolate or separate social animals from conspecifics. Conversely, the captive situation can result in abnormal crowding, and generate abnormal agonistic behaviors. Subordinate animals may not be able to flee or avoid dominant animals. Crowded conditions in captivity can spread disease pathogens, ectoparasites, and endoparasites.

Captive non-human primates face another complication. Because of their genetic relatedness to humans, human diseases can spread to captive primates from human caretakers and handlers. Mild human respiratory diseases such as the common cold can cause non-human primate deaths from unmanageable respiratory disorders. Alternatively, deadly human diseases such as polio are equally deadly to non-human primates. In addition, captive primates can serve as reservoirs of unknown new diseases that can affect unsuspecting human caretakers that come into contact with non-human primates. There are incalculable dangers in developing experimental species that are closely related to humans. Pathogenic exchange is reciprocal. In addition, non-human primates may harbor unknown pathogenic agents. Human personnel live in dread of contracting dangerous, unknown simian viruses. The primatologist K. R. L. Hall, for example, died of a novel strain of hepatitis after being bitten by a captive patas monkey. Even if the rate of infection is low, if one-third or more of the infected die, extraordinary prudence is necessary. Non-human primates are literally biohazards, yet animals in colonies are continually having blood drawn, and swabs taken of bodily orifices. Hence, human caretakers must be constantly vigilant, and must eliminate any potential for contact with non-human primate bodily fluids.

These considerations of mutual pathogen exchange also affect human/non-human primate interactions in the wild. For example, hundreds of humans in the Congo have died from the deadly and untreatable Ebola virus, but African great apes in the Congo and Gabon also die from Ebola. Within the Lossi Sanctuary in the Congo, up to two-thirds of the gorillas may have already died from Ebola (Kaiser 2003). Updated analysis confirms this pessimistic projection. When population densities of gorillas and common chimpanzees within the Lossi Sanctuary are compared from before and after the 2003 Ebola outbreak, gorillas appear to have suffered a 56% reduction and chimpanzees an 89% reduction in numbers (Leroy *et al.* 2004: Table 2). Recovered carcasses confirmed Ebola infection, and indicated a failure to develop immunoglobin G responses. This failure was similar to that of infected humans who died from the infection (Leroy *et al.* 2004). Attempts to control or stem Ebola outbreaks among humans and great apes through eliminating inter-species contacts or creating physical barriers to movement are problematic, because of the difficulties of establishing such controls, and the possibility that other sympatric primate species are the ultimate source of the contagion (Vogel 2003).

In addition, human AIDS undoubtedly had its origins in simian immuno-deficiency viruses (SIVs) that were transferred to humans through African bushmeat hunting (Petersen 2003). Non-human primates, including apes, are hunted and consumed by rural people. The meat may be traded across long distances, and it may ultimately appear in urban markets. The hunting of non-human primates and the subsequent skinning, preparation, and consumption of carcasses, allows people to be exposed to contaminated body fluids. In the case of Ebola, it is now established that dead and infected chimpanzees and gorillas appear in forested regions before a human outbreak occurs, and that hunters and villagers who process infected carcasses spread the disease to other humans (Leroy *et al.* 2004). Ebola and SIV/AIDS are two cases that illustrate how non-human primates can serve as reservoirs of human disease, or as the source of emerging novel diseases.

Truly aberrant behaviors

The array of abnormal behaviors displayed by animals under normal captive conditions becomes accentuated to an amazing degree among non-human primates raised as pets or surrogate human children. The behaviors that these animals exhibit are truly aberrant. A sampling of the strange, disturbing array of these behaviors can be glimpsed in Hooton (1942), Sanderson (1957), Morris & Morris (1968), and Hahn (1988). Although these authors concentrate on the early twentieth century, primates continue to be acquired as exotic pets in Western countries. Rural people living in tropical countries can easily acquire young primates, and may keep them as pets (Figure 6.1). A recent survey of residents of Mexico City noted 179 non-human primates kept as pets, distributed across 11 taxa (Duarte-Quiroga & Estrada 2003). Two-thirds of the animals were black-handed spider monkeys (*Ateles geoffroyi*), and the overwhelming majority of these were juveniles or young adults. Twenty-seven percent of the reported deaths were caused by the owner battering an adult animal that had become intractable or destructive (Duarte-Quiroga & Estrada 2003: Table III). As animals mature, they grow large adult canine teeth, achieve adult muscle strength, or grow to their natural, alarming adult size. Owners usually abandon or destroy animals at this point. Attempts have been made to rehabilitate or re-educate captive animals when their owners abandon them. The sad failure of these rehabilitation attempts can be seen in facilities such as Sweetwaters in Central Kenya. Chimpanzees and other non-human primates that are

Figure 6.1. A young black-handed spider monkey (*Ateles geoffroyi*) kept as a pet in Costa Rica. It is usually tethered to a post outside the house, but here moves freely along the verandah using a bipedal gait. Note the elevated prehensile tail.

maintained here and at similar facilities are truly equivalent to institutionalized humans, and there is no hope of these animals regaining any vestige of normal behaviors.

Risk factors

A broad survey of captive primate situations has identified particular factors that affect the likelihood that animals will develop abnormal behaviors. Housing animals in a solitary condition at an early age and for prolonged periods, as well as invasive research protocols (e.g., repeated blood drawing) are some of these factors (Lutz *et al.* 2003). Once established, abnormal behaviors tend to endure. However, individual animals may be predisposed to develop abnormal behaviors. The sex of captive animals has been documented to affect these behaviors. Males are significantly much more predisposed to develop abnormal behaviors than are females (Lutz *et al.* 2003). This indicates some physiological buffering system in females.

Alleviating abnormal behaviors

In the future, drugs may be administered to captive great apes that suffer from abnormal behaviors. Because many of these behaviors appear similar to human behavioral pathologies, it is possible that these pongid behaviors are homologous to human behavioral abnormalities. Hence, drugs used to treat depression, anxiety, repetitive and stereotyped movement disorders, and cognitive decline in humans may someday be used to alleviate these conditions in captive great apes (Brûne *et al.* 2004). However, the high cost of these psychiatric drugs and the necessity for long-term administration would probably prohibit their widespread use among captive species.

Sometimes abnormal behaviors can be alleviated through the construction of appropriate enclosures. The scaling of enclosures to body size and species-specific natural home range sizes may reduce the occurrence of abnormal behaviors. Certainly, small enclosures relative to body mass, natural home range size, and median distance of daily travel adversely affect the mortality of captive infants, and increase the frequency of stereotyped pacing in captive mammalian carnivore species (Clubb & Mason 2003).

Non-human primates are notorious for their abilities to destroy enclosures. This is due partly to their hands – catarrhine primates with truly opposable thumbs are capable of fine manipulation – and partly to boredom experienced under captive conditions. "Monkey-proof" cages are made of metal or concrete. These cages are impervious to destruction, and have no paint or wood that can be chipped or splintered off and ingested. Metal or concrete cages are easy to keep clean, because they can easily be hosed down. Because non-human primates do not maintain latrine areas, and defecate at random, the ability to clean cages easily is important. Sanitation also prevents the spread of disease pathogens and parasites. Given the fragile nature of non-human primate health, this is an important consideration. Hygiene is also compromised when non-human primates throw or hurl food, feces, or other objects, which they may do with stunning accuracy. Glass or plexiglass around cages can prevent unfortunate accidents to zoo tourists, while avoiding contagion to both species. However, simple cages or enclosures that are easy to clean conflict with the necessity of constructing cages that are designed to provide stimulation to captive animals. Branches, ropes, ladders, and trees allow captive primates to exercise, while simultaneously utilizing their arboreal specializations. Captive primates can climb, swing, and hang, as expected for arboreal animals. These structures develop normal perception and good motor control as captive animals exercise, as well as offering sensory stimulation. Yet, these complicated structures may be difficult to clean, and artificial structures that mimic natural foliage offer still more of a challenge to hygiene. Furthermore, these structures are targets for destruction by bored primates. Electrified wire sometimes protects trees in outdoor enclosures, while it also seals off potential escape routes.

Good laboratory or colony management requires understanding the factors that contribute to the origins of abnormal behaviors and attempting to alleviate them. Animals reared by their mothers, and young animals in contact with conspecifics, tend to develop more normally. Even animals that are housed alone may not demonstrate the full range of abnormal behaviors, although animals raised in a solitary condition tend to exhibit more self-directed and self-injurious behavior than animals raised with conspecifics (Lutz *et al.* 2003). This agrees with the general lack of social behavior in individually housed animals.

It is thought that a number of abnormal behaviors are displacement behaviors that arise because of an inability to display species-specific behavior under captive conditions. Classical ethologists might then consider these abnormal behaviors, such as stereotyped pacing or rocking, to be displacement activities caused by a frustration of normal goals or behaviors. Because of this hypothesis, attempts have been made to create captive conditions where normal behaviors such as species-specific locomotion or foraging can occur. Clever additions to the captive primate environment have been invented. For example, an artificial "foraging/grooming" board has been shown to ameliorate abnormal behavior in captive rhesus macaques that are housed alone (Bayne *et al.* 1991). In the wild, non-human primates spend much of their time foraging and eating when awake. Zoo curators have designed clever ways in which captive primates can search for or discover food, or be rewarded with food after performing some activity. The wonderful "Think Tank" facility at the U.S. National Zoo was created in 1995. It features primate species that demonstrate their cognitive abilities to zoo tourists while obtaining food rewards. Orangutans and Sulawesi macaques (*Macaca nigra*) are the two primate species exhibited at the facility, along with archer fish, hermit crabs, and leaf-cutter ants.[1] Orangutans can utilize their species-specific mode of slow, quadrumanous climbing during arboreal locomotion as they travel along a high network of cables and towers that joins the Think Tank to the Great Ape House. Yet, captive enrichment programs that emphasize conspecifics prevail over the design of special structures and exhibits that counteract problems associated with captivity. A topical breakdown of papers published between 1990 and 1998 in the online bibliography

"Environmental Enrichment for Nonhuman Primates" demonstrates that 65% of these papers deal with animate enrichment (Howell 1999: Fig. 3).

Biomedical primatology

In Europe, as well as in North America, there is significant political pressure against the use of non-human primates in biomedical research. This distress far exceeds the degree of apprehension expressed about the use of other mammals or other vertebrates in biomedical experimentation. Nevertheless, a survey of both Swedish and Kenyan medical and veterinary students reveals that over 90% of the students in both countries accept the necessity of animal experimentation, and over 75% of the students in both countries accept the use of non-human primates in research (Hagelin *et al.* 2000). It is noteworthy that no significant differences about the use of non-human primates in biomedical research separate perceptions in the two countries, in spite of the fact that Kenyan students view baboons and vervets as pests and crop raiders, not as rare, threatened, or exotic species (Hagelin *et al.* 2000).

In spite of concerns about the use of non-human primates in medical research, there are details of anatomy and physiology that can be gleaned only from captive animals. An example of this is the revelation that common marmosets (*Callithrix jacchus*) have the potential for a far greater reproductive output than observed in the wild. The records of 12 colonies show that, in captivity, triplets can be more frequent than twins are, and quadruplets and quintuplets can also be born. Improved nutrition in captivity causes increased ovulation. As revealed by ultrasound and hysterotomies, a strange long pause occurs between blastocyst implantation and the main embryonic growth phase (Windle *et al.* 1999). Litter size can thus be flexibly reduced during pregnancy. This reproductive potential can only be observed in captivity, because more than two offspring cannot be reared under normal natural conditions. Yet, rare, optimal natural conditions allow the common marmoset to expand its numbers quickly. Furthermore, gestational details underlying this strange reproductive physiology would remain unknown without access to laboratory and medical technology.

The importance of primates as model species for human disease and therapeutic research can be assessed from the existence of specialized journals such as the *Journal of Medical Primatology*. *Current Primate References* was a monthly bibliographic amalgam of all published material featuring primates. It was issued by the Primate Information Center, Regional Primate Research Center, University of Washington.[2] Although studies of primate behavioral ecology in the wild or under naturalistic conditions capture the attention of the lay public, these studies constitute only a small fraction of the research and publications related to primates. A perusal of *Current Primate References* shows a very different picture from the general perception, and illustrates the dominance of biomedical research. I would estimate that, during any given month, about 70–80% of the primate publications related to biomedical primatology. The obvious explanation for this is that non-human primates are models for research on the many health problems that plague humankind. Non-human primates also substitute for human subjects in physiological experiments or pharmaceutical research when ethical considerations prevent the use of humans. Howell (1999) examined *Current Primate References* between January 1990 and December 1998, in order to assess trends and predict likely future areas of research. She discovered that the "anthropological categories" behavior and ecology and conservation accounted for only 17.5% of the citations, whereas physiology and microbiology accounted for 72% of the total number of citations ($n = 57\ 305$).

It also appears that the rate of publication devoted to primate behavioral ecology is declining. Howell (1999) discovered that the rate of publication in primate behavior, ecology, and conservation declined during the 1990s, as documented in *Current Primate References*. In contrast, biomedical topics are strong areas of future growth, indicated by a rise in publication rate through the 1990s. Immunology, nervous system, genetics, and pharmacology and therapeutics are some of these areas; immunology represents the fastest area of growth. The nervous system alone accounts for 18.6% of the total number of citations, and represents the category with the largest number of publications (Howell 1999: Table 5). These results are not surprising, given the use of nonhuman primates as model animals in biomedical research on human nervous system function, or in PET imaging analysis of brain activity during various activity states or during drug uptake.

Non-human catarrhine primates predominate in biomedical research. Given the physiological similarities of all catarrhines, it is understandable that these species are used in place of humans when ethical considerations prevent the use of human subjects in biomedical and pharmaceutical experimentation with infectious agents that have high human mortality rates. Non-human catarrhines are also used to establish baseline and stress levels of hormones like cortisol, and so are fundamental to studies of the physiological bases of human behavior. Primatology is integrated with medical research. The provisioned, free-ranging rhesus macaques on Cayo Santiago Island off the coast of Puerto Rico quickly increase their numbers because they suffer no problems with food supply. Yet, space is limited on Cayo Santiago Island, and the macaques are periodically culled down to about 600 animals. The "excess" animals are supplied to biomedical researchers. In Japan, The Tsukuba Primate Center for Medical Science has been established at Tsukuba Science City.

Chimpanzees were utilized in research on yellow fever as long ago as 1907. Monkeys were used to isolate the poliovirus in 1909, and the poliovirus was reproduced in chimpanzees in 1940. Karl Landsteiner[3] and his colleagues studied ABO blood group types in non-human primates during the early twentieth century, and constructed a primate phylogenetic tree from the ABO data (Snyder 1927: Fig. 3). ABO blood group typing of non-human primates continues to be an important area of investigation (Corvelo et al. 2002). Species of Macaca, Papio, and vervets (Cercopithecus aethiops) are routinely used in biomedical research. Their impact on the history of medicine has been profound. For example, the Rhesus blood group in humans takes its name from the rhesus macaque (Macaca mulatta), the species in which the blood group was first discovered. Understanding that a maternal Rhesus negative status affected the survivability of a Rhesus positive fetus was based on macaque research. This explained a mysterious medical condition that necessitated immediate transfusion of human newborns. The condition was known as erythroblastosis fetalis, and human newborns that suffered from it experienced life-threatening red blood cell destruction. Discovery of the Rhesus blood group therefore had immediate benefits for the survival of human newborns. Knowledge of a patient's Rhesus status (Rh- or Rh+ status) is absolutely necessary in current human blood-typing work. Reproductive research yields another example of the use of a catarrhine model in important medical research. Details of gestation, embryo implantation, and techniques of embryo collection that have implications for research into human fertility and reproduction were first investigated in female baboons (Hendrickx 1971).

Congenital malformations and diseases in non-human primates that also occur in human newborns continue to be studied. However, in recent years, non-human primates have been fundamental to research on the transmission of pathogenic viruses and possible treatment strategies. These viruses include herpesvirus, and simian and human immunodeficiency virus. For example, SIV-infected macaques, like humans, can show either a rapid or a slow progression of the disease. They can therefore be used to investigate variables that determine the course of AIDS infection in humans. Furthermore, captive SIV-infected macaques showed signs of the disease at least a decade before the onslaught of the human AIDS epidemic, and could have been used to predict the sudden appearance of the human epidemic, if primate researchers had been canny enough to discern similarities between human and macaque symptoms (Gardner 2003).

Because the exportation of rhesus macaques from India has now been halted, other model species have been introduced in biomedical research. New World monkeys (especially Saimiri spp., Aotus spp., and callitrichids) are starting to figure largely in biomedical research. Their smaller size allows them to be maintained with relative ease, in contrast to larger bodied catarrhines. The high rate of increase in callitrichids, and the discovery of behavioral tests in newborn callitrichids that predict infant survivorship (Tardif & Layne 2002), create conditions for rapid colony build-up. The threatened status of pongids, as well as their cost, currently prohibits much use of these species in biomedical research, in spite of their desirability in central nervous system and AIDS research.

Endnotes

1. Information about the Think Tank facility can be obtained from the following URL: http://natzoo.si.edu/Animals/Think Tank/. A range of animal tool behaviors is also exhibited in the facility.
2. Hardcopy versions are no longer available. An online primate citation database of over 165 000 references dating to 1940 is maintained by the Primate Information Center via the Wisconsin National Primate Research Center: http://primatelit.library.wisc.edu.
3. In 1900, Landsteiner began investigating the basis of blood agglutination during transfusions. Landsteiner received the Nobel Prize for Physiology or Medicine in 1930 for his discovery of the ABO blood group, and its application to transfusion technology.

7 What can non-human primate anatomy, physiology, and development reveal about human evolution?

The catarrhine substrate

This chapter is a linchpin between the earlier and later portions of this book. It briefly synthesizes points made in the previous chapters, and introduces topics that will be discussed in detail in the remainder of the volume. It presents a basic substrate or blueprint for catarrhine anatomy, physiology, and development, and it also discusses catarrhine behavioral ecology. Twenty-nine features comprise the catarrhine substrate. They are enumerated below. The end of this list briefly highlights differences between the behavioral ecology of hominids and non-human primates, which foreshadows discussions of sociality and interpretation of the archeological evidence.

1. All catarrhine species are diurnal.
2. Catarrhines emphasize certain senses, and downplay or lose others. Vision is more important than smell: trichromatic color vision exists, with three peaks of color sensitivity. Olfaction is downplayed. The vomeronasal organ, which lies in a cartilaginous cavity in the lower anterior part of the nasal septum, and which opens into the mouth by a small passage just behind the upper central incisor teeth, is absent. This organ is important in virtually all mammals. It receives scent information relating to reproduction. Its absence in catarrhines implies that chemical signals influencing reproduction are not as important as in other mammals.
3. The cerebral hemispheres of the brain, which form the neocortex, are asymmetrical. This asymmetry creates a visibly larger left cerebral hemisphere even when measured with the naked eye. Cerebral lateralization is linked to processing of species-specific vocalizations in the left cerebral hemisphere.
4. A larger body size exists than in platyrrhine anthropoids.
5. Pronounced sexual dimorphism exists in body size and sometimes other traits (e.g., canine size, coat color, and pattern), as well.
6. A sexual skin exists in the genital region of adult females in many species. Depending on the species, the sexual skin can even extend to the root of the tail and the lower thighs. This skin undergoes periodic changes in color and tumescence related to the reproductive cycle. Given its distribution in living species, a sexual skin is probably the primitive condition. Menstruation exists.

141

7. Very restricted mating periods (estrous) occur in female mammals. Modern human females are often thought to be unique because they have lost estrous cycles. However, some wild catarrhine primate females have an extended period of estrous, and it is apparently continuous in pygmy chimpanzees. In captivity, many more catarrhine species also extend estrous, so that much more sexual activity occurs under zoo or laboratory conditions than in the wild.

8. Truly opposable thumbs exist. True opposability depends on the ability to rotate the thumb about its own long axis. The thumb is able to make pulp-to-pulp contact with the other digits. This is explained by a saddle-shaped joint at the base of the thumb or the first carpometacarpal joint. Many researchers once thought that true opposability explained the remarkable tool behavior of hominids. Yet, if tool behavior were only dependent upon this joint anatomy, all catarrhines would show a similar degree of tool behavior.

9. In a trend that is documented back to 21 million years in East Africa in the genus *Morotopithecus*, hominoid catarrhines (great apes, lesser apes, hominids) acquire some specialized skeletal traits. The following traits are acquired in a mosaic fashion: a short, laterally broad thorax and sternum; short, broad pelvic bones; loss of the tail; shifting of the scapula to the dorsal side of the thorax; lengthening of the clavicles; and the developing of a remarkably mobile forelimb by increasing mobility at the shoulder, elbow, and wrist joints. The complete suite of characters does not appear until the late Miocene (9.5 mya) in European *Dryopithecus laietanus* from Can Llobatares, Spain.

10. There is a general primate trend to carry more body weight on the hind limbs. This is shown by a longer femur, heavier hind limb muscles, and a center of gravity that is located closer to the tail. Laboratory experiments show that primates use their hind limbs not only to accelerate, but also to brake, and they have a gait sequence (diagonal couplets), which is highly unusual among mammals. In catarrhines, the trunk is held upright during sleeping and resting postures (orthogrady). In combination with the general primate trend to carry more body weight on the hindlimbs (and perhaps accentuated by arboreal suspensory postures or vertical climbing in the hominoids), this ensures that some form of bipedality is probably an inevitable development in catarrhines. This idea will be explicated in Chapter 13.

11. Ischial callosities are the hairless, keratinized pads of skin bonded to the ischium of the catarrhine pelvis. The existence of ischial callosities appears to be the primitive catarrhine condition. These pads probably relate to widespread orthograde sleeping and resting postures, because resting and sleeping upright are more comfortable when ischial callosities are present. Hominids lose ischial callosities, because the form and orientation of the gluteus maximus muscle that are adaptations to bipedality create another kind of self-contained cushion or sitting pad.

12. Terrestriality or semi-terrestriality is widespread in catarrhines, and some terrestrial specializations can be detected in fossil species as far back as the middle Miocene. Terrestriality may be the primitive condition for Old World monkeys. Terrestriality is not simply caused by the spread of grasslands, because some open-country habitats can be detected in North Africa as far back as the late Eocene, without causing the appearance of terrestrial or semi-terrestrial primates. Furthermore, there are no terrestrial or semi-terrestrial monkeys in the New World right now, although open-country habitats exist there now. In addition, in spite of widespread grasslands in ancient South America, no fossil evidence of terrestrial or semi-terrestrial New World monkeys has ever been discovered.

13. Seasonality of breeding among catarrhines may be pronounced. The length of the breeding season is inversely related to the number of adult males within the group, but there is some argument that phylogeny or evolutionary history overrides this phenomenon. An increase in dietary fat in young females may determine an earlier onset of menarche (first menstruation) and a shorter length of inter-birth intervals. This effect is experimentally shown to be mediated through the release of the hormone leptin, produced by adipose tissue. Female reproductive fitness is therefore associated with dietary quality. The greatest nutritional demands on females occur when they are lactating. Dietary constraints are most severe during this period.

14. Observations of free-ranging provisioned and captive species suggest that food (which is a density-dependent factor) is a major limiting factor on population size. Female–female food competition is probably important.

15. Singleton births are the normal condition: one newborn is delivered per pregnancy.

16. Neonates and infants are carried by adults, and not left at a den or sleeping nest – i.e., no "infant parking" takes place. Contact with adults and other conspecifics is important for normal socialization, which is confirmed by systematic deprivation experiments on captive animals.

17. Weaning occurs when juveniles are approximately four times their birth weight.

18. Home ranges exist. Yet, home ranges have been found in all animal species investigated, whether solitary or social.

19. Sociality exists, but not complex sociality. Complex sociality is eusociality: communal rearing of young, reproductive division of labor, and overlap of generations. Food is probably the major factor in limiting population (a density-dependent factor), and hence is the principal focus of natural selection affecting sociality. Females are more affected by access to food than are males.

20. Males and females differ in terms of philopatry, or remaining within their natal groups. Female philopatry is ancient: females generally do not

disperse from their natal groups. Transfer between troops may occur several times over the lifespan of a male. Dispersing males are responsible for gene flow. Matrilines (female descendants of a common female ancestor and their offspring) are important. Matrilines within the same social group may compete with each other. As detailed in Chapter 15, many of these traits – female philopatry, groups with matrilines, female dominance hierarchies, female–female competition, female–female cooperation (e.g., grooming associations, allomothering) – appear to be traits associated with cercopithecoids, and not with hominoids. They are also traits that appear to be derived in comparison to other primate groups.

21. Dominance is an important factor in the organization of catarrhine society. Dominance is expressed overtly, generally by agonistic behaviors or by presenting and mounting behaviors that are non-reproductive in function. Dominance may also be expressed covertly through displacement. Male and female dominance hierarchies are separate. The female dominance hierarchy is more stable, and relative status within it is more firmly linked to reproductive success, because female tenure in a dominance position within the hierarchy is longer.

22. Mothers high in dominance tend to produce offspring high in dominance. Because of female philopatry, dominant daughters will remain in their natal group. Hence, the past history of a matriline can affect current demography and social dynamics.

23. Rearing (socialization) and grooming are the most important factors creating social cohesion. Only in pygmy chimpanzees (bonobos) does another factor develop: non-reproductive sexual behavior.

Some major differences between hominids and other catarrhines will now become apparent. Early hominid behavior is documented through the archeological record.

24. Non-human catarrhines take shelter at night in sleeping trees or sleeping cliffs, even if the animals are wholly or largely terrestrial during the day. There is no construction of artificial shelters. Only the modern pongids construct sleeping nests, and these are very crude platforms of vegetation. There is no control of fire. There is no alteration of the landscape.

25. There is minimal cooperation among non-human catarrhines. There is a lack of care of sick and helpless individuals; there is no cooperation in acquiring or preparing food; there is no food-sharing; there is no communal rearing of young; there is no division of labor or evidence of social "roles."

26. There is abundant evidence of individual status-striving, deception, and "political" maneuvering among non-human catarrhines. There is evidence that individuals recognize kinship relationships of other animals within the social group. "Machiavellian" behavior on the part of individuals does not

seem to generate the level of cooperative behavior seen in other species (e.g., mammalian social carnivores).

27. Non-human catarrhines exhibit no food-hoarding, or evidence of resource management.

28. The foraging behavior of non-human catarrhines is the "routed foraging" discussed originally by the archeologist Lewis Binford. "Routed foraging" is a catch-as-catch-can mode of foraging, with no evidence of forethought or planning in resource exploitation. Mental mapping probably exists, but it appears to be widespread in the animal world. However, there is no evidence of elaborate foresight or planning with regard to food foraging, resource manipulation, or resource preparation, except on a rudimentary scale (e.g., nut-cracking, or killing scorpions and removing the inedible exoskeleton).

29. Non-human catarrhines exhibit minimal tool behavior. Tool behavior is unimpressive compared with many non-primate taxa.

It is apparent from the foregoing list that hominids greatly resemble their fellow catarrhines in many anatomical and physiological traits. There are, of course, distinctive differences in hominid anatomy and physiology, such as specializations for bipedal locomotion, or the hominid ability to cool the body by evaporating copious sweat from a naked body surface. However, in the realm of behavioral ecology, many distinctive hominid differences appear. These very clearly separate hominids from their fellow catarrhines. These differences will be discussed in the following chapters, and examined in terms of the possible forces of natural selection that might underlie their appearance.

8 Natural history intelligence and human evolution

Introduction

The human brain is generally perceived to be the summit of animal creation, irrespective of whether one thinks that it is the supernatural result of direct intervention by an intelligent designer or the product of known evolutionary processes. In the middle of the nineteenth century, before the advent of natural selection theory, Sir Richard Owen advocated that humans be formally classified as separate from other mammals. He contended in 1858 that humans should be put in their own taxonomic category, a subclass of the mammals that he proposed calling the Archencephala ("ruling brain"). The brain was the seat of consciousness, reason, and volition, and was the generator of peerless human behaviors. Much to the fascination of Victorian society, Owen later regularly dueled with Thomas Henry Huxley over whether humans possessed unique anatomical structures that differentiated their brains from those of other primates. Owen lost these arguments, because he mistakenly claimed that only humans had enlarged cerebral hemispheres that covered the cerebellum, and that only humans possessed a hippocampus minor. Other than unconscious bias, there is still no explanation for Owen's mistakes, because he had access to anatomical preparations that clearly demonstrated his errors (Desmond 1984).

Sometimes the perception of human uniqueness is implicit. Studies of brain evolution, for example, regularly include illustrations of animal brains shown in a ladder-like array, with the human brain placed neatly at the top of the sequence (Figure 1.4). Among anthropologists, interest has always been intently focused on the first appearance of a larger hominid brain size (presumed to signify the advent of intelligence). There has also been a great concerted effort by anthropologists to explain the origins of hominid intelligence. In terms of evolutionary processes, the explanation of first appearance and origins lies in discerning the selection pressures that are responsible for the appearance of higher cognitive functions. Because the interior of the neurocranium preserves some of the complex topography of the living brain, natural and artificial endocasts of hominid fossils have been used to establish the existence and estimate the relative size of brain regions with different functions (Figure 8.1).

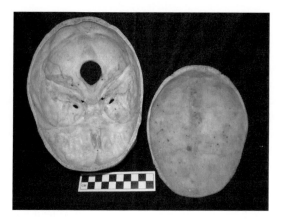

Figure 8.1. Sectioned human neurocranium, illustrating the complex internal topography that preserves gross features of the living brain. The scale is in centimeters.

How does the human brain compare with those of other mammals?[1] The human brain is far larger than expected for a typical mammal of the same body size, but the primate order as a whole is not remarkably encephalized in comparison with other mammalian orders (Stephan 1972, Jerison 1973). Magnetic resonance imaging (MRI) has recently been used to reconstruct the brains of living and dead hominoid species in three dimensions. Using small data sets, the authors of these studies argue that human frontal lobes are not enlarged, but are isometric. They are only as large as expected for a human-sized hominoid, because the volume of the human frontal lobes falls on the regression line generated when frontal lobe volume is compared to the total volume of both cerebral hemispheres (Semendeferi *et al.* 1997, Semendeferi & Damasio 2000). However, the mammalian neocortex is extraordinarily varied in both histology and neural connections, and this renders any gross structural comparisons essentially trivial. When the frontal lobe is examined in terms of its components, the hominid prefrontal is enlarged relative to primary sensorimotor elements, as are parts of the posterior association cortex (Preuss 2001). When gene expression is examined in leukocytes, liver, and brain in humans, common chimpanzees, orangutans, and rhesus macaques, expression is markedly pronounced in the human brain. When three outbred mouse species are used for comparison, quantitative changes in gene expression between human and chimpanzee brain appear much more striking than in the rodent species (Enard *et al.* 2002).

Experiments with engineered transgenic mice have recently demonstrated that the protein β-catenin, which regulates the production of neural precursors, may have a fundamental effect on cell number during brain development

(Chenn 2002, Chenn & Walsh 2002). The population of precursor cells is greatly increased when β-catenin is over-activated, because precursors re-enter the cell cycle without differentiating. The brains of engineered mice are grossly enlarged, when compared with wild type littermate controls (Chenn & Walsh 2002: Fig. 3). The cortical surface area of an engineered brain increases without a corresponding increase in volume. This results in a convoluted cerebral cortex with sulci and gyri that resemble those of other mammals, rather than the smooth cerebral cortex of a normal rodent brain. It is clear from these experiments that cortical cell number and decisions about cortical cell fate can be affected by protein activity in the developing brain, rather than through a fundamental neural reorganization in the ratio of cortical surface area to volume.

One fact is irrefutable. Hominid brain size has tripled over the span of the last four million years. The first hominids emerge with a brain size that is no different from that of a pongid. During the time of the first hominids, the australopithecines, the average endocranial capacity was about 400 cm^3, and the average endocranial capacity of living humans is about 1200 cm^3. Because gross endocranial capacity has long been taken as a proxy indicator of higher cognitive functions, or "intelligence," the measurement of endocranial volume in fossil primates has long been a staple of primate and human paleontology. To many researchers, this tripling of endocranial capacity has either been outside the realm of known selection pressures, or else it is something that requires a quantum advance over the non-human condition. Nevertheless, if one tries to estimate the degree of selection intensity that is necessary to triple brain size during the course of hominid evolution, the selection intensities do not seem aberrant (Table 8.1). Gingerich (2001) has tried to link laboratory data on microevolutionary rates of change to the gross macroevolutionary rates of change discernable in the fossil record. He conservatively estimates that a rate of at least 0.2 standard deviation of change per generation can yield the patterns discernable at the scale of hundreds of generations.

In fact, the estimated selection intensities fall within the realm of those that are documented under field biology conditions or in experimental laboratory settings. The increase in hominid brain size might therefore be reasonably explained by the existence of a known evolutionary process – natural selection. However, the factors that account for such selection pressures remain to be elucidated.

During the course of this chapter, I will emphasize two themes that will identify me as an apostate from primatology. The first theme is that many animals other than primates – especially other mammals and birds – can be useful in examining selection pressures involved in the origins of higher cognitive functions, or complex behaviors like tool use and manufacture or food storage. The second theme is that, among primates, the great apes are not necessarily the most

Table 8.1. *Selection pressures and human brain size*

Human brain size has tripled during the last four million years. Since the time of *Australopithecus afarensis* to living humans, endocranial capacity has increased from 400 cm^3 to 1200 cm^3. It is possible to estimate the selection pressure necessary to achieve this result, and see if it is beyond the range of selection pressure detected experimentally or under field biology conditions.

If one assumes that:

The heritability of volume is 0.5.
The average generation time is 15 years.
The coefficient of variation (SD/mean) is 0.1.
The response to selection equals the selection differential (actually, the response is likely to be less).

Then the observed change is calculated by

$$I = S/SD_p$$

I is the intensity of selection; S is the selection differential (the difference between the mean value of selected parents and the population as a whole); SD_p is the phenotypic standard deviation of the population before selection; the number of generations is 2.67×10^5 [$4 \times 10^6/15$ years].

Assuming that the increase is *proportional* (i.e., volume increases by a factor of x per generation), then

$$(1 + x)^{2.67 \times 10^5} = 3 \text{ [tripling of brain size]}$$
$$x = 0.4115 \times 10^{-5}; S = 0.4115 \times 10^{-5}(0.5); I = 0.20575 \times 10^{-5}/0.1$$

- **I = 0.20575 × 10^{-4} [if average generation time is 15 years]**
- **I = 0.165 × 10^{-4} [if average generation time is 12 years]**

Assuming that the x increase is a *constant*, with n a fraction of the original volume, then

$$(1 + nx)^{2.67 \times 10^5} = 3 \text{ [tripling of brain size]}$$
$$x = 0.749 \times 10^{-5}; S = 0.749 \times 10^{-5}(0.5); I = 0.3745 \times 10^{-5}/0.1$$

- I = 0.3745 × 10^{-4} [if average generation time is 15 years]
- I = 0.3 × 10^{-4} [if average generation time is 12 years]

These selection intensities are within the range of intensities detected today in field or lab. Selection pressures necessary for a tripling of endocranial volume are therefore biologically reasonable. Note that selection intensities decline as generation time shortens. Faster maturation in earlier hominids would mean lower selection intensities.

useful animals to study when examining the origins of hominid intelligence or complex sociality. The great ape bias in cognitive studies is so pervasive that one major researcher even coined the neologism "chimpocentrism" when arguing that studies of tool behavior and learning had been skewed by a mind-set that automatically distinguishes chimpanzee intelligence, behavior, and sociality as the *ne plus ultra* of the animal world (Beck 1982). For example, the existence of chimpanzee boundary patrols, warfare, and long-term anticipation and planning (Byrne 1995: 156–158) are routinely accepted by primatologists. Yet, many palaeoanthropologists deny that such behavior exists in fossil hominids until the advent of anatomically modern humans. Do chimpanzees really trump Neanderthals in terms of intelligence and sociality? I believe that "chimpocentrism" is at work here.

But I have been strongly making the point elsewhere in the book that hominids are good catarrhine primates, and that all of the living catarrhine primates form a relatively homogenous group (Schultz 1969). Am I now backtracking on this point? No. What I am saying is that the evolution of hominid morphology and physiology can be illuminated by the study of other catarrhines. However, in my opinion, hominid behavior and sociality represent a clean break from that of other catarrhines, and this break can be identified with the first appearance of an archeological record between 2.6 and 2.5 mya. As a result, the evolution of hominid behavior and sociality might be better illuminated by studying non-catarrhine primates like marmosets and tamarins, or social carnivores, or birds. Of course, one can continue to study great apes, but, to be brutally pragmatic, all of the great ape species will almost certainly go extinct in the wild within the next decade. Of more fundamental importance is the fact that about 40 years of studying the behavioral ecology of wild pongids has not yielded any special insight into the origins of the kind of behavior or sociality that leaves behind an archeological record. This archeological record, especially when it becomes more dense and variable at about 1.6 mya, documents the arrival of obligate, habitual stone tool use and manufacture, complex foraging for food and lithic resources, complex ranging behavior, and an understanding of the intricate behavior and ecology of other animals.

Ideas on the origins of hominid intelligence

Among the many ideas that have been advanced to explain the origin of hominid intelligence are primate extractive foraging (Parker & Gibson 1977, Gibson 1986) and primate foraging for dispersed and unpredictable food items (Milton 1981, Byrne 1995). In spite of the fact that foraging has been used to explain primate intelligence, other animals exhibit more complex foraging behaviors than primates do. Food hoarding is an example. Animals hoarding food may use scatter or larder hoards, dry food prior to caching, retard decay through a variety of measures, and fine-tune their caching response according to environmental variables, the amount of food stored, and the quantity of body fat accumulated by the maker (Vander Wall 1991, Gendron & Reichman 1995). Experimental evidence documents that caches are located through sophisticated cognitive mapping (e.g., Kamil & Jones 1997). No wild non-human primate species caches food; neither do these species engage in elaborate cooperative foraging.

Furthermore, comparative study of animal foraging demonstrates that there is no necessary correlation between foraging and intelligence.

The concept of a special relationship between tool behavior and human uniqueness can be traced back to the late eighteenth century (Oakley 1957). The manufacture of objects, especially tools to make tools, has long been implicated in the origin of intelligence[2] (Bergson 1907: 150). Washburn (1959) argued that tool behavior had a catalytic effect on human evolution ("autocatalysis"), so that evolutionary rates, particularly the rates of brain evolution, were speeded up. Tool behavior continues to be studied by archeologists who use stone implements to assess the nature and evolution of human thinking and knowledge during the Paleolithic (Wynn 1993). Animal tool behavior has been intensively investigated in a comparative fashion by Alcock and Beck (Alcock 1972, 1989, Beck 1980, 1982), and there seems to be no necessary relationship between tool behavior and the origin of intelligence. In any case, wild non-human primate tool behavior, the construction of artifacts or structures like nests, and extensive modifications of the external environment are not impressive when compared to the rest of the animal world.

However, tool behavior has been newly resurrected as a factor that increases brain size in primates. Reader & Laland (2002) emphasize that "technical intelligence" generates an increase in the size of the "executive brain," the neocortex and striatum. Tool behavior and the capacity for innovation underlie "technical intelligence." "Technical intelligence," in turn, stimulates the growth of the "executive brain."

Much consideration has been given to the existence of culture in great apes, especially in common chimpanzees. A major review article (Whiten *et al.* 1999 [and supplementary online file]) argued that tool behavior and other behaviors in common chimpanzees show a variability that exceeds that of any other non-human species, and this variability is probably the result of learned behavior transmitted through social interaction – i.e., it is the result of cultural differences between chimpanzee communities. However, I believe that the 65 behaviors examined in this paper do not seem to exceed the range of behavioral variability exhibited by other social mammals or birds. Furthermore, the list of non-primate species exhibiting cultural behavior continues to grow (e.g., Hunt 1996, Smolker *et al.* 1997, Heinrich 2000). The seven long-term chimpanzee studies synthesized by Whiten *et al.* (1999) totaled 151 years of field time – a number far surpassing the human labor expended in the systematic study of other animal species. Anthropologists and primatologists have devoted much attention to cataloging and analyzing great ape (particularly chimpanzee) foraging behavior, food sharing, nest construction, and tool behavior (e.g., Tuttle 1986, McGrew 1992, Sept 1992). The implication is that phylogeny or genetic relatedness is the main determinant of these behaviors, and that their existence in

Figure 8.2. A chimpanzee nest from the Ishasha River region, Kivu Province, Eastern Congo. Note the rough structure. There is no intertwining or weaving of materials. Tree branches are simply bent back and flattened to form a crude resting platform.

great apes reveals something about their origin in hominids (Figure 8.2). Yet, no wild non-human primate engages in complex cooperative foraging, food sharing, food caching, habitual tool use and modification, artifact creation, or landscape modification to the extent that hominids or other animals do (Alcock 1989, Vander Wall 1991, Gendron & Reichman 1995, Kamil & Jones 1997, Emery & Clayton 2004). Corvid birds, for example, exhibit habitual tool use and manufacture, intricate tool modification under novel circumstances, food caching, an arms race of strategies between conspecific food storers and thieves, complex recovery of cached and perishable food items in fluctuating temperatures, and recaching of food items upon observation by potential thieves (Emery & Clayton 2004). This range of complex behaviors rivals that of great apes, and has been noted both in the wild and under captive, experimental conditions (Figure 8.3). Tool manufacturing occurs spontaneously in hand-reared naïve juvenile crows, demonstrating that the ability is hereditary – it does not need to be learned, and is not dependent on cultural transmission (Kenward *et al.* 2005). In particular, the routine food caching behavior of corvids appears to exceed the ability of great apes to plan for the future. Yet, forethought and planning must underlie complex hominid ranging and foraging behavior documented from archeological evidence. Furthermore, the hominid record of complex, long-distance ranging behaviors to acquire food and lithic resources, habitual tool use and modification, artifact creation, and landscape modification, extends deep into the archeological record.

Hence, there is a disjunction between hominids and other primates. Beck (1982) discusses the bias against recognition of this disjunction as a form of "chimpocentrism": "The emphasis on chimpanzees as subjects of cognitive

Figure 8.3. Animal tool behavior in non-pongids. A. A captive, naïve New Caledonian crow spontaneously creates and uses a tool to acquire insects. Modified from Kenward *et al.* (2005: Fig. 1). B. A wild tufted capuchin uses a large hammerstone to crack open a palm nut on a stone anvil. Modified from Fragaszy *et al.* (2004: Fig. 1A).

ethology has not greatly retarded the study of cognition in the other great apes but . . . it has diverted attention from other taxa" (Beck 1982: 4). Whiten *et al.* (1999) laboriously document the existence of chimpanzee cultures – a good example of how "chimpocentrism" drives research. I contend that if an equivalent amount of time were expended studying other social mammals, equivalent behavioral variability would be discovered.

The current consensus opinion is that complex sociality itself selects for intelligence in vertebrate animals. The social cognition hypothesis holds that "[S]ocial integration and intelligence probably evolved together, reinforcing each other in an ever-increasing spiral" (Jolly 1966: 504). The "chief role of creative intellect is to hold society together" (Humphrey 1976: 307). Status-striving within the group emphasizes "political" maneuvering and develops reasoning abilities (de Waal 1982). Many researchers stress the existence of Machiavellian Intelligence: intelligence in primates and other vertebrates evolves from the assessment of social relationships, as well as the prediction of behavior, and the occurrence of deceit and manipulation within a social group (Byrne & Whiten 1988, Whiten & Byrne 1997). According to this model, primate intelligence arises in social cognition through individual competition, tactical status-striving, and "political" maneuvering. In fact, the social cognition of great apes may not be significantly different from that of monkeys (Tomasello &

Call 1994, 1997), although this is an implicit assumption of the Machiavellian model.

Further experimental study of what chimpanzees know about the psychological states of other animals demonstrates that chimpanzees can understand something about the goals and desires of other animals, particularly in competitive interactions. However, domesticated dogs outperform chimpanzees in using social cognition to interact with humans in cooperative tasks (Tomasello *et al.* 2003). Chimpanzees can understand something about the intentions of other animals, and how sensory information influences intention. However, chimpanzees do not appear to have a human-like Theory of Mind, which would allow them to understand that others may be attending to different objects within the same visual field, or to comprehend a wider array of psychological states understood by human children (Tomasello *et al.* 2003). Povinelli & Vonk (2003) argue that interpreting the social world in terms of mental states is a unique, species-specific human trait. They suggest abandoning experiments based on testing for a Theory of Mind, and instead advocate experiments creating unpredictable experiences. The ability of an animal to generalize from these novel experiences can test for whether the animal formed a mental abstraction of the experience (Povinelli & Vonk 2003).

A recent refinement of the social cognition model downplays the Machiavellian factor, and suggests that selection pressure for large cohesive social groups in primates causes an increase in relative neocortex size, and the concomitant evolution of complex communication systems (Dunbar 1992, Aiello & Dunbar 1993). Byrne (1995, 1996) now believes that Machiavellian intelligence may not explain the origins of mental representation and planning. Byrne (1995: 177–194) now presents a "food for thought" model, which is explicitly based on anthropoid primates. He argues that anthropoid primates experience substantial selection pressures via nutritional constraints, because they are largely herbivorous, yet lack the body size and gut specializations to digest low-quality plant food. They respond to these selection pressures by developing sophisticated foraging and extraction techniques, which ultimately generate intelligence, and, in the case of the great apes, even "insight." Yet specialized herbivory can occur in primates without a substantial increase in body size taking place. Some examples are found in members of the genera *Hapalemur* and *Lepilemur*, howler monkeys, and colobine monkeys. Specialized herbivory without a considerable increase in body size is accomplished in a variety of ways: a decline in metabolic rate, coprophagy, the appearance of large guts and relatively complex sacculated stomachs, and the maintenance of symbiotic bacteria to detoxify plant secondary compounds. Furthermore, if an adequate food supply is a problem, the most obvious adaptive response is to reduce body size. This is a response that has independently occurred in lineages across many mammalian orders. Body size is, in fact, very labile, and can quickly increase, decline, or reverse in a mammalian lineage, as explained in another chapter.

Hominid attention to natural history

The first stone tools are found in the Pliocene of East Africa, and now date back to 2.6–2.5 mya (Semaw *et al.* 1997). These tools are identified as representative of the Oldowan stone tool industry, the oldest known stone tool industry. Unequivocal archeological localities occur at 1.8 mya. FLK 22 ("the Zinjanthropus floor") at Olduvai Gorge, Tanzania, is an example. The early archeological record becomes more complicated (i.e., more dense and variable) about 200 000 years later. Archeological localities dating back to about 1.6 mya document unique hominid behaviors involving transportation of stone, carcasses or partial carcasses, alteration of stones (tool behavior), alteration of animal bones through the removal of flesh or marrow, and knowledge of the behavioral ecology of other species, including interaction with sympatric carnivore species (Cachel & Harris 1995, 1996, 1998). At the Olduvai MNK chert factory site, dated to 1.6 mya, chert cobbles were flaked and then transported at least 8–10 km away. In the Koobi Fora region, east of Lake Turkana, Kenya, sites dating between 1.64 and 1.39 mya demonstrate alternative foraging strategies and complex foraging for dispersed resources. In areas where suitable raw material for stone tools is not available, hominids do not discard tools – tools are carefully retained or curated (Cachel *et al.* 2000, Cachel & Harris in press). Raw material for stone tool manufacture may have been a crucial resource for hominids. In fact, the proximity of lithic raw material sources of a suitable clast size may have profoundly affected hominid ranging and foraging behavior. Complex foraging for dispersed resources implies intimate knowledge of the regional environment, and the ability to locate and predict the abundance of resources that fluctuate widely in space and time. A number of hominid species coexist during these time ranges in East Africa. While more than one species may have engaged in tool behavior, the taxon *Homo erectus* was probably responsible for the archeological evidence at 1.6 mya and afterwards.

These complex behaviors necessitate an attention to natural history, which appears documented at an ancient date. At these early time ranges, inferences of hominid mental activity are also made from degree of artifact symmetry, utility of the working edge of the artifact, careful use of raw material or tailoring of specific raw material to specific uses, and repeated use and transport (curation) of specific artifacts. Far later in time, beginning about 40 000 yrs BP during the European Upper Paleolithic, first portable and then parietal art documents the attention of its creators to fine details of the morphology and behavior of contemporary non-human animals (Bahn & Vertut 1988). By 12–13 000 years BP, portable and parietal art even documents human awareness of ontogenetic processes within other species, and an apparent symbolic manipulation of this awareness, so that the birth, growth, and death of other species can represent events in the human world or signify eternal recurrence. With the advent of a written record of human thought, there is undeniable evidence of human attention to and manipulation of the external environment. A Sumerian cuneiform document, for example, referred to as "The First Pharmacopeia" details mineral, animal, and botanical sources for a physician's materia medica (Moore 1993).

Changes in hominid cognition through time must certainly have occurred, if only because of the changes in brain size that took place. However, these changes in brain size are not outside the realm of known rates of natural selection (Table 8.1). The appearance of an archeological record by itself indicates that a qualitative break in intelligence occurred between hominids and other primates at least as far back as 2.5 mya. Intricate hominid symbolic manipulation remains invisible, if art is absent from the archeological record, or if a written record does not exist.

Other researchers downgrade natural history intelligence

Although I have just argued that even the earliest stone tools and archeological sites indicate the presence of a distinctive type of hominid intelligence, many researchers consider that such qualitative differences in behavior and intelligence occur very late in time. These researchers associate such differences with humans who are anatomically modern, although there may be a significant lag in time between the first appearance of such humans and archeological evidence of art or symbolic manipulation. Neanderthals, for example, who precede and coexist with modern humans in the late Pleistocene of Europe, are frequently (e.g., Binford 1989, Klein 1999) considered to be cognitively and behaviorally more primitive than any normal living humans, and lacking in any capacity for planning or forethought. Mithen (1996), who uses archeological evidence to infer cognition, argues that Neanderthals engaged in tool behavior and foraging without any conscious awareness of what they were doing, and lacked an ability for self-reflection, or introspection. Mithen (1996: 148) maintains that Neanderthals had three types of intelligence, which remained unintegrated as separate cognitive domains in a "Swiss-army-knife-type mentality": social, technical, and natural history intelligence. He contends that the modern human mind arrives when these three components merge and function together. Unlike

Mithen, I am not separating technical intelligence (modules for tool behavior) and natural history intelligence (modules for understanding the biological or physical worlds). I subsume both of these under the rubric of "natural history intelligence," following my earlier work (Cachel 1994).

A more fundamental difference with Mithen (1996) is that I contend that complex interaction with the biological and physical worlds demands not merely behavioral flexibility, but an integration of data from multiple senses, and a reorganization of neuroanatomical pathways that organize attention. The ability to monitor, to predict, and to manipulate objects and events in the natural world result from this complex interaction. As a consequence, natural history intelligence by itself contributes to the formation of general or fluid intelligence by expanding attentiveness and awareness away from the tyranny of the social world that generates the social intelligence identifiable in non-human primates.

Animal behavior and artificial intelligence

A second line of evidence indicates hominid attention to natural history and its distinctive emphasis on modeling the environment, manipulating abstract symbols, and planning future behavior. This involves research in artificial intelligence. Traditionally, such research targeted the construction of robots with modules for collecting data about the external environment, processing the input, mapping the external world, and planning activities within that world. The result was bulky structures that needed a great deal of memory, and processed constantly, but produced very little activity. This was a mystery to researchers, who could observe common, annoying insects engaging in constant activity with very little nervous system to underpin it. These insects explored their environment, fed, eluded human pursuit, and lived to fly another day. How did they do it? They did it through a nervous system that emphasizes activity, and not processing or "thinking." The solution to the artificial intelligence puzzle was to construct robots like insects, creating modules for producing activity, instead of modules for processing and planning.

Recent experiments in artificial intelligence have consequently resulted in the invention of a new type of artificial intelligence, an "insect" robot created by researchers at M.I.T. (Brooks 1991, McFarland & Bösser 1993). Researchers working with insect robots explicitly treat these robots as if they were animals, and state that they use design principles similar to those exercised by natural selection. This new type of artificial intelligence can duplicate many complex kinds of animal behavior. The mimicking of animal behavior is startlingly realistic. In fact, the new artificial intelligence is the basis of novel toys that are currently being marketed as "companions." These are robots that explore the

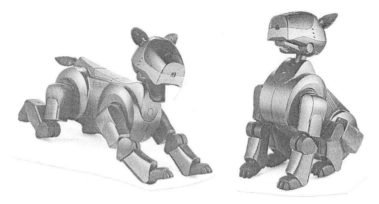

Figure 8.4. AIBO© entertainment robots marketed by Sony. The copyrighted name is an acronym for Artificial Intelligence Robot, but also fortuitously means 'companion' in Japanese. A cat-like version is also available. The toys move freely and actively, explore their environment, learn to modify their behavior, and interact with humans in a life-like fashion. Similar machines can function in industry, and swarms of them may serve in the future as emergency response teams during catastrophes or industrial accidents.

environment, modify their behavior, and interact realistically with people as if they are living pets. In fact, they are constructed to look dog-like or cat-like (Figure 8.4). Besides being playthings, the robots have a profound scientific utility. Sony's AIBO© dog robots mimic canine activities so faithfully that they can interact with real dogs, allowing researchers to study questions about species recognition.[3] The new generation of robots has already been used in the field to study complex questions about animal behavior, such as communication in lions (West & Packer 2002) and sexual selection in bowerbirds (Patricelli *et al.* 2002). The impact of information from different senses or reactions to conspecific behavior can be quantified, because field experiments with wild animals are conducted non-invasively through human manipulation of robots. It therefore appears that the new generation of robots, the insect robots, represents a new frontier in animal behavior studies, because these machines allow researchers to achieve new insights into animal behavior and physiology while leaving animals intact (Webb 2001, Kenneally 2003). Ironically, although the study of behavior is now being transformed in the twenty-first century as robots illuminate animal behavior (Machine as Animal), nineteenth century biology was similarly transformed when animal behavior was compared to the activity of complex machines (Animal as Machine). A revolution occurred in nineteenth century physiology when animal behavior was compartmentalized into a series of simple reflexes. Behavior was not caused by unknown and unknowable forces. Instead, animal behavior was reduced to muscle contractions and nerve reflexes that imitated the activity of machines. Huxley himself (1874) synthesized pioneering physiological studies by comparing animals to automata (robots, in modern parlance).

However, the new insect robots do not model the environment, create symbolic representation of the environment, or engage in abstract manipulation of that representation. The old type of robot processed sensory information, modeled the environment, and planned activity. The old type of robot exhibited very little activity at all, and little complex activity that duplicated animal behavior. Yet, it mirrored the type of intelligence found in its hominid creators, and was an obvious first step in the human creation of artificial intelligence. In fact, classical artificial intelligence research was explicitly based on inferences about human intelligence (McFarland & Bösser 1993). Evidence from both the archeological past and modern human experiments with artificial intelligence thus demonstrates hominid attention to natural history: symbolic representation of the external environment and abstract manipulation of that environment.

Natural history intelligence

Modern humans possess both natural history and Machiavellian intelligence. Their separate status is indicated by the consensus opinion that recognizes a difference in modern humans between fluid intelligence and social intelligence (e.g., Eysenck 1993). Machiavellian intelligence corresponds to social intelligence, though with an emphasis on competition and status-striving, rather than cooperation and empathy. Natural history intelligence corresponds to general or fluid intelligence, highlighting the importance of planning, predicting, and manipulating items in the non-social environment. Because the constitution of awareness is a model of the world, an organism must be highly aware of the world outside of its social group and social relationships before it can model, create symbols for, predict, or manipulate the external environment (Cachel 1994). Natural history intelligence is based on awareness of and attention to the world outside of the social world. Pre-attentive neural mechanisms for processing sensory information are doubtless crucial to its functioning. Natural history intelligence probably exists in other vertebrate species, but it does not seem well developed in non-human catarrhine primates, which have received most attention in studies of intelligence, and which provide the bulk of the evidence for Machiavellian intelligence. With the advent of an archeological record at 2.6–2.5 mya that preserves hominid behavior (and certainly with the advent of a dense archeological record at 1.7–1.5 mya), it becomes clear that hominids are behaving differently from any other primate. Paleoanthropologists routinely study Plio-Pleistocene hominid habitat exploitation through a survey of site locations, the selection and transport of raw materials for stone implements, lithic technology and function, the acquisition of vertebrate meat and marrow through hunting and/or scavenging (including confrontational scavenging), the transportation of carcasses and partial carcasses, the alteration of animal bones to obtain meat and marrow, the patterning of bone and stone within a site, and differences between sites caused either by different activities, seasonality, or length of time over which the materials accumulated.

Natural history intelligence incorporates awareness of and attention to vegetation beyond the realm of day-to-day foraging or predator shelter; the morphology and behavior of other animal species; and the inanimate world (for example, rock types and their distribution, and functional properties of these rock types). Natural history intelligence is based on complex, symbolic representation of the non-social environment and its abstract manipulation. The ability to generalize about objects and events in the outside world, to recognize patterns through the use of metaphor, to generate rules and laws about processes in the physical and biological world (outside of the social world), to predict events, and to plan behaviors that anticipate or control these events is predicated in natural history intelligence. This intelligence functions in cognitive problems that involve anticipating properties of objects outside the social world, once the object is recognized as possessing certain other properties, or being able to reason about cause and effect in the world outside the social group.

Some developmental psychologists have recently argued that learning in human infants and children is organized precisely like scientific learning through data collection and hypothesis-testing (Gopnik *et al.* 2001). A contrasting situation exists in the great apes. As might be expected, common chimpanzees have been the focus of study. Povinelli *et al.* (2000) conducted intricate research to determine the nature of chimpanzee understanding of the material world that lies behind their tool behavior. They also reviewed studies of the other apes. Povinelli and his colleagues used the label "folk physics" to describe an understanding of the material world. "Folk physics" comprises knowledge of energy, movement, objects, and object interactions. Povinelli *et al.* (2000) ultimately conclude that humans are different from the apes. Humans possess species-specific differences that allow them to interpret, explain, manipulate, and predict the material world, as well as the social world, in ways not found among the great apes. Chimpanzees must observe phenomena: "their central reasoning systems do not reason about things which have the status of being 'hypothetical'. In our view, this is because the chimpanzee does not form such concepts to begin with" (Povinelli *et al.* 2000: 298). On the other hand, humans, who possess the same sensory systems as chimpanzees, have a cognitive system that can deal with unobservable, hypothetical phenomena and spatial relationships. Povinelli *et al.* (2000: 337–340) conclude that 30 years of modern research on the mentality of apes – 75 years if one goes back to the work of the Yerkes and Wolfgang Köhler – was devoted to challenging human uniqueness, only to have human uniqueness emerge at the beginning of the twenty-first century. And this contradicts the embellished expectations of both primatologists and the lay public.[4]

Thus, natural history intelligence is not well developed in non-human primates, even if they are closely related to humans, as all Old World higher primates are. Habitual tool behavior, material and symbolic culture, control of the physical environment based on culture, knowledge and manipulation of other species, and interests in the secrets of nature are the result of natural history intelligence. A reciprocal interaction may occur between understanding and technology: human understanding of structures and events in the natural environment may be influenced by the course of human technology. For example, Maynard Smith (1992: 231) argues from the history of biology that understanding of the mechanisms of flight and the genetic code could occur only after Western science and technology had invented machines to duplicate flight and process information.

Because it constrains much non-human primate attention and awareness to happenings in the social world, Machiavellian intelligence may militate against the development of natural history intelligence, rather than contributing to its appearance (Cachel 1994). Competitive social life may enhance social intelligence, but, unless awareness of and attention to the non-social world occurs, this social intelligence will not be applied to the world outside the social group. Cognition is ordered, probably at the neuroanatomical level, to emphasize conspecifics and social dynamics. Competitive social life consumes attention, as animals constantly strive for status. A major researcher on primate social dynamics (Silk 2002) argues that high-ranking female cercopithecine monkeys launch randomly timed attacks against random subordinates in order to induce insecurity and stress in all subordinate animals. There are continual contests over status. This behavior benefits the dominant animals, but the physiological underpinnings of this behavior must be costly, even to the dominant animals. Furthermore, dominance is an inherently unstable condition. It is only established with reference to other individuals, and the playing field is subject to change. Natural history intelligence is based on neuroanatomical substrates that order awareness of and attention to the world outside the social group. Complex, symbolic representation and abstract manipulation of this world can then occur. Alertness to the social world appears to be a deterrent to the perception of and reaction to the world outside the social group.

After concluding a primatological literature review, Reader & Laland (2002) argue that there is no visible tradeoff between individual innovation or asocial learning and social intelligence or social learning. This appears to contradict the argument that I have been making about Machiavellian intelligence hindering natural history intelligence. However, Reader & Laland (2002: 4440) specifically distinguish between Machiavellian intelligence and social intelligence, which they define as social learning – the ability to acquire new information from other members of the social group. Furthermore, Reader & Laland emphasize that "technical intelligence" generates growth of the "executive brain." This type of intelligence is based on tool behavior, technical skills, and the ability to behave in a complex, novel, and flexible way as an adaptive response to ecological variables. Selection pressure for this type of intelligence favors "[i]ndividuals capable of inventing new solutions to ecological challenges, or exploiting the discoveries and inventions of others . . ." (Reader & Laland 2002: 4440). Another noteworthy aspect of this work is that there is no correlation between the size of the primate social group and the frequency of social learning behavior, as defined above. Hence, group size in primates has no relationship to the ability to acquire new information or spread innovations (Reader & Laland 2002).

Implicit in the Machiavellian intelligence theory is the idea that social intelligence is transferable to other spheres. It implies that the field of focus is immaterial: if an animal demonstrates social cognition, manifested through tactical status-striving, deceit and manipulation of conspecifics, and political maneuvering, then these abilities should be transferable outside the social realm and be manifested in tool use and manufacture, and knowledge of the behavioral ecology of other species. It is assumed that marked abilities for innovation and imitation in the social realm reflect abilities for imitation and innovation outside the social realm, thus affecting, for example, the possibilities for tool behavior. Yet this is not the case for primate tool behavior, which has been extensively investigated.

"Among primates, imitation of novel [tool] behaviors seems to be a rarer event than has been assumed in the past. The lack of imitation of new complex behaviors decreases the chances that new skills will catch on in a group or in a population. In short, cultural evolution is perhaps more strictly a human phenomenon than we have been led to suppose . . . Whereas in *Homo sapiens* . . . and to an unknown degree in the ancestral hominids, any tool-using behavior can disseminate quickly by imitation, it is plausible that the imitative spread of innovative behaviors occurs only in circumscribed situations in apes, and may be almost absent in monkeys" (Visalberghi 1993b: 147). Some researchers (e.g., McGrew 1992) argue that culture does not necessarily require imitation and teaching; hence, the dissemination of innovations may still occur in apes and monkeys that are admittedly deficient in these behaviors. But Visalberghi (1993a) questions the probability of culture existing and diffusing without the occurrence of imitation and teaching.

If social cognition cannot be transferred outside the social realm, there would then be no unbroken line of development between non-human primate sociality and the appearance of hominid intelligence. If this is the case, then selection pressures responsible for the origin of hominid intelligence cannot be readily inferred from the behavioral ecology of other catarrhines.

Problems with the social cognition model

Criticisms of the social cognition model are not yet generally found in anthropology or primatology texts, but they do appear in cognitive psychology (Heyes 1993) and in European primatology (Kummer *et al.* 1990, Kummer 2002). Beck (1982) and Povinelli *et al.* (2000) direct specific arguments against the social cognition model when it relies largely or heavily on great apes, particularly chimpanzees.

Animals form representations of their environment that allow complex decisions about resource acquisition to occur (Real 1991, 1992). Natural selection acts on the individual organism. Within that organism, behavioral, anatomical, and physiological components of the phenotype that relate to sociality may compete with phenotypic components that relate to the external environment. From this I infer that intense natural selection on vertebrates for the development of competitive social individuals may develop Machiavellian intelligence, but detract from the development of natural history intelligence. The intense focus on catarrhine sociality, which has been present in anthropology since the mid-1960s, was supposed to yield insights into the origins of hominid behavior and sociality. A focus on catarrhine sociality is a reasonable first approach, given that, for most traits, hominids are morphologically and physiologically typical catarrhines (Schultz 1969). It is ironic that hominid conformation to normal catarrhine morphology and physiology may have blinded researchers to divergent evolution in hominid behavior and sociality. Furthermore, there is serious reason to doubt that social cognition in the great apes (a particular focus for researchers of human evolution) differs qualitatively from that seen in other catarrhines (Tomasello & Call 1994, 1997). In short, granted the existence of a typical catarrhine morphological and physiological Bauplan, the origin of hominid intelligence may have been linked to a divergent sociality.

The mystery of the disappearing Kenyan vervets

Several avenues of investigation raise doubt that Machiavellian social cognition in non-human primates is a necessary substrate for the origin of hominid

Figure 8.5. An adult female vervet monkey (*Cercopithecus aethiops*) foraging for food in Nairobi Game Park, Kenya.

intelligence. The first line of evidence involves non-human primate behavior, both in the field and under captive conditions. The vervet monkey (*Cercopithecus aethiops*) is an abundant species of African monkey that occurs in a great number of different habitats (Figure 8.5).

Unlike many other primate species, vervets are not greatly affected by human alteration of the environment. In fact, vervets thrive in the presence of humans, happily raiding farm fields and garbage dumps for food, like a diurnal, African version of raccoons. Furthermore, the vervet mind has been studied under natural conditions. Vervets at the Amboseli Reserve, Kenya, are the subjects of the most intensive study among wild land mammals of non-human cognition (Cheney & Seyfarth 1990).[5] Non-human primate cognition is said to be based on social relationships and characterized by Machiavellian intelligence (Cheney *et al.* 1986). Yet, notable surprises can be discerned among vervets in "how monkeys see the world" (Cheney & Seyfarth 1990). Vervets are knowledgeable about inter-personal dynamics within their social group, and are knowledgeable about vervets outside their social group. But vervets do not seem to be aware of the natural world. One intuitively expects that a species' predators would be the focus of intense scrutiny and attention. Vervet mortality at Amboseli is estimated at 65% in 1987; selection pressure exercised through suspected leopard predation accounts for about 70% of the vervet mortality seen in 1987 (Isbell 1990). Leopard predation thus accounts for about 46% of the 1987 Amboseli vervet mortality. Yet, this predation by leopards is not associated with vervet awareness of leopard natural history (or natural history in general).

Leopards routinely cache carcasses or partial carcasses of their prey in trees, keeping the food away from other carnivores, and returning to feed later (Figure 8.6). Even the most naïve human tourist easily learns to recognize the presence of leopards by noting the occurrence of tree-stored carcasses. Some paleoanthropologists have even argued that early hominids might filch food from such tree-stored leopard caches, as part of a general repertoire of scavenging behavior, which they consider critical for the initial incorporation of meat and marrow into the hominid diet (Cavallo & Blumenschine 1989). Yet, vervets are unable to associate carcasses cached in trees with leopard presence. Cheney & Seyfarth (1990) even deliberately placed a stuffed carcass in a tree in order to elicit some vervet response (Figure 8.7). There was no vervet reaction. Given that leopards are suspected to account for significant vervet mortality at Amboseli, one might expect that the vervets would react to this carcass in the same way that humans would react to an iconic skull and crossbones figure on a bottle of poison – as a symbol of menace and danger. If early hominids had the minds of vervets, they could not have scavenged meat and marrow from a leopard cache, and would often have been leopard prey, ending up cached in trees themselves.

Vervets in another region of Kenya also suffer from intense leopard predation. Three species of catarrhine primate (vervets, patas monkeys, and olive baboons) have been studied through time by a number of researchers at Segera Ranch, Laikipia District, Central Kenya. A female leopard with her cubs recently set up residence at Segera. She began to specialize in eating vervets, and eventually completely wiped out the study sample of vervets by 2000 (fortunately after one of our Rutgers graduate students had finished collecting dissertation data on vervet endoparasites and sociality). The elimination of the Segera vervets is a consequence of not possessing natural history intelligence. Yet a specific vervet alarm call, documented by Cheney & Seyfarth (1990), exists for leopards.

There are other documented examples of vervet lack of awareness of natural history (Cheney & Seyfarth 1990): (1) the inability to associate python tracks (another predator) with python presence; (2) the inability to recognize during playback experiments that hippopotamus and black-winged stilt calls were coming from inappropriate dry woodland habitats; and (3) the inability to distinguish dust clouds raised by wildebeest from those raised by domesticated cattle. In sum, despite possessing Machiavellian intelligence that is based on social awareness, vervet monkeys appear to be unaware of much of their external environment, and are peculiarly obtuse about making associations and predictions about the external world, sometimes despite intense positive selection pressure for them to do so. This is important to note, because a reasonable first response to concerns about vervet lack of natural history intelligence is that vervets must not experience selection pressure to develop such intelligence. But, in fact, such selection pressure does exist at both Amboseli and Segera.

Figure 8.6. Typical signs of leopard presence and predation. A female Grant's gazelle was killed by a leopard the night before. Tracks and dung confirm leopard predation. The major portion of the gazelle's carcass remains on the ground (right). Besides the axial skeleton (shown here), the cranium and mandible lie scattered nearby. The leopard has carried one limb up into a fever tree (left) immediately adjacent to the carcass remnants on the ground. Photos taken at Segera Ranch, Laikipia District, Central Kenya.

Figure 8.7. The skin of a Thompson's gazelle that has been artificially stuffed to look more life-like, and hence to look like a very recent kill, virtually untouched, and cached in a tree by a leopard. Humans would instantly infer the close presence of a hungry leopard. Cheney and Seyfarth strategically placed this "carcass" in a tree where vervets would encounter it. There was no vervet response. From Cheney & Seyfarth (1990: Fig. 9.9).

Further primatological evidence against social cognition as a generator of intelligence

I have argued (Cachel 1994, and above) that competitive social life and Machiavellian interaction weighs against attentiveness to events outside the social group. In contrast to birds and other mammals, living primates seem to exhibit inefficient and impoverished anti-predator vigilance behavior; anti-predator vigilance is relaxed when conspecific members of the primate social group are nearby (Treves 2000, Treves *et al.* 2001, Treves & Pizzagalli 2002). Only dominant animals and the mothers of neonates remain vigilant when conspecifics are near. However, dominant animals and the mothers of newborns monitor the behavior of conspecifics, as well as predators, in the effort to protect their kin and social allies. Treves & Naughton-Treves (1999) argue that ancient hominids in environments with high densities of dangerous predators must have needed a high degree of anti-predator vigilance, and therefore ancient hominids must have had a type of social organization that was much more cohesive and less volatile than that of common chimpanzees or bonobos.

Adult wild female chimpanzees at Taï National Park, Ivory Coast, are more proficient and persistent than males at opening coula and panda nuts with stone tools. The argument is made that males, concerned with male–male relationships, monitor the behavior of other group members, and move away from nut sites in response to other animals – relatively non-social, individual females are not impeded by the social milieu (Boesch & Boesch 1984). Measurement of the energetics of nut-cracking by the Taï chimpanzees demonstrates that benefits outweigh costs by a ratio conservatively estimated as 9:1 (Gunther & Boesch 1993). Given the advantages of a nine-fold energy gain – a wonderful energetic advantage – natural selection should favor this tool behavior in both chimpanzee sexes, if other factors were not operating against it. At Gombe National Park, female chimpanzees engage in termite-fishing significantly more than males, and at a younger age (Lonsdorf *et al.* 2004). No active teaching occurs. Yet, young females spend more time observing their mothers, and eventually employ their mothers' tool behavior techniques. Mothers interact equally with both male and female offspring, but males play more at the termite mound, observe their mothers less, and demonstrate less termite-fishing behavior (Lonsdorf *et al.* 2004).

Captive studies on non-deprived and "learning-enhanced" or trained non-human primates often demonstrate an increase in abstract reasoning abilities or tool behavior, especially when animals are raised with humans in a relatively normal human environment (Savage-Rumbaugh 1986, Toth *et al.* 1993, Savage-Rumbaugh & Lewin 1994, Tomasello & Call 1994, 1997). Under these circumstances, subjects have no contact or a lessened degree of contact with conspecifics. It is significant that competition between conspecifics exists even under laboratory conditions. Even if human experimenters train animals for cooperation (e.g., cooperative chimpanzee communication using a computer keyboard), cooperative behavior between conspecifics is difficult to elicit, and requires a long period of training. The mere presence of conspecifics inhibits attention to things outside of social interaction. New World capuchin monkeys (*Cebus* spp.) have long been noted for tool behavior (Figure 8.8). The tufted capuchin (*Cebus apella*) is the taxon that exhibits the summit of tool behavior in non-human primates. Extensive use of tools or substrate anvils to pound and crack open thick husked fruit (Family Lecythidaceae, the Brazil nut family) is documented among wild tufted capuchins in Central Amazonia, and tool behavior is so common among this species that the stresses of captivity can be

Figure 8.8. An adult male white faced capuchin (*Cebus capucinus*) in Corcovado National Park, Costa Rica. He was foraging for insects in dead *Cecropia* leaves, when he was disturbed by my presence on the forest floor. His reaction was carefully to position himself above me, bite through *Cecropia* limbs, and wait intently for the crashing branches to discourage my observation. I followed the animal for half an hour, while dodging branches. Beck (1980) would not classify this activity as tool behavior, because merely letting a branch fall is too passive a behavior – the animal may have accidentally dislodged a branch by passing nearby. Nevertheless, because of the length of the interaction and the number of branches involved, I am convinced that this is a genuine example of spontaneous tool behavior in a capuchin monkey.

alleviated if tufted capuchins are provided with objects to manipulate (Boinski *et al.* 2001). Tufted capuchins use tools nearly on a daily basis in the Caatinga dry forest of northeastern Brazil (Moura & Lee 2004). These animals even forage terrestrially – a rare behavior for platyrrine primates – during the prolonged Caatinga dry season, when food is very limited and the risk of starvation is high. Tufted capuchins use branches, twigs, and stones as tools. The stones are used to dig into soil, and to crack or break open food items (Moura & Lee 2004). At another dry woodland site in Brazil, tufted capuchins have been observed for about 30 years using stone or log anvils and hammerstones to crack palm nuts (Fragaszy *et al.* 2004). Animals transport river pebbles to anvils, which are covered with stone dust, multiple depressions, and palm residues. Experimental observations conducted from a blind document that animals may sit or stand bipedally when using hammerstones, and that the hammerstones require significant muscular effort to use: the hammerstones weigh between 19% and 25% or 37% and 50% of the animal's body weight (Fragaszy *et al.* 2004). However, despite the frequency of tool behavior in the wild, long-term laboratory studies of tool behavior in the tufted capuchin monkey confirm that most subjects engage in more exploratory behavior when left alone with the experimental apparatus. Exploratory behavior is inhibited when a conspecific is present. This holds true even if the other animal is engaged in tool behavior (Fragaszy & Visalberghi 1989).

Captivity may artificially increase contact with conspecifics. In chimpanzees, the high population density possible under captive conditions has an adverse effect on social behavior. Social behavior is inhibited, and even agonistic behavior is lessened, while individual animals show behavioral signs of anxiety (Aureli & de Waal 1997).

Chimpanzees have been trained in the sophisticated use of mathematical symbols under laboratory conditions; they therefore demonstrate numerical competence (Boysen & Berntson 1989, Boysen *et al.* 1993). Two female chimpanzees were also subjects in a successful demonstration of Machiavellian deception. Yet, they were unable to act upon abstract mathematical rules when given food items. Alternating rounds of experiments with Arabic numerals and food items showed no improvement: the animals were unable to transfer abstract manipulation learned with symbols to actual food items (Boysen 1993). Even under laboratory conditions, animals are unable to act upon abstract rules when confronted with food items – they could act upon these rules only when using the symbol system. This suggests that brain mechanisms underlying foraging and food acquisition in the wild are buffered from abstract symbolic manipulation (Fischman 1993b). Even laboratory conditions of enhanced learning are unable to penetrate this buffering. This implies that the processes of natural selection that generated the foraging abilities of chimpanzees did not concomitantly generate abstract reasoning abilities related to food acquisition (although such was obviously the case for Plio-Pleistocene hominids). In short, these experiments confirm that the field of focus is important in chimpanzees, and that abilities for abstract reasoning and understanding cannot be easily transferred from one frame of reference to another. If abstract reasoning and manipulative abilities exist in a social field, this does not imply an inevitable transfer to realms outside the social world. Vigilance or sustained attention may be a problem, but the principal problem appears to be overriding neurological mechanisms that relentlessly prompt the animal to select for larger quantities of food items. Another interesting factor is that experimenters were unable to elicit cooperative food-sharing with symbols signifying more (rather than less) food items.

Experiments with tool use in captive hamadryas baboons (*Papio hamadryas*) and pig-tailed macaques (*Macaca nemestrina*) demonstrate that social dynamics involving status and sexual behavior impede tool behavior in these species (Beck 1980). Experiments with captive tufted capuchins (*Cebus apella*), whose tool behavior in the wild is the most pronounced among non-human primates, also repeatedly demonstrate minimal ability of animals in a social group to acquire tool use through imitation, although imitation of social behaviors occurs (Fragaszy & Visalberghi 1989, Visalberghi 1993b). Not only does imitation play no particular role in the acquisition of novel tool behaviors, but the presence of conspecifics also hinders exploratory behavior with the experimental apparatus (Fragaszy & Visalberghi 1989).

Finally, experiments with captive bonobos or pygmy chimpanzees (*Pan paniscus*) seem to demonstrate that bonobos possess greater abilities in human language use, abstract reasoning, and possibly tool behavior than common chimpanzees possess. This might be the result of species-specific differences in behavior: wild bonobos show more affiliative behavior and less dominance interaction than wild common chimpanzees do. The reduction in status-striving and competition among bonobos might result in more natural history intelligence. In a later section, I will make the argument that such conditions generate complex sociality in New World callitrichids.

Brain mechanisms underlying natural history intelligence

I equate natural history intelligence with fluid intelligence. Some researchers dispute the reality or significance of general or fluid intelligence, measured by Spearman's g factor in intelligence tests across cognitive domains. Nevertheless, there seems to be a neuroanatomical basis for high g. In human twin studies, researchers discovered (Plomin & Kosslyn 2001, Thompson *et al.* 2001) that the quantity of frontal gray matter is under significant genetic control (although heritability could not be estimated), and this quantity is related to g. Furthermore, PET scans of living humans exhibit localized frontal lobe activation when subjects are engaged in tasks requiring high g (Duncan *et al.* 2000).

The dichotomy between Machiavellian intelligence and natural history intelligence discussed above is much like the dichotomy between the "emotional" brain and the "executive" brain that is discussed in studies of primate cognition and genomic imprinting. The genomes of each parent may contribute differently to brain development by silencing autosomal alleles contributed by the other parent. Experimentally duplicated mouse paternal genomes contribute significantly to the development of the hypothalamus and septum (emotional brain); duplicated mouse maternal genomes contribute significantly to the development of the neocortex and striatum (executive brain) (Keverne *et al.* 1996). These results have been imported into discussions of primate sociality – e.g., different male versus female reproductive strategies, group size, adult sex ratios, and philopatry (Falk 2001). In terms of the present analysis, the hormonally driven emotional brain underlies Machiavellian intelligence expressed by political maneuvering and dominance interactions. The executive brain underlies natural history intelligence, which is more flexible, relatively released from hormonal influences, and attentive to events outside the competitive arena of the social group.

The divorce between natural history intelligence and social intelligence is confirmed in modern humans by two congenital conditions that severely affect normal existence. These conditions are autism and Williams' syndrome,[6] and both have strong genetic substrates. Autistic individuals interact freely with the non-living world, and may exhibit great mechanical and numerical competence, but they experience difficulty in interacting with other humans, because they cannot assess the minds, emotions, or motives of their fellow humans (Rodier 2000). Autistic humans apparently experience a kind of blindness to other minds. In Williams' syndrome, affected individuals are facile and charming in social communication, exhibiting great skill in small-talk and the superficial, milling social interactions that occur on the level of party conversation and activity. Yet, Williams' syndrome people appear to be handicapped in understanding and interpreting the physical world (Bellugi *et al.* 1994, Mervis & Bertrand 1997, Hoogenraad *et al.* 2004). As children, they are slow at exhibiting tool behavior, object categorization, understanding how objects affect each other, and appreciating biological processes like birth, growth, and death. As adults, they are affected by a blindness to spatial relationships that inhibits their safe movement and return even within their own neighborhood.

Functional magnetic resonance imaging (fMRI) of the brain of normal living humans shows characteristic activity when simple experimental tasks are performed. The same neural systems are activated when subjects are asked to anticipate the action of another subject doing the same task (Ramnani & Miall 2004). These results suggest that normal humans predict the behavior of other individuals by activating the same neural pathways in their own brain – that is, forming a "Theory of Mind" about the mental states of others by stimulating the same areas in their own brains.

What brain mechanisms underlie attention and attentiveness? Studies of the living brain in normal human subjects using non-invasive PET and functional MRI imaging show that changes in brain activity unfailingly occur during goal-directed activity. In fact, there is apparently a uniform baseline or default state of human brain function that is changed during goal-directed behavior that demands attention (Raichle *et al.* 2001). The default state of brain activity involves areas such as the medial prefrontal cortex, the posterior cingulate, and the medial parietal cortex; if focused attention is necessary, brain activity in these areas is diminished (Raichle *et al.* 2001). Areas involved with focused attention and goal-directed behaviors are then briefly activated to support attention and behavior.

Weak, subliminal sensory stimuli still invoke an EEG response in the human brain, but are functionally inhibited to buffer the cortex from meaningless noise (Blankenburg *et al.* 2003). A major debate in cognitive studies concerns whether the processing load that requires attentiveness can affect the perception of irrelevant stimuli. A human study has shown that high loads of attention can reduce the perception of even biologically important stimuli that may be especially difficult to ignore (Rees *et al.* 1997). This study might explain the proposed trade-off or inverse relationship between Machiavellian and natural history intelligence. It might also suggest tests of the hypothesis. For example, electroencephalograms or functional imaging analysis of brain activity in relevant areas can record visual-evoked and auditory-evoked responses from non-human primates being introduced to novel stimuli from inanimate objects or different species. If stimuli from conspecifics are introduced, and if attention is then focused on these conspecific stimuli to the detriment of the other experimental stimuli, the evoked responses will then be reduced. The reduction of the evoked response from the introduction of stimuli from conspecifics should be greater than when other novel stimuli (e.g., birdsong, a new wooden toy) are introduced. Conspecific stimuli should produce a statistically greater effect because of the attentional load demanded to process these stimuli.

I do not mean to imply that hominid brain evolution occurs independently of trends taking place in other mammals. In fact, there is clear evidence for convergent evolution in the mammalian brain (Jerison 1973, Clark *et al.* 2001, de Winter & Oxnard 2001). When broad mammalian comparisons are done, brain size per se is the overriding factor accounting for differences, with an influence ranging from 98 to 90% (Oxnard 2004). But when ratios using functional relationships between different brain structures are examined using multivariate statistics (de Winter & Oxnard 2001, Oxnard 2004), the size influence is reduced. Primates, insectivores, and bats are widely separated in multivariate space, demonstrating the differences in brain organization between the three orders. Three primate subsets are differentiated by locomotion: hindlimb dominated leapers, forelimb dominated arboreal hangers, and quadrupeds. These functional subsets cut across phylogenetic divisions. Canonical variates analysis, performed on primate species assigned to families and superfamilies, allows for a quantitative assessment of brain differences. Humans are well separated from all other primates on two axes measuring standard deviation units (Oxnard 2004: Fig. 11). Chimpanzees are encompassed within the non-human catarrhines, and the degree of difference between humans and chimpanzees is equivalent to the greatest distance separating all non-human catarrhines. Oxnard (2004) infers that this difference reflects not only neocortex size, but also unique internal differences in the human brain. He argues that the human brain evolved 4–6 times faster than the brain of chimpanzees since divergence from their last common ancestor. Yet, considering brain size alone, normal rates of selection pressure can account for the tripling of hominid brain size over 4 million years (Table 8.1). No unique or aberrant evolutionary mechanisms need be invoked to account for hominid brain size increase.

Mental abilities are subject to convergent evolution. Similar selection pressures operating on phylogenetically distant groups can create similar modes of cognition. Complex cognition in corvid birds is equivalent to that of apes, but is based on completely different brain mechanisms (Emery & Clayton 2004). The nidopallium and mesopallium, brain areas analogous to the mammalian prefrontal cortex, are relatively enlarged in corvids. A recent interpretation of vertebrate brain evolution uses connections in the nervous system and anatomical regions that express gene products to argue that a revised nomenclature should be used for the avian brain (The Avian Brain Nomenclature Consortium 2005). Birds appear to have a relatively large pallium that processes information like the sensory and motor cortices of mammals. Complex cognitive functions are based not only in avian nuclear pallial areas, but also in a mammalian six-layered cortex. There is no single neural architecture responsible for complex cognition, and intricate cortical folds are not necessary for sophisticated cognition. Thus, convergent evolution can occur in mental abilities, in spite of very divergent trends in neural anatomy (The Avian Brain Nomenclature Consortium 2005).

Because the hippocampus plays a role in filtering sensory data committed to memory, it might be expected to show a relative increase in size if complex environmental data were crucial to survival. In fact, the relative size of the hippocampus does increase in bird species that cache food (Harvey & Krebs 1990, Krebs *et al.* 1996, Clayton 1998). New neurons are formed in the hippocampus of adult birds and mammals, including humans, when environmental stimuli or exercise take place under experimental conditions. In mice, these new hippocampal neurons are functional (van Praag *et al.* 2002).

Neuroanatomical bases for attending to, analyzing, sequencing, or predicting natural history data may be particularly important. For example, in humans the earliest measurable physiological responses to novel auditory events occur in the auditory cortex, and are governed by a pre-attentive sensory memory mechanism (Tiitinen *et al.* 1994). In humans it is well known that subliminal perception, pre-attentive perceptual processes, and implicit memory all exist and influence mental function and behavior, even though they are not consciously perceived or remembered (Kihlstrom 1987). Therefore, the detection of events and pattern recognition do not need to be conscious in humans.

In humans, a common semantic system for processing words and pictures lies in regions within the left cerebral hemisphere (Vandenberghe *et al.* 1996). However, language-based knowledge of the physical properties of animals, and residual knowledge of the properties of animals has a special neural subsystem in humans (Hart & Gordon 1992). This specialization supports natural history intelligence in humans, and indicates that novel neural systems may evolve to maintain it.

In humans, brain activation caused by emotion does not interfere with cognitive activation in the same brain area – instead, cognitive activity is enhanced, indicating that emotion can adjust activity within the same area (Gray *et al.* 2002). Recent human tests using functional MRI in many subjects establish that cognition and emotion are divisible in humans. However, cognition and emotion are integrated in the lateral prefrontal cortex, and the integrated signal influences thought and goal-directed behavior (Gray *et al.* 2002). Human lateral prefrontal cortex activity selectively mirrors emotional loads or demands, in order to maintain goal-directed behavior at a certain level (Gray *et al.* 2002).

The anterior principal and lateral dorsal limbic nuclei have an absolutely and relatively greater number of nerve cells in modern humans than in investigated species of great ape or lesser ape. Some researchers suggest that the larger size of these features modulates the integration of emotion and cognition, or relays a larger emotional component into the posterior cingulate gyrus (attention) and the posterior association region (Armstrong 1982, 1985, 1990, 1991, Simmons 1990). Instilling experiences with emotion would enhance attentiveness and augment memory. This might help to focus attention toward the external environment, and heighten cause and effect determinations. Such mechanisms might then serve to control Machiavellian intelligence. If expansion of the human limbic system occurred before expansion of the hominid neocortex, control of Machiavellian intelligence could have existed in the earliest known members of the hominid family, the australopithecines, whose fossil record extends back well before 4 mya. However, evidence for a general trade-off effect between emotion and focused attention during cognition comes from imaging techniques on the brains of normal living human subjects. These techniques show decreases in activity of the medial prefrontal cortex during focused attention (Raichle *et al.* 2001), and may support my arguments for an inverse relationship between emotional states and cognitive states demanding focused attention.

There is evidence that intense competition for space occurs in the brain, simply because brain tissue is so metabolically expensive. Brain tissue, in fact, accounts for about 20% of a mammal's oxygen and glucose consumption – this is true for humans, as well as for smaller mammals (Diamond 1996, Raichle *et al.* 2001). As an example of the competition between brain structures for space, the somatosensory cortex and the visual cortex have a reciprocal relationship in animals that experience a natural selection trade-off effect in these neural areas (Diamond 1996). In the human brain, functional MRI studies show that memory systems with different substrates may compete with each other during experimental learning tasks (Poldrack *et al.* 2001). Studies in rats also show "zero-sum" competition between brain areas controlling learning behavior. Such competition may be an adaptive response – it can mediate between the contrary demands for learning both flexible and automatic responses.

I infer from such evidence that cognitive differences based in different neuroanatomical substrates should show a reciprocal relationship if an animal's energy consumption were significantly affected. This might be the neuroanatomical explanation for the empirical evidence that Machiavellian intelligence militates against the appearance of natural history intelligence in primates. Furthermore, sex differences in adult human cerebral glucose metabolism occur while the brain is in a resting state (Gur *et al.* 1995). If these sex differences (higher male metabolism in the temporal-limbic system, higher female metabolism in the middle and posterior cingulate gyrus) are upheld by further experimentation, they imply cognitive differences between the sexes in emotional processing and symbolic ability or tasks involving abstraction. For humans, these differences might indicate a neuroanatomical basis for the trade-off between Machiavellian intelligence and natural history intelligence that I suggested above. Because similarities in brain metabolism between the human sexes outweigh sexual differences (Gur *et al.* 1995), a species-specific human baseline for cognitive processing may exist. This underscores the appearance of a distinctive human neuroanatomy that evolved from a non-human primate substrate.

In mammals, the cerebellum occupies a relatively constant percentage of about 10–11% of total brain volume across various orders (Jerison 1991, Clark *et al.* 2001). This implies that cerebellar function is deeply implicated in general brain function. There are certain animals, however, whose specialized senses mandate additional cerebellar increase for sophisticated processing of sensory input (Clark *et al.* 2001). Initial comparisons of the relative size of the cerebellum across the primate order apparently demonstrated that modern humans have an enlarged cerebellum (Stephan 1972, Stephan *et al.* 1981). The apparent cerebellar enlargement is traditionally explained by the fact that the unique bipedal locomotion of hominids mandates extensive motor control. Yet, tool behavior is also invoked, as well as cognitive advances, because the cerebellum integrates both movement and thinking during tool behavior.

Jerison (1991: 42–47) reanalyzes the data in Stephan *et al.* (1981) and argues that the human cerebellum is not enlarged. These arguments are confirmed by

Clark *et al.* (2001). Jerison contends that the size of the human cerebellum is exactly that expected for anthropoid primates. In comparison to other orders, primates, in fact, have a slightly smaller cerebellum, and anthropoid primates show even more reduction than prosimians. Jerison explains this by claiming that expanded visual processing centers in the anthropoid brain involve the cerebellum only in a peripheral fashion. "Anthropoid primates are not under-cerebellarized. Rather they are overvisualized, as it were, so their gross brain size is greater than one would expect, other things being equal" (Jerison 1991: 47). Yet, he remains mystified by what he perceives as a lack of cerebellar advance in living humans.

As was shown in the discussion of frontal lobe size (Introduction), the variety of histological and neural connectivity differences that exist between species renders trivial any straightforward comparison of the size of brain components. The smaller than expected human cerebellum might be explained by qualitative differences at a histological level. When the primate cerebellum is examined in detail, certain components appear species specific (Matano *et al.* 1985, Matano & Hirasaki 1997). In humans, the dentate nucleus of the cerebellum is larger than in other hominoids.

Recent neuroanatomical investigation certainly implicates the cerebellum in cognitive functions. Experiments with the tufted capuchin monkey (*Cebus apella*) demonstrate that neurons in the basal ganglia and cerebellum project to the dorsolateral prefrontal cortex (Walker's area 46) via the thalamus (Middleton & Strick 1994). This fact implies that these cerebellar neurons also have cognitive, rather than strictly motor, functions. Furthermore, human diseases that result in cerebellar destruction (e.g., Huntington's disease, Parkinson's disease) affect working memory, rule-based learning, and the planning of future behavior (Middleton & Strick 1994). Magnetic resonance imaging reveals that the dentate nucleus of the cerebellum is active during problem-solving, and is therefore presumed to be involved with higher cognitive functioning independent of the control of eye and limb movement (Kim *et al.* 1994). Functional magnetic imaging discloses that the cerebellum integrates and coordinates sensory input in humans, which agrees with micromapping studies of cerebellar activity in other animals (Bower & Parsons 2003). Autopsies and new imaging techniques reveal anomalies of the cerebellum, cranial nerves, and motor nuclei in autistic modern humans, whose cognitive abilities are clearly abnormal (Rodier 2000). These anomalies occur during fetal life, and result in a range of deficits, the most obvious being an inability to understand the minds of other humans. Even in very early infancy, this can create problems involving imitation of and interaction with other people.

Because the cerebellum in living humans has cognitive functions, there are implications for paleoanthropology. A marked enlargement of the cerebellar fossae occurs in the occipital bones of some robust australopithecines (Aiello & Dean 1990: 168–169). Using an intense LED flashlight beam to reveal topographic contrast, I have observed huge, deep cerebellar fossae within the basioccipital of a female *Australopithecus boisei* specimen (KNM-ER 407A) in the Palaeoanthropology Section, Kenya National Museums, Nairobi. The cerebral fossae are much smaller. In fact, when one compares the occipital bones of hominids and chimpanzees, the deep cerebellar fossae of hominids account for the greatest difference between the species (Mowbray 2002). Later hominids expand the cerebral hemispheres. In the original Nariokotome (KNM-WT 15000) *Homo erectus* specimen, for example, topographic contrast reveals that the cerebellar fossae are smaller than the cerebral fossae. What does the early increase in the size of the hominid cerebellum signify? Paleoneurology is debatable. However, it is possible that this increase affects cognitive activities related to understanding the minds of conspecifics, learning new motor skills, or planning complex activities. Based on hand anatomy, one major researcher (Susman 1994) is convinced that robust australopithecines, as well as members of genus *Homo*, were capable of tool behavior. Robust australopithecines are certainly often found in association with stone-tools in East Africa (e.g., the Koobi Fora region, Olduvai Gorge) and in South Africa. Tool behavior using both stone and bone as a raw material has been inferred for robust australopithecines at the South African site of Swartkrans.

What brain regions are active during stone-tool making? Brain activity in one human subject was recently examined during stone-tool manufacture using positron emission tomography (PET) brain imaging (Stout *et al.* 2000). The subject, Dr. Nicholas Toth from Indiana University, is an experienced stone-tool knapper. Through this non-invasive technique, researchers monitored increased localized blood flow caused by augmented activity in several areas of Dr. Toth's brain. Three states were investigated: a control or baseline state, a cognitive state involving examining a core and imagining or planning stone flake production, and a state when active knapping took place. Many brain regions were active during both the mental imagery and the knapping states, although activity was much more intense during knapping. There was cerebellar activity during both these states. During the imagery state, areas of the parietal and occipital lobes, and the fusiform gyrus were active, signifying spatial cognition, motor imagery and processing, and visual association and processing. During knapping, areas of the parietal lobe, central sulcus, postcentral gyrus, and fusiform gyrus were active, signifying spatial cognition, primary motor and somatosensory processing, and visual association (Stout *et al.* 2000). Cerebellar activity is expected in movement, posture, and motor learning. Significantly, activation of the superior parietal lobe in both the imagery and knapping states implies cognition involving attention, representation, and modeling of external space from diverse sensory inputs. Hence, even the most rudimentary knapping to produce Oldowan stone flakes necessitates neuroanatomical structures for attentiveness, representation, and modeling. These brain mechanisms underlie the natural history intelligence that makes stone tool behavior possible.

Other tests of the social cognition theory

Natural history intelligence may be tested by playback experiments or other non-invasive techniques performed on wild non-human primate species that exhibit Machiavellian intelligence. Cognition in the solitary orangutan (*Pongo pygmaeus*) should be investigated, and comparison made with cognition in other pongids, which are social. As predicted in Cachel (1994), orangutans should exhibit natural history intelligence. The recently published inventory of wild orangutan cultural behaviors (van Schaik *et al.* 2003) rivals that found in chimpanzees (Whiten *et al.* 1999). Some elements of natural history intelligence may be discerned in wild orangutan behavior, although they need detailed confirmation. These elements are tool behavior, including the use of tools for extractive foraging both in the trees and on the ground, the building of bridge nests that connect trees separated by a river, and hiding from predators or humans behind detached branches, which may document a Theory of Mind (van Schaik *et al.* 2003: Table 1). The degree to which orangutans consume insects, and the effort expended in their consumption, can be inferred from this inventory. Also documented in this repertoire is orangutan hunting and consuming of slow lorises (*Nycticebus coucang*), which are presumably caught while they are asleep during the day. Chimpanzees are therefore not the only pongids that hunt and eat their fellow primates. Because non-human primates form a significant part of the vertebrate biomass in tropical rainforests, it is not strange to find apes consuming other primates.

Van Schaik (2004) argues that orangutans in the swamp forests of western Sumatra offer particular insights into the origins of human tool behavior. Tool behavior is common among these orangutans, and it is learned from adults. Tools are made for specific uses, and saved for later use. Van Schaik believes that tool behavior arises for two reasons: arboreal orangutans are safe from predation, and the swamp forest environment offers rich food resources for orangutans. Relative safety and leisure generate tool behavior, which appears when selection pressures from predation and dietary stress are ameliorated.

Among non-human primates, callitrichids (the New World marmosets and tamarins) show the most complex sociality. Intense investigation of cognition in wild callitrichid species is also warranted. The behavioral ecology of other mammals may be informative. The diet of mammalian social carnivores, for example, is based upon intimate knowledge of the behavior and ecology of other species. Social carnivores (rather than non-human primates) also exhibit elaborate cooperative foraging, food sharing, and food caching behaviors. If Machiavellian intelligence is not apparent in a species, natural history intelligence may be more developed than in the non-human primate world. Neuro-anatomical substrates that order attention outside of the social realm might also be investigated. In social species with natural history intelligence, one would expect to find relative enlargement of brain areas to control or focus sexual, feeding, or fighting emotions, or areas that contribute to integration of information from several sensory modalities.

One might also identify selection pressures that could contribute to the origin of hominid natural history intelligence. These would be factors diminishing status-striving or the internal competition between matrilines in a social group. Increasing genetic relatedness or a reduction of food competition through the availability of new food resources might accomplish this. (Yet captive, provisioned non-human primates still demonstrate status-striving and Machiavellian intelligence.)

Another trigger for hominid sociality might be selection for predator detection and defense in a small-bodied species, analogous to the way that complex sociality is thought to have evolved in the dwarf mongoose. Brain (2001) argues that the threat of constant predation by large mammalian carnivores fueled both an increase in hominid brain size and the origins of tool behavior. (Yet both documented aerial and terrestrial predation have not caused the evolution of eusociality in Old World monkeys.)

The diminishment of status-striving through the establishment of rigid dominance hierarchies or positive selection pressure for cooperative behavior would also create natural history intelligence in hominids. Factors reducing dominance interactions and increasing cooperative behavior need to be examined, and methods of identifying these factors in the paleoanthropological record need to be studied.

Natural history intelligence over the course of human evolution

During the last 4 million years, hominid brain size tripled from about 400 cm^3 to about 1200 cm^3 (Table 8.1). Cranial capacity scaled to body size yields a value of neurocranial volume relative to body size that has been widely studied in vertebrate paleontology (Jerison 1973, 1975, 2001). The investigation of relative brain size is also a staple of paleoanthropology (Tobias 1971, 1991, Falk 1987). In both cases, relative brain size is used to infer the intelligence of fossil organisms, although this is admittedly only a proxy indicator of intelligence.

The decrease in relative brain size subsequent to domestication in non-human animals, and hominid brain size decrease over the last 10 000 years (Henneberg 1988) both imply significant lability in relative brain size.

Jerison (1973) argued that different functional modules of the mammalian brain evolve independently from each other in a mosaic fashion. He also contended that similar selection pressures create convergent brain proportions and structures in distantly related mammalian groups that are undergoing similar adaptive responses. Subsequent investigation by Finlay & Darlington (1995) appeared to disprove this. The scaling of structures in mammalian brains relative to absolute brain size seemed to demonstrate a highly conservative, non-linear linkage, with greater increases occurring in the neocortex. If this were true, Finlay & Darlington (1995) argued that strong pleiotropic effects might result from selection for a single behavioral trait, and an increase in neocortical size could concomitantly influence many other behavioral variables. However, more recent investigation (Barton & Harvey 2000, Clark *et al.* 2001, de Winter & Oxnard 2001) supports Jerison's (1973) observation of mosaic evolution in the mammalian brain, and multiple mammal radiations that convergently develop similar proportions in different brain structures. Similar selection pressure on variables of niche structure like sensory modalities and locomotor adaptations can create species whose brain organization is similar, in spite of the fact that their lineages have been separated by 30 to 60 million years (de Winter & Oxnard 2001).

Brain asymmetry and left hemisphere dominance unequivocally appears to underlie language and speech in modern humans. Tobias argued (1991) that endocasts of *Homo habilis* exhibit Broca's Area for motor speech in the left cerebral hemisphere, and that members of this taxon were therefore capable of speech. However, MRI images of 20 chimpanzees, 5 bonobos, and 2 gorillas have recently documented the existence of Brodmann's area 44, a part of Broca's Area, in these animals (Cantalupo & Hopkins 2001), although they obviously lack speech. In addition, Falk (1987) documents asymmetry in the cerebral hemispheres of Old World monkeys that are also clearly speechless. Brain asymmetry may therefore be an ancient feature of catarrhine primates. Although asymmetry is now linked with areas supporting language in modern humans, asymmetry may have originally been associated with gestures or species-specific vocalizations in ancient catarrhines.

Modern humans certainly possess both natural history and Machiavellian intelligence. If, as I have argued above, Machiavellian intelligence deters the development of natural history intelligence, this implies the existence of two phases in the evolution of higher cognitive functions in hominids. In the first phase, the general anthropoid primate emphasis on social cognition is downplayed, and natural history intelligence emerges. The human ability to interact with the physical and non-social biological worlds in an abstract fashion (distinctive among primates) arises at this time. Its origin might come about through the diminishment of status-striving, either through the establishment of rigid dominance hierarchies or the reduction of matriline competition, or positive selection pressure for cooperative behavior in hominids. Unequivocal evidence of natural history intelligence is first seen at complex East African archeological sites that appear at 1.8 mya. Shortly thereafter, at 1.6 mya, the record becomes even denser, and more variable. Because the record of hominid manufacture of stone tools dates back to 2.6–2.5 mya, natural history intelligence probably dates to a time before this. Human natural history intelligence (identified as fluid intelligence in modern humans) is initially divorced from a social context, and thus more nearly resembles the intelligence of social carnivores.

What might hominid sociality have been like 2.5 mya? Tool behavior already occurs at 2.5 mya, but what type of social interactions might have taken place? Chapter 17 presents some speculation about early hominid sociality, and the selection pressures contributing to growing social complexity.

Conclusions

Researchers interested in the origin of human intelligence have traditionally focused on non-human primate tool behavior, foraging behavior, and sociality. Non-human primate sociality is widely believed to be the necessary factor in the evolution of human intelligence. Human intelligence is capable of attending to the non-social world, symbolically representing that world, and abstractly manipulating it ("natural history intelligence"). The "Machiavellian intelligence" recently emphasized as typical of non-human primates may not be a necessary step in the development of natural history intelligence. Because the structure of awareness is a model of the world, Machiavellian intelligence may even militate against the development of natural history intelligence. The arguments elaborated above imply that the social intelligence seen in non-human primates and humans is partly convergent or non-homologous. Applying human social terminology to the description of non-human primate social behavior (e.g., "politics," "political," "Machiavellian") thwarts appreciation of the fact that human and non-human primate intelligence related to social behavior may have disparate sources.

In conclusion, I emphasize five points:

(1) Modern human intelligence is capable of attending to the non-social world, symbolic representation of that world, and its abstract manipulation. I call this natural history intelligence. It is predicated on awareness of and

attention to the world outside of the social world. Although Upper Paleolithic art certainly demonstrates natural history intelligence, it can also be documented at archeological sites dating back to 1.8 mya. This intelligence is not well developed in non-human catarrhine primates, which have been the focus of most attempts to discover the origin of hominid intelligence.

(2) Non-human primate sociality is thought to generate Machiavellian intelligence. If this is true, there may be a break between non-human primate social intelligence and the appearance of hominid intelligence. Because the composition of awareness models the world, Machiavellian intelligence may even militate against or impede the development of natural history intelligence. Several lines of evidence from field and captive studies of non-human primates and neuroanatomical investigation were examined to show that Machiavellian intelligence may deter natural history intelligence.

(3) No single living species can serve as a model for the behavioral evolution of early hominids. Rather, a theoretical composite of many species (a "conceptual model"; Tooby & DeVore 1987) is probably of greater heuristic value. The behavioral ecology of callitrichid primates, as well as other mammals (especially social carnivores) may reveal more about the development of natural history intelligence than the study of non-human catarrhine primates in general and of great apes in particular.

(4) If one assumes that hominid sociality emerged from a substrate similar to that of other catarrhines, then factors responsible for the origin of natural history intelligence might include reduction of status-striving or the internal competition between matrilines which appear to typify non-human catarrhines today. An increasing degree of genetic relatedness, reduction of intra-group food competition, and the diminishment of status-striving through the maintenance of rigid hierarchies may have been crucial in the evolution of natural history intelligence. In addition, positive selection pressure for cooperative behavior unlike that seen in other catarrhines (such as elaborate cooperative foraging and food sharing) may have occurred.

(5) Finally, anthropologists have traditionally emphasized the link between language and intelligence in human evolution. This inevitably implies that hominid intelligence evolves from sociality based on complex communication. However, sociality in other mammals does not necessarily generate complex communication. This idea was formally tested using alarm calls of closely related rodent species – there was a general positive relationship, but a considerable amount of variation was unexplained by social complexity (Blumstein & Armitage 1997). If the model developed here has any validity, and hominid intelligence was initially divorced from a social context, then a radically different implication must be considered. It is possible that hominid intelligence had its origins apart from social communication, and the close links now apparent

between hominid language and intelligence only develop at a later stage of human evolution. In terms of neuroanatomy, a pre-existing non-human primate object-recognition system in the left cerebral hemisphere may have been altered to yield the common semantic system for both objects and words that occurs in modern humans (Vandenberghe *et al.* 1996).

Endnotes

1. A compendium of facts and illustrations of mammalian brains can be found at the Website called Comparative Mammalian Brain Collections: http://brainmuseum.org/.
2. "... l'intelligence ... est la faculté de fabriquer des objets artificiels, en particulier des outils à faire des outils, et d'en varier indéfiniment la fabrication" (Bergson 1907 [1945]: 150).
3. Hungarian researchers are studying factors involved in dog recognition of conspecifics by observing the interaction of dogs with AIBO© dog robots at the Sony Computer Science Laboratory in Paris: http://www.csl.sony.fr/Research/Experiments/DogAIBO/index.php.
4. "Determined to demonstrate similarity, many researchers train on, without ever seriously considering the possibility that the very extent of the efforts required to produce human-like behaviors in their animals undermines the very claims they wish to make in the first place. Nonetheless, the final results of such projects seem impressive – so impressive that the layperson may be excused for thinking that the only thing left to settle on is the degree of similarity between the minds of humans and apes. The visual rhetoric of *National Geographic* and *BBC* documentaries on chimpanzee social organization, tool use, and cooperative hunting has already paved the way, preparing the general public to be persuaded that the remarkable behavioral similarity between humans and apes is a sure guide to a comparable degree of psychological similarity" (Povinelli *et al.* 2000: 337–338).
5. The minds of some bird and small-toothed whale species have received a similar degree of study (e.g., Heinrich 2000).
6. Williams' syndrome is listed as OMIM 194050 in the Online Mendelian Inheritance in Man catalog (http://www.ncbi.nlm.nih.gov/Omim).

9 Why be social? – the advantages and disadvantages of social life

Why be social?

There are disadvantages to social life. These disadvantages are not immediately apparent to anthropologists, who unconsciously assume that existence is impossible without the support of a social group. This bias was confirmed by early reports of chimpanzee sociality, in particular, which tended to depict an idyllic existence. Only later were reports of competition, cannibalism, and infanticide allowed to cloud the picture of a chimpanzee Eden (Goodall 1977, 1979, 1986). E. O. Wilson (1975) discovered that the pinnacle of cohesive social life existed among the invertebrates. He marveled that cohesive social life could ever be achieved among vertebrate animals, where distinctive individuals, pursuing their own gain, either vie with each other directly for resources or vie for social status that ultimately confers pragmatic benefits. Intra-group competition exists for food, access to mates, and other scarce resources. Animals within the vertebrate group may constantly jockey for position within a dominance hierarchy. In primates, males and females have separate dominance hierarchies. The utter misery of life that exists in a primate social group is well expressed in the title of a recent paper examining status interactions in primate social groups: "practice random acts of aggression and senseless acts of intimidation" (Silk 2002). This is a Hobbesian world.

However, complex sociality in mammals is not necessarily predicated upon competitive interactions in a hierarchical society. Female lions (*Panthera leo*), for example, are egalitarian in their interactions within a pride (Packer *et al.* 2001). Thus, the nature of interactions within primate social groups is not the only form of cooperative mammalian life.

Scholars have tended to emphasize long-term associations ("friendships"), conflict resolution ("peacemaking"), and reconciliation among non-human primates (e.g., Smuts 1985, de Waal 1989, 1996, Aureli & de Waal 2000). Yet, the existence of these peaceable behaviors highlights the behavioral norm for primate societies. The norm is intense social competition that may indirectly affect access to food, space, and mates. It is telling that a recent report of female reproductive success in yellow baboons (*Papio cynocephalus*) emphasizes that close social bonds between females improve infant survivorship (Silk *et al.*

185

2003). Female baboons that form close long-term bonds with other females, as measured by maintaining proximity and grooming interactions, had relatively higher infant survival. However, emphasis on the positive effects of social integration, as demonstrated by the title of the paper ("social bonds of female baboons enhance infant survival"), tends to downplay the starkest truth: female–female competition within a social group creates differential reproductive success among female baboons. Mammalogists should find this intensely interesting, because one of the basic differences between the sexes is supposed to lie in the fact that males compete with other males for access to mates, and thus have much greater variability in reproductive success than females do. This difference between the sexes is maintained in yellow baboons. Yet, female baboons also show differing reproductive success, and this is the direct outcome of female–female competition. The effect of social integration on reproductive success is independent of female rank and ecological variation (Silk *et al.* 2003). However, because female baboons remain within their natal group and associate mainly with kin, matriline affiliation is important, and interaction between matrilines can affect female reproductive success. Thus, another potential arena for intra-group competition appears in primate social life – competition between matrilines. A mother's close social associations affect her infant's survivorship, but, if her matriline is low-ranked, she may fare no better than a female who is relatively isolated.

Male–female "friendships" in Chacma baboons (*Papio ursinus*) may be a female counterstrategy to infanticide, which is documented to occur in this species. Females that cultivate male "friends" may ward off potential infanticidal attacks by other males (Palombit *et al.* 1997). It is noteworthy that female Chacma baboons compete with other females for male "friends" (Palombit *et al.* 2001). This highlights the degree of competition that exists within free-ranging catarrhine primate groups – male "friends" that are necessary to counteract potentially infanticidal males become the object of female–female competition.

Proximate causes of primate sociality examine the nature of long-term associations between group members (Mendoza *et al.* 2002). Researchers investigating the ultimate causes of primate sociality focus on food, predation risk, access to mates, and protection from infanticidal conspecifics. The ability of juveniles to learn about edible food items and increasing efficiency in locating food are thought to underlie much of non-human primate sociality. Female primates often engage in contest and scramble competition over food. Besides competition between individuals, food competition can also occur between matrilines (Su 2003). Yet, increased foraging efficiency and safety from predation are often invoked as the principal factors molding primate sociality (van Schaik 1989). A high risk of predation is believed to force primates to live in cohesive groups for protection. Two recent platyrrhine examples demonstrate that the benefits of group living can yield more than a simple visual deterrent to predation. Group mobbing and cooperative defense of a subadult moustached tamarin (*Saguinus mystax*) saved the potential victim from consumption by a boa constrictor (Tello *et al.* 2002). Similarly, white-faced capuchins (*Cebus capucinus*) rescued a 3-year-old female from a 2 m long boa constrictor. All 38 group members rescued the potential victim by biting, mobbing, alarm-calling, and shaking tree branches at the snake (Perry *et al.* 2003). Nevertheless, active cooperative defense of a group member threatened by a predator is rare in primates, in comparison to some other social mammals. The rarity of such defense among primate species accounts for the publication of such events. A similar level of active defense by elephants, musk-oxen, or dwarf mongooses, for example, would not warrant the appearance of a special paper.

Contagious parasites are a problem among social animals, and parasitism consequently has the power to mold social behavior. Alexander (1974) first discussed parasitism as a disadvantage of social life. Female birds selecting mates are thought to assess the relative health and parasite load of males by assessing the color and radiance of male plumage. Blue jays anoint themselves with ants, which produce formic acid, to deter or kill ectoparasites, such as mites – a spectacular use of living tools. Similarly, capuchin monkeys (*Cebus* spp.) and owl monkeys (*Aotus* spp.) anoint themselves with millipedes, which produce the foul chemical benzoquinone, and owl monkeys also use a variety of plants to deter ectoparasites (Zito *et al.* 2003). Gilbert (1994, 1997) examines how endoparasites mold the behavior of red howler monkeys (*Alouatta seniculus*) in Amazonia. These highly folivorous primates use specific defecation sites that stand clear of underlying foliage. This reduces the probability of later ingesting leaves contaminated with parasite ova. This strategy succeeds, because red howlers carry relatively small parasite loads. The problem of parasitism as a cost of social life may be accentuated for animals with relatively small home ranges, such as folivores. Because Brazilian red howlers also face the additional problem of living in fragmented rainforest (Gilbert 1994, 1997), where the best habitat occurs in limited patches, one might expect that endoparasite infection would be a powerful cost of sociality in this taxon.

When many mammal species are surveyed, it becomes clear that body size positively affects parasite load. Because sexual selection creates sexual dimorphism, male mammals tend to be larger than females, and thus suffer from statistically greater parasite loads and higher mortality than females. In species where reverse sexual dimorphism exists, parasite loads and mortality are biased in favor of females. Hence, body size alone affects the degree to which mammals suffer from ectoparasites and endoparasites (Moore & Wilson 2002). Using 63% of the living anthropoid primate species, Nunn *et al.* (2003) examined variables that affect parasite load under natural conditions. Besides helminths, arthropods, and protozoa, the authors use an expanded definition of "parasite" to include viruses, bacteria, and fungi. Body mass was associated with total parasite diversity, but closely related host taxa resembled each other in total parasite diversity, indicating a phylogenetic component to parasite susceptibility. The population density of anthropoid primate host taxa was positively correlated with total parasite diversity, but the size of social groups and promiscuous mating behavior had no significant effect on parasite species richness. The researchers expected that sociality would be important. They attribute their failure to find a relationship between sociality of primate hosts and parasite species richness to the fact that their analysis collapsed together parasites with different modes of transmission and different host specificity (Nunn *et al.* 2003). For example, this analysis did not separate parasites with direct or indirect modes of transmission, and thus group size, which affects the physical association of host animals, showed no impact.

How to become social

During the first half of the twentieth century, the origin of social life was thought to be embedded within group selection, under which animals behave selflessly "for the good of the group" or "the good of the species." Animal ecologists like W. C. Allee, Alfred E. Emerson, and Thomas Park at the University of Chicago wrote of how "natural cooperation" worked against competition, harmful tendencies, and "disoperation" (Allee *et al.* 1949: 395–397). This viewpoint had been developed through the earliest formal study of animal ecology, which was practical work conducted on rodents and other crop pests. The goal of this early pragmatic work was to investigate the factors that determined population size. This early ecological research had enormous economic consequences, because factors that determined population size might be manipulated to control the abundance and distribution of vermin. Many species appeared to have relatively stable population sizes in the wild, which implied the existence of natural population controls, such as territoriality. Social animals were especially characterized by stable populations. Thus, the idea arose that individuals within a social group might forgo reproducing or simply die outright, in order to limit a striking rise in numbers under which all would suffer and starve.

The ecologist V. C. Wynn-Edwards (1962) later presented the classic argument for group selection, emphasizing that group selection promoted population regulation. George Williams (1966: Chapter 4) finally silenced the argument for group selection by outlining how it conflicted with natural selection, particularly with regard to "beneficial death." Williams demonstrated how some unknown evolutionary process would need to be invoked if group selection were to operate. Edward Wilson persuasively demonstrated how natural selection and kin selection could be used to explain specifics of social organization first among social insects (Wilson 1971), and then among both invertebrates and vertebrates in a magisterial comparison of the origins of social life (Wilson 1975).

Since the demise of group selection theory, a number of suggestions have been made about the generators and determinants of social life. I list eight of these suggestions below. These suggestions are not mutually exclusive. Combinations of these factors probably operate to create sociality in any given species.

(1) William D. Hamilton (1964) hypothesized that kin selection was the fundamental generator of social life. Genetic relatedness created sociality through inclusive fitness, as animals favored kin, or behaved altruistically towards kin who shared their genes. A meta-analysis of kin selection was performed on 18 cooperatively breeding bird and mammal species, where non-breeding subordinates care for the young (Griffin & West 2003). This analysis demonstrates that animals discriminate between kin, and the finest level of discrimination occurs when the benefits of helping kin are the greatest. Female Chacma baboons (*Papio ursinus*) can simultaneously distinguish both kinship and dominance rank within their social group (Bergman *et al.* 2003). Kin within a matriline are ranked, and the matrilines themselves are ranked. Chacma baboons thus experience a world in which social interactions occur not only at the level of individuals, but also at the level of a hierarchical classification of the society (Bergman *et al.* 2003). Note, however, that the Chacma baboons do not exhibit cooperative breeding, unlike the bird and mammal species that also discriminate finely between kin (Griffin & West 2003). Furthermore, hierarchy is a major structural feature of baboon society, which emphasizes the argument about the competitive nature of primate social life made at the beginning of this chapter.

Although kin selection is revealed to be important both theoretically and under field biology conditions, the mechanisms through which animals recognize kin are still being investigated. Kin may be known through simple association or proximity. Yet, because animals in the wild or in experimental conditions appear able to discriminate between their half-siblings whom they have never seen and unrelated individuals, mechanisms for recognizing kin other than familiarity and socialization must exist. Olfactory cues mediated through the major histocompatibility complex (MHC) will probably prove important in most mammal species. In humans, this complex is known as the human leukocyte antigen system (HLA). Unconscious cues – probably olfactory – influence human mate choice. Among the Hutterites, who are a religious isolate, mates are chosen for their lack of HLA resemblance. Marriages where the mates closely resemble each other in HLA type are relatively infertile, caused by fetal loss (Ober *et al.* 1998). Phenotype matching is another mechanism for kin recognition (Lacy & Sherman 1983). In this mechanism, animals behave differently toward other animals whose phenotype resembles their own. A variety of senses (e.g., vision, olfaction) can provide the cues that are being matched.

(2) In vertebrates, social groups are often formed by relatives that do not disperse after they mature. Hence, kin selection can create eusocial behavior such as cooperative breeding through delayed dispersal of the young from their natal group. However, most social vertebrate species consist of relatives living in stable social groups, and yet these animals do not breed cooperatively, or exhibit other elements of eusocial behavior. Old World monkey species such as guenons, macaques, and baboons are typical in this regard. Hence, the conditions under which vertebrate cooperative breeding evolves remain obscure. Furthermore, a pattern of low dispersal from a natal group can promote competition between relatives, rather than promoting the altruism between relatives expected with kin selection (West *et al.* 2002). West and his colleagues argue that the most striking examples of kin selection may occur when competition between relatives is slight or insignificant. In catarrhine primates where females are philopatric and do not disperse from their natal group, competition may also occur between matrilines formed by descendants of a single female.

Another problem that has emerged with the study of known genetic relatedness among wild primates is that complex social interactions may occur with low genetic relatedness. Adult male common chimpanzees are philopatric and demonstrate complex male–male interactions. If kinship alone were the generator of social life, one would expect a high degree of relatedness among these males, in comparison to the virtually asocial females. Nevertheless, male common chimpanzees show only a low degree of genetic relatedness in spite of male bonding (Vigilant *et al.* 2001, Di Fiore 2003). Even when dyadic male interactions are examined, affiliative male dyads are no more related than non-affiliative dyads (Di Fiore 2003). In bonobos, relatedness between adult males is virtually nil; this is also true for adult females, and for male and female relatedness (Di Fiore 2003).

(3) Robert Trivers (1971) suggested that reciprocal altruism was a generator of complex social relationships. He applied it to human social interactions. Trivers uses reciprocal altruism to explain Good Samaritan behavior that occurs between unrelated individuals who may never have met each other before. Interactions between cooperative individuals resemble a game of The Prisoner's Dilemma, where both actors gain if both cooperate. Human honor or personal reputation, sympathy, guilt, and a moral code combine powerfully to assure reciprocity. Cheaters suffer a loss of reputation, ostracism, or more severe punishment. The spread of a genetic basis for reciprocal altruism is affected by population size and the extent of the cooperative network (Boorman & Levitt

1973). Reciprocal altruism is based on the existence of long-term relationships and detailed memory of complex interactions. Interactions are not isolated, anonymous events. Interactions tend to occur in dyads, and individuals that are known to each other interact with each other repeatedly. Reciprocal altruism has largely figured in analyses of human sociality, but has also appeared in descriptions of coalition behavior in non-human primates.

Game theory was originally used by John von Neumann to examine human economic behavior. Promoted by the theorist John Maynard Smith (1982), game theory has now become a fundamental component in the study of the evolution of behavior. In particular, it has been used to determine the circumstances under which cooperation can evolve, given the presence of individuals who always defect from the system. Defectors receive the benefits of altruistic cooperation, but they do not cooperate. Evolutionary game theory demonstrates that cooperation can emerge even in the presence of individuals who always defect. A single cooperative individual using a "tit for tat" strategy can invade a population of defectors if the population is finite, and the cooperative mutation is favored by natural selection (Nowak & Sigmund 2004, Nowak *et al.* 2004). The minimum and maximum population size is important, because if there are only two individuals in the population, a defector always has higher fitness than a "tit for tat" cooperator; in a very large population of defectors, a single "tit for tat" cooperator cannot achieve the threshold for invasion. At intermediate population sizes, however, natural selection can favor the invasion and replacement of defectors by "tit for tat" cooperators (Nowak *et al.* 2004).

(4) Punishment of non-cooperators can also contribute to social life. Many species of social insects exhibit "policing" – workers detect and prevent illicit reproduction by other workers, who may try to evade policing (Ratnieks & Wenseleers 2005). Because bees, ants, and wasps independently evolve eusociality, policing has also evolved recurrently. It may be a common and effective solution in the animal world efficiently to resolve conflicts between the interests of individuals and the interests of society. That is, "effective policing can induce individuals to act in ways that are better for society" (Ratnieks & Wenseleers 2005: 56).

Boyd *et al.* (2003) present a model in which human punishment of non-cooperators explains great levels of cooperation in human societies. Punishment of defectors within a group acts to increase altruistic cooperation. The costs of altruistic punishment decrease as defectors become rarer, because punishment becomes increasingly infrequent. In fact, cooperation may not be possible in large groups if there is no punishment of non-cooperators (Fehr & Fischbacher 2003). Computer simulations of the evolution of human cooperation show that groups larger than 16 individuals cannot exist without altruistic punishment. If punishment of defectors occurs, a group of 32 can maintain an average cooperation rate of 40%. However, if both defectors and non-punishers can be punished, groups of several hundred individuals can achieve an average cooperation rate of 70–80% (Fehr & Fischbacher 2003: Fig. 4). Lest this punishment of malefactors be thought a rarified component of human sociality alone, it resembles the "policing" behavior that occurs among social insects. Furthermore, experiments with New World tufted capuchin monkeys (*Cebus apella*) demonstrate that animals have expectations of equitable outcomes during experiments involving resource division (Brosnan & de Waal 2003). Hence, "fair play," turn-taking, and other normal components of human social interaction may also exist among other social mammals.[1]

It is noteworthy that punishment may underlie mutualism generally, not only among animals. For example, symbiotic interactions in plants may be maintained by punishment of defectors. Soybeans punish symbiotic bacteria that do not fix nitrogen inside the root nodules of the host (Kiers *et al.* 2003).

(5) W. D. Hamilton (1971) also suggested that mere aggregation itself might generate benefits that contribute to sociality. Hamilton called this concept the "selfish herd." It is exemplified by herds of ungulates or schools of fish or tadpoles in which the probability that an individual animal suffers predation is lessened by its membership in a group. Clutton-Brock (2002) terms this concept "group augmentation," and discusses it within the range of behaviors that are categorized as mutualism.

(6) Autonomous agents obeying simple behavioral rules and making simple behavioral decisions can spontaneously create social complexity. Hans Kummer, one of the founders of modern field-oriented primate studies, makes the point that many primatologists incorrectly assume that social complexity cannot exist without social intelligence. "Many researchers connect social complexity with social intelligence and interpret anecdotes accordingly. Where experimental support is lacking, we may remember that the complexity of living systems is generally not produced by intelligence, but by emergence based on complex systems of unconscious simple rules. This applies not only to the building activities of termites . . . Focusing on the single explanatory hypothesis that monkeys or apes generate complex behavior by intelligence is not justified" (Kummer 2002: 74).

The emergence of complexity is a fundamental topic in Artificial Intelligence (AI) research. Lest this be thought of as merely a boring sidebar to thrilling field observations of primate social interactions, computer simulations can, in fact, produce typically primate-like social activities. A computer simulation of chimpanzee social structure has been designed with software called ChimpWorld (Povinelli *et al.* 2000). The software is written so that each animal behaves as an autonomous agent and takes turns with other agents. Simple motivations are present like hunger, fatigue, and aggression. These motivations are ranked to determine the outcome of events. "Animals" in the program have motivations, goals, plans, and knowledge, but they have no representations of the plans and knowledge of other "animals." Thus, the software is written so that the "animals" have no Theory of Mind. Nevertheless, complex, sophisticated behaviors are generated in a dynamic interplay. Povinelli *et al.* argue (2000: 61–65) that computer simulations such as those generated by ChimpWorld falsify the notion that natural selection must necessarily evolve animals with a Theory of Mind in order for complex sociality to appear. Similarly, computer modeling of group movement or foraging demonstrates how a small number of knowledgeable individuals can influence the behavior of a large group. In fact, the larger the group size, the smaller the ratio of knowledgeable individuals needs to be (Couzin *et al.* 2005). This abstract study of information transfer is also important for showing that coordinated group activity is not necessarily dependent on individual recognition, higher cognitive abilities, or the dominance of specific individuals. Rather, naïve members of a group can spontaneously respond to informed individuals (Couzin *et al.* 2005).

(7) Penalty-free altruism is another suggestion for the origins of sociality. It was first suggested by Bednekoff (1997), who pointed out that researchers always assumed that altruism was hazardous or costly without assessing its costs. He argued that sentinels, for example, might be actually safer than their unwary conspecifics. Penalty-free altruism has been confirmed during elegant field experiments on free-ranging meerkats (*Suricata suricatta*). These are South African desert mongooses less than 1 kg in weight. Meerkats are easily habituated to human presence and their behavior can be manipulated in the field (Clutton-Brock *et al.* 1999, 2002, Clutton-Brock 2002, Russell *et al.* 2003). Sentinel behavior and care taking of the young ("babysitting") have been demonstrated to arise from penalty-free altruism.

(8) Repression of competition has also been proposed as a generator of cooperative social life. Mathematical modeling explores repression of competition at a number of levels, from genes in the genome, to cells in metazoan bodies, to hierarchical social groups (Frank 2003). Individuals within a social group differ in competitive ability or, in the case of humans, command more resources. Punishment or policing tends to shift to these individuals, because they have more to gain, and they experience lower costs associated with monitoring and punishing (Frank 2003). In human social interactions, group resources held in common are often overexploited if monitoring systems do not exist and sanctions are not imposed. So frequent is the overuse and abuse of public resources that it has famously been termed the "Tragedy of the Commons" (Hardin 1968) – the tragedy being that these resources appear doomed to extinction. However, research with anonymous human subjects in a public goods game shows that continuing interaction between loners, defectors, and cooperators sets up an

oscillating succession of strategies if players can refuse to participate, or can volunteer to defect or cooperate (Semmann *et al.* 2003). Thus, besides sanctions, the punishment of defectors, or the besmirching of reputations, the freedom to volunteer may contribute to the maintenance of cooperation.

Explanations of primate social complexity

Explanations for increasing complexity in primate societies began to be generated soon after the first field-oriented behavioral studies appeared in the 1960s. The initial assumption was that the great apes possessed the most complex social life among non-human primates. This assumption was shattered by the first field studies of the orangutan, which demonstrated that this species, contrary to expectations, was solitary (MacKinnon 1974a, 1974b). Another early explanation was that increasing terrestriality in primates inevitably leads to social complexity (Crook & Gartlan 1966). Later, a single variable was focused upon – the number of adult males in the social group. An increasing number of adult males involved in the social group generated social complexity (Eisenberg *et al.* 1972). However, Ridley (1986) surveyed 33 primate species, and discovered that the number of adult males in a group is affected by the length of the breeding season. The relationship is an inverse one: single-male groups have a long breeding season, and multi-male groups occur when the breeding season is short. Obviously, the adult sex-ratio of groups is also correlated with length of the breeding season. The operational sex ratio – the number of available mates – is also an important variable. John Eisenberg (1981) later abandoned the idea that a single variable could account for social complexity in mammals.

What is the catarrhine substrate for sociality?

Can any outlines be discerned of the general catarrhine substrate for sociality? The outline that looms up is a grim and gloomy one, and not one conducive to the origins of complex sociality found among hominids. Components of this outline include kin selection, infanticide, parental investment, parent–offspring conflict, possible different male and female strategies of reproduction developed by sexual selection, intrasexual competition, and competition between matrilines.

Intra-group competition is an important aspect of catarrhine sociality. Silk (2002) notes that status contests in primate groups occur among individuals that live together, know each other, and remember previous interactions. She argues that random aggression and intimidation of subordinates by dominant animals generates constant low-level stress in the subordinates that must have injurious consequences, but the dominant animals benefit. "Dominants benefit because they are able to inflict these costs on subordinates but are able to minimize the risks associated with escalated aggression. Thus, randomly timed attacks on randomly selected targets may be favored by natural selection because this strategy is both effective and efficient. Cercopithecine females seem to adopt this strategy, launching unprovoked attacks on unsuspecting subordinate targets" (Silk 2002: 225). Competition between matrilines also occurs within catarrhine social groups (Su 2003). This is a Hobbesian world, diametrically opposite to the idyllic Eden originally portrayed among chimpanzees of the Gombe Stream Reserve. One searches in vain for the intense cooperative efforts seen in modern humans, documented from historical records, or inferred from complexities of the archeological record dating back to 1.6–1.5 mya. The question then becomes, What evolutionary processes led to the break between the basic catarrhine substrate for sociality and the beginnings of complex hominid sociality at 1.6–1.5 mya?

Endnote

1. Some animal behaviorists argue that the origins of a basic human moral sense involving "fair play" and turn-taking originate from a social milieu that contains multiple interactions between known individuals, an equitable division of resources, and the identification and punishment of malefactors (non-cooperators or defectors).

10   *Evolution and behavior*

Proximate and ultimate factors in behavioral evolution

The methodology for studying behavior in an evolutionary framework was established at the beginnings of the science of ethology. However, most textbooks on animal behavior (e.g., Slater 1985, Alcock 2001a) include a detailed explanation of this framework. The evolutionary approach is best exemplified by the search for proximate versus ultimate causes of behavior. John Alcock (2001a: Chapter 2) lucidly discusses these causes in terms of hypotheses that pose How (proximate) versus Why (ultimate) questions. For example, an analysis of arboreal suspensory posture and arm-swinging locomotion in lesser apes could involve description of the musculoskeletal structures that underlie such posture and locomotion. These would be proximate causes of the behavior. Alternatively, an analysis of this posture and locomotion could involve selection pressures for feeding and foraging in the high canopy, or selection pressures for agile, three-dimensional display in order to maintain the boundaries of a territory. These would be ultimate causes of the behavior.

However, most of the fiery polemics about evolution and behavior do not involve locomotor behavior or feeding behavior. Instead, controversy involves interactions between conspecific animals, such as mating behavior or agonistic interactions. Empirical documentation of behaviors such as rape, unprovoked aggression, or infanticide has caused many social scientists to develop a distaste for evolution and for studies of animal behavior. Biology offends their sensibilities. The idea that evolution can produce such behaviors leads to a rejection of evolutionary ecology. A schism is developing in disciplines like anthropology. At least in the U.S.A., evolutionary anthropology (physical anthropology, archeology, primatology, human behavioral ecology) is now cleaving away from other anthropological sub-disciplines. Sociobiology, in particular, has become the subject of opprobrium. Hence, it is important to note that sociobiological studies are not mere adventures in story-telling, as they are characterized by those who are dubious about natural selection and adaptation (Chapter 5). Instead, sociobiological hypotheses about animal behavior are rigorously framed and tested (Alcock 2001a, 2001b). Furthermore, the distaste for genetic bases for behavior and the infamy of sociobiology are generated mainly by the perception that natural behaviors imply a natural basis for morality (Alcock 2001b). Nothing could be further from the truth, as Thomas Henry Huxley argued in *Evolution and Ethics*, which was written in response to Herbert Spencer's concept of "social Darwinism." Over 100 years have passed since the publication of *Evolution and Ethics*, but the same arguments are continuing.

Factors limiting population size

A number of factors limit the population size of organisms. Natural selection acts through these factors. They consequently drive adaptations. These factors can be divided into density-independent and density-dependent categories (Table 10.1). These factors have been studied in animal ecology since the early twentieth century, and have been the focus of attention by researchers interested in population dynamics and major population cycles, including the amplitude of population fluctuations, and their periodicity (Allee *et al.* 1949, Andrewartha & Birch 1954, Andrewartha 1961). Food, predation, and parasites have all been investigated as factors limiting non-human primate populations. Food appears to be the principal factor influencing population size in non-human primates, as documented by population increase in many species under zoo or colony conditions, or when free-ranging animals are provisioned. Provisioned Japanese macaques (*Macaca fuscata*) or rhesus macaques (*Macaca mulatta*) at Cayo Santiago Island show a pronounced rise in numbers. Food and contagious pathogens certainly have played a major role in limiting human populations during the course of human history. Hence, it can be seen that both human and non-human primate populations tend to be limited by density-dependent factors.

If food is the major limiting factor affecting primate populations, what is the relative contribution of other factors, such as predation and parasites? As noted in Chapter 9, all three of these factors have been thought important in molding primate sociality.

Ungulate species ($n = 28$) and large carnivore species ($n = 10$) were examined in the rich savannah mammal community of the Serengeti ecosystem, extending through Tanzania and Kenya in East Africa (Sinclair *et al.* 2003). The largest predators focus upon the largest prey, but also consume a large range of prey species. Hence, the smallest ungulates have many more predators than the largest ungulates. Ungulates above a body size of 150 kg experience a significant reduction in mortality caused by predation, and instead become food-limited. Why are primates apparently food-limited at a size far smaller than this critical threshold for ungulates? The general arboreality of primates saves them from

Table 10.1. *Factors that limit population size*

Density-independent factors	Density-dependent factors
(These factors operate regardless of the number of animals)	(These factors are affected by the number of animals)
• Climate (temperature, precipitation, light, wind)	• Food
• Degree of seasonality	• Intrasexual competition (especially female–female competition)
• Natural catastrophe (e.g., windstorms, floods, volcanic eruptions)	• Predation
• Non-contagious pathogens	• Contagious pathogens
	• Parasites (ectoparasites and endoparasites)

many potential predators, although arboreal mammalian carnivores, snakes, and raptorial birds may still afflict them. Only when they become terrestrial do primates enter the sphere of diverse mammalian predators. However, terrestrial primates are diurnal, and are therefore separated temporally from the activities of the major nocturnal predators. Terrestrial primates are also rarely far from trees as they forage during the day, and they hurriedly take shelter in trees or rocky sleeping cliffs as night begins to fall.

Diet and foraging behavior

In one of the first monographs produced during modern field-oriented research on primate behavior, George Schaller (1963), a student of the University of Wisconsin zoologist John T. Emlen, Jr., emphasized the link between mountain gorilla ecology and behavior. However, ecology remained relatively neglected by primatologists until the 1980s. A variety of reasons accounts for this, but a principal contributing factor in America is the emergence of behavioral primatology from anthropology (Chapter 2). The anthropological background of researchers assured that initial studies would be oriented toward social interactions, to the detriment of ecology. To anthropologists, much of the ecological literature appeared abstruse, and strangely focused on the factors that influenced population numbers (Richard 1981). Furthermore, simple behavioral observations and the quantification of social interactions do not require rigorous preparation in tropical ecology, vegetation analysis, or an understanding of evolutionary processes like natural selection or co-evolution. Many of the first modern field-oriented primatologists entered their research sites with no prior training in biology, and lacking a comparative perspective on animal behavior (Goodall 1971). In addition, because primates breed relatively slowly and are long-lived, the changing demographic patterns of primate groups or the reproductive success of individual animals are difficult to discern without establishing a long-term research project to monitor the fate of certain primate groups over decades.

K. R. L. Hall, who was one of the founders of field-oriented primate studies in the 1960s, argued that food type, and its distribution and abundance, was a major determinant of primate sociality (Hall 1963, 1965). He studied both Chacma baboons and patas monkeys in the wild, and experimented with captive animals. Hall demonstrated that adult male–male tolerance was a species-specific behavior through manipulation of captive animals. The capacity for male–male tolerance was well developed in baboons, and quickly appeared after animals were housed together. However, male–male tolerance was non-existent in patas monkeys, and strange males that were introduced to each other would not stop fighting. Hall interpreted these species differences as ultimately being determined by the type and distribution of major dietary components. Hall unfortunately contracted a virulent simian virus from the bite of a captive patas monkey. He died as a martyr to science, near the beginning of the Baboon Renaissance during the 1960s (Chapter 2). Hans Kummer (1968, 1971) further developed the idea of ecological determinants to primate sociality, based on his original fieldwork with hamadryas baboons, and comparative analyses of other baboons and cercopithecoid monkeys. Continuing study of hamadryas baboons has led to ideas about the forces of natural selection that created their characteristic small, female-based foraging units (Kummer 1995).

Katharine Milton (1980) later documented that the foraging behavior of even folivorous howler monkeys was affected by food availability and quality. Hence, folivores are also influenced by dietary quality. Foraging behavior involves not merely the ability efficiently to locate and exploit a food source. The food search itself may be hazardous. A current topic in primatology is how predators and predation risk affect the search for food (Miller 2002).

Cultural traditions

Inspired by a suggestion made by K. Imanishi in 1951, individual innovation and transmission of new behaviors within primate groups was noted at an early date by Japanese primatologists among Japanese macaques (Kawamura 1959, Kawai 1965). Free-ranging, provisioned troops of this endemic species (*Macaca fuscata*) were observed to engage in sweet potato washing, wheat washing, stone handling, and other distinctive behaviors. Some of these behaviors have been observed for over 40 years. They can occur with very high (sometimes 100%) frequency within social groups. An inventory of these and other cultural traditions in Japanese macaques, vervets, chacma baboons, and two species of *Cebus* is given in Perry & Manson (2003: Table 1). Japanese macaques currently dominate the inventory, probably as a simple result of the long-term investigation of social groups by Japanese primatologists. The platyrrhine species *Cebus capucinus* also exhibits a long list of traditions, including squirrel hunting, and *Cebus apella* uses stones to crack open nuts. It is important to note the existence and frequency of these cultural traditions among monkey species, because the documentation of such traditions among common chimpanzees and orangutans is now often used to presage the origins of tool behavior and culture among the earliest hominids (Chapter 16).

Hunting of small vertebrates, including relay hunting, has also been observed in the olive baboon, *Papio anubis*. Animals learn to hunt from other members of the social group, and the size of prey items is related to the size of the baboon hunter. Strum (1975) ingeniously experimented with a troop of these baboons, trying to elicit scavenging behavior among animals that hunted by presenting them with freshly killed carcasses. However, she was not successful in eliciting scavenging by baboons. She speculated that scavenging under natural conditions might be so dangerous that baboons would be unwilling to feed even on the sort of small, freshly killed carcasses that she presented them with (Strum 1975). Yet, baboons certainly are not loath to scavenge from human garbage dumps. They do this enthusiastically, in fact. However, any animal food that baboons scavenge under these circumstances is long dead, or highly processed, and not a freshly killed carcass. Additional field experimentation with scavenging in baboons and other species seems appropriate, especially in light of broad questions about the cultural transmission of behavior, the adoption of novel food items, and the mode of acquisition of vertebrate meat and fat.

Hunting by common chimpanzees was first investigated by Teleki (1973), who reported ten cases among the chimpanzees of Gombe. Teleki reported the prolonged dismemberment and deposition of carcasses in detail, and noted instances of successful begging and tolerated scrounging. Baboons were the prey items. Because intense competitive interactions between chimpanzees and baboons were occurring at Gombe over jackpots of provisioned bananas, it was possible that the killing of baboons was a result of an increased general level of inter-species aggression. The lengthy and sportive treatment of a carcass also did not resemble the rapid feeding that occurs when hunger alone drives a kill. Stanford (1998) later published details of chimpanzee predation on red colobus monkeys. With no human interference, the killing of colobus monkeys here appeared to be part of the natural repertoire of chimpanzee behavior. However, the attacks and confrontations with red colobus monkeys seemed to be generated by chimpanzee male interactions, and the attacks were thought to promote social cohesiveness among the predatory males. Wrangham & Peterson (1996) previously argued that complex male–male relationships in common chimpanzees simultaneously forged cohesiveness and triggered aggression.

Phylogenetic inertia and phylogenetic constraint

The investigation of behavior sometimes reveals mysteries that have no explanation in terms of the modern world. In these cases, the concept of phylogenetic inertia may be invoked (Wilson 1975). The concept argues for a lag-time between a past world, in which a certain morphology or behavior was adaptive, and the present world, in which the morphology or behavior apparently has no function. One of the best examples of phylogenetic inertia is provided by the American pronghorn antelope (*Antilocapra americana*), which exhibits a suite of characters that are inexplicable in modern terms (Byers 1997). Pronghorn antelopes can run at speeds of 100 km/hour, and have the stamina of Thomson's gazelles – yet, no modern American predator exists that would warrant such speed and stamina. Pronghorn mothers hide their young, and remove all trace of scent by drinking the young's urine and consuming the young's feces. It seems probable that these pronghorn locomotor and behavioral specializations evolved in a late Cenozoic world replete with dangerous predators that are now long extinct (Byers 1997).

The concept of phylogenetic inertia implies some degree of mismatch between an animal's phenotype and the present world. It further implies the existence of a vigorous phylogenetic signal within a phenotype. A search for major ecological patterns or rules that could explain ecological traits in many temperate and tropical insect species has revealed that phylogeny can often create a stronger signal than habitat variation (Price 2003). Phylogeny overrides ecological differences caused by habitat. The adaptive radiation of a taxonomic group can be explained by phenotypic traits that appear at the base of the radiation, and that are distinctive for the group. Price (2003) has hypothesized that "Strong Phylogenetic Constraints" can explain variables studied by classical animal ecologists of the early twentieth century, such as the distribution and abundance of animals and their population dynamics.

Some researchers prefer to use the term "phylogenetic signal," arguing that "inertia" and "constraint" are not easy to demonstrate (Blomberg *et al.* 2003). These researchers also argue that some degree of phylogenetic signal will probably always occur when groups are being investigated. Most traits have less phylogenetic signal than body size does, and thus bivariate plots of body size versus other traits can be used to investigate trait evolution. Even physiological traits, when body size was accounted for, showed less phylogenetic signal than body size. In general, behavioral traits showed less phylogenetic signal than body size, physiological traits, morphological traits, or life-history traits (Blomberg *et al.* 2003). The importance of these observations is that behavior exhibits more evolutionary plasticity than do these other traits. Hence, when investigating a group of organisms, one could expect a lability and plasticity of behavior to occur, although body size, morphology, physiology, and life-history traits would be more conservative. This expectation will be utilized in later chapters, when I model hominization and early hominid behavior, as inferred from the paleontological and archeological records.

11 *The implications of body size for evolutionary ecology*

Introduction

Size is a fundamental variable in the life of animals. Size is, in fact, probably the most important single variable that a biologist can learn about a species. Size affects such basic parameters of animal existence as temperature regulation, diet, locomotion, ranging, foraging, activity patterns, predator pressure, reproduction, growth and development, mating systems, intra-group competition, and sociality itself.

Body size ranges enormously during the course of mammal evolution. Aquatic mammals, whose body weight is supported by water, can obviously grow to larger sizes than terrestrial mammals. Bats, which fly and therefore experience aerodynamic constraints, tend to be small. One of the two smallest living mammals is the bumblebee bat (*Craseonycteris thonglongyai*). With an adult body size of 2 g, it is literally the size of a large bumblebee (Nowak 1999). The other contender for smallest mammal is a terrestrial species, the Etruscan shrew (*Suncus etruscus*), which also has an adult body size of about 2 g (Churchfield 1990). This is equivalent to the very ancient early Jurassic *Hadrocodium wui*, only 2 g in weight, with a head and body length the size of a common paperclip (Luo *et al.* 2001; also see cover illustration). There is an impressive range of body size within land mammals. Extinct species represent the alpha and omega among land mammals. The smallest known land mammal was an early Eocene fossil insectivore (*Batodonoides vanhouteni*) with a body weight estimated at about 1.3 g (Bloch *et al.* 1998). The largest known land mammal was the rhinoceratoid *Indricotherium*, with a shoulder height of approximately 5.5 meters (Gregory 1951: Fig. 21.80) and an estimated body weight of 11 000 kg (Burness *et al.* 2001).

Body size is a crucial variable for paleontologists who are attempting to reconstruct the lifeways or niches of extinct species. How can one infer the paleoecology of a fossil species? Through the best of luck, three variables – body size, diet, and locomotion – account for most of the information about niche structure in living animals. Hence, if body size, diet, and locomotion can be estimated in a fossil species, the broad outlines of a fossil niche can ultimately be roughed out. Sometimes one can also infer the degree of nocturnal

203

or diurnal activity by examining the evidence of a fossil's dependence on senses like olfaction, touch, or sight. But body size affects locomotion – note the difference between bats and whales. And body size affects diet. More food is needed to support a larger body size, but a larger body size can accommodate low-quality dietary items (grass, mature leaves, bark), while a smaller body-size necessitates high-quality items (meat, blood, nectar). The smallest living mammals are insectivorous or carnivorous, the largest land mammals are herbivorous, and the largest aquatic mammals are filter-feeders on plankton suspended in the open ocean.

The primacy of body size is abundantly documented in the biology of living animals (e.g., Van Valen 1973b, Calder 1984, Schmidt-Nielsen 1984, West *et al.* 1999). Living animals must take in oxygen, oxidize food, and expend energy in calories simply to maintain those physiological processes that are necessary for life. This fundamental energy expenditure is the basal metabolic rate. However, most measurements of oxygen consumption are done under sedentary or resting conditions, giving the resting metabolic rate (RMR). Energy expenditure relating to normal activities is the field metabolic rate (FMR). Energy expenditure may obviously fluctuate dramatically with different behaviors (e.g., lounging at rest, routinely searching for food, frantically escaping from a predator, doggedly digging a den, or quietly grooming).

Max Kleiber first explored in depth the relationship between animal body size and metabolic rate during the 1930s. Kleiber's studies eventually culminated in the classic publication *The Fire of Life* (1961). Mammalian body mass scales to metabolic rate with a slope of 0.75, a number now referred to as Kleiber's law or the Kleiber relationship. The regression line itself is sometimes known as the Kleiber line. Because small body size is associated with a rapid metabolism, some researchers contend that mammals smaller than 2.5 g should experience a fundamental biological limit – they should not be able to eat fast enough to support a high metabolism and avoid starvation. They would need to lower their body temperature, periodically enter a state of torpor, or evolve some alternative strategy to cope with small body size. Some living shrews confound these expectations: they have high metabolic rates, well above the Kleiber line predictions for their body size (Calder 1984, Churchfield 1990). The ambient temperature, as well as an animal's internal heat production, appears to be important. All of these matters impinge on discussions of the niches of fossil mammals – what diet could be associated with a certain body size? And these matters impinge on discussions of life-history evolution – how are growth and development or reproduction affected by diet, metabolism, and body size? At any given size, animals with higher metabolic rates have a greater reproductive output (McNab 1980, Calder 1984). Food appears to underlie the rate of reproduction and reproductive success, and hence food ultimately determines the outcome of competitive interactions between species of equivalent size.

Larger mammals have a lower metabolic rate relative to body mass (Calder 1984, Schmidt-Nielsen 1984). Scaling in animals is described by the allometric power function $y = ax^b$, where y is a biological variable, a is a constant, x is body mass, and b is a scaling exponent. The area of mammalian postcanine teeth (those teeth that process food items) generally scales isometrically with body size (Fortelius 1990), although tooth area in suids and early hominids shows positive allometry (Cachel 1996b). Masticatory muscles in primates also scale isometrically with body size (Cachel 1984). This isometric scaling implies that occlusal stress is independent of mammalian body size, and raises questions about whether food intake rate, volume processed per chew, food comminution, or chewing rate scale isometrically as well, independently of body size (Fortelius 1990, Shipley *et al.* 1994). Isometry is generally explained by arguing that neither teeth nor masticatory muscles are weight-bearing structures. But isometry contradicts expectations for positive allometry in teeth and masticatory muscles with traditional assumptions of geometric scaling. That is, under the classical biomechanical scaling model, used since the time of Galileo in 1638, size increase in teeth and jaw muscles should be greater in larger bodied animals. This expectation is generated by the higher caloric base demanded by larger body size. Lucas (2004) explains the unexpectedly smaller molar size of large mammals by the texture and fracture mechanics of different foods. Because larger food particles fracture at lower stresses than smaller particles do, larger mammals ingesting larger food particles can have unexpectedly smaller molar teeth. However, it is also possible that scaling factors relating to gut absorption, digestion, and nutrient transport are involved.

Classical biomechanics leads one to expect geometric scaling, so that the scaling exponents should be multiples of 1/3. However, the 3/4 -power scaling of metabolic rate in organisms (i.e., the Kleiber slope of 0.75) demonstrates that the scaling exponents are multiples of 1/4. The expected geometric scaling is not occurring. This is because metabolic capacity is maximized when areas relating to nutrient and energy uptake (e.g., gut surface or absorption area, total vascular network area distributing resources) are scaled in a fractal manner (West *et al.* 1999). These areas scale like volumes. Area effectively scales like a 3-dimensional structure through the operation of fractal geometry, because the fractal dimension of fractal biological area is 3. The unexpected isometric scaling of mammalian teeth and jaw muscles is explained if cellular determinants of digestion and the internal transport of digested foods are fractal in nature. Because the scaling exponent is a simple multiple of 1/4, this demonstrates that fractal geometry of gut absorption area or total capillary area is involved. This implies that the processes of digestion, nutrient uptake and distribution, and energy utilization or minimization of energy expenditure in living mammals are apparently of more functional or adaptive significance than any mechanical processing of food with the postcanine teeth. In any case, mechanical processing of food with the postcanine teeth occurs through a variety of different crown morphologies that evolve independently in different mammalian orders (Jernvall *et al.* 1996). Once a crown morphology that can adequately triturate food items evolves within a group of herbivorous mammals, or one that can adequately shear meat evolves within a group of carnivorous mammals, this morphology tends to be maintained within the lineage. Crown morphology can then be used by taxonomists as a character that differentiates lineages. For example, disparate crown morphologies distinguish artiodactyl subgroups that independently evolve grazing or browsing specializations. This is also true for other mammalian orders (e.g., perissodactyls, condylarths) that independently evolve specializations for grazing or browsing diets. Occlusal stress generated by the masticatory muscles is also apparently subordinated to the processes of digestion, nutrient distribution, and energy utilization. In short, if mammalian subsistence is considered within the framework of functional morphology, fractal geometry trumps positive allometry.

The relationship between body size, diet, metabolic rate, and life-history variables has not escaped the attention of primatologists and paleoanthropologists. Robert Martin (1983) interjected much discussion of mammalian evolutionary ecology into the primatological literature, and argued that major questions of reproduction, infant growth, and brain evolution were ultimately affected by diet and energetics. Humans evaluate these factors from the perspective of a large animal species. When the totality of living animals is considered, the average animal species has a body size that is less than 10 mm in length. The average mammal is the size of a rodent, because nearly half of all living mammal species are rodents. Primates are large even when accepting a higher cut-off point of ≥ 2 kg for inclusion into the category of large animals. The mean body weight of living primates is 8.120 kg (Table 11.1). Thus, primates in general can be classified as large animals. The size of living primates ranges from a mouse lemur species (*Microcebus myoxinus*) of 24–38 g to the gorilla of 70–275 kg (Nowak 1999). The smallest living primates are unable to maintain a constant body temperature, even in captivity, with constant control over ambient temperatures. In the wild, they may experience periods of true hibernation, especially under conditions of marked seasonality. The fat-tailed dwarf lemur (*Cheirogaleus medius*) hibernates in hollow trees for 7 months of the year, even though they are tropical mammals. The body temperature of these hibernating animals fluctuates over a range not observed in other mammals – most subjects have a body temperature that fluctuates daily over a range of 20 °C (Dausmann *et al.* 2004). In fact, the hibernating lemurs are ectothermic through the

Table 11.1. *Body size in living primate genera*

Calculated from data in Conroy (2003: Appendix 1), which gives the mean body weight in kilograms of 62 living primate genera
All living primates: mean weight of 8.120 kg (n = 62)
Prosimians (including *Tarsius*): mean weight of 1.265 kg (n = 23)
The range is from 0.065 kg (*Microcebus*) to 6.335 kg (*Indri*); 5 genera are above 2 kg
Anthropoids: mean weight of 11.548 kg (n = 39)
The range is from 0.116 kg (*Cebuella*) to 124.683 kg (*Gorilla*); 30 genera are above 2 kg
Platyrrhine anthropoids: mean weight of 3.041 kg (n = 16)
The range is from 0.116 kg (*Cebuella*) to 8.841 kg (*Lagothrix*); 8 genera are above 2 kg
Callitrichid platyrrhines (including *Callimico*): mean weight of 0.418 kg (n = 5)
The range is from 0.116 kg (*Cebuella*) to 0.583 (*Leontopithecus*)
Catarrhine anthropoids: mean weight of 20.055 kg (n = 23)
The range is from 1.750 kg (*Miopithecus*) to 124.683 kg (*Gorilla*); 22 genera are above 2 kg
Hominoid catarrhines: mean weight of 57.755 kg (n = 5)
The range is from 6.793 kg (*Hylobates*) to 124.683 kg (*Gorilla*)

duration of the Malagasy winter, when daily temperatures are high and vary widely.

Using molar size to estimate body weight, the fossil prosimian *Altanius orlovi* weighed 21–23 g. *Eosimias*, a possible anthropoid genus, weighed 67–179 g (Gebo *et al.* 2000). However, two adult primate calcanei from the Eocene of China are less than half the size of the calcaneus of *Microcebus*, the smallest living prosimian genus. Using absolute calcaneal size and other body size estimates, Gebo *et al.* (2000: 587) argue that these Chinese fossils come from the smallest known primates. Body size may have been 10.6 g for the smaller specimen, and 17.2 g for the larger specimen. This is within the size range of living shrews. For this reason, I am inclined to think that these 45 mya Chinese calcanei represent fossil prosimians.

Anthropoid primates tend to be larger than prosimians, although the callitrichid anthropoids have secondarily reduced their body size. A decline in global temperatures and more pronounced seasonality at the Eocene/Oligocene boundary created selection pressures for increasing body size in the first anthropoids. This body size increase initiated a cascade of morphological changes associated with anthropoid origins, including a broadening of diet to include larger fruit with tougher pericarp (Cachel 1979a, 1979b). The relationship between primate body size and diet has been studied for a long time. Kay and Couvert (1984) noted that body size alone can sometimes be used to infer the diet of extinct primates, because living primates smaller than 300 g are constrained by energy expenditure to be insectivorous or carnivorous.

Daily energy expenditure in wild mammal species is now often investigated with radioactively labeled water, a technique that is routinely employed in nutritional research on humans. A comparison of data from 10 human samples, 2 wild non-human primate species, and 73 other wild mammal species demonstrates that the scaling of body mass to metabolic rate is the same: the slope of the regression line is 0.75 – the Kleiber relationship (Snodgrass *et al.* 2001). Why does this matter? It matters because humans and non-human primates obey the rules for other mammals. They are not a special case. As a result, the ecological impact of body size, and forces affecting the evolution of body size are the same in humans and non-human primates as they are in other mammals.

Measuring body size in fossil species

Given the great body of data from the present, one might confidently assume that body size would exercise an equally fundamental role in the life of fossil animals. The great nineteenth century American paleontologist Edward Drinker Cope suggested that a pattern (Cope's rule) existed in mammal evolution through the last 65 million years: lineages first originate at a small size and then independently evolve larger sizes (Cope 1871). This pattern of directional selection is explained by such advantages of larger size as safety from predators, larger brain size, and increased longevity. Alroy (1998) recently studied body mass in over 1500 fossil mammal species from North America and confirmed that a consistent pattern of size increase occurred from older to younger in matched species pairs. This directional selection for size increase in mammal lineages may not be true for other animals; for example, it does not hold true for mollusks, whose fossil record is dense and well-studied (Jablonski 1997). Jerison (1973) examined mammalian brain evolution through the Cenozoic, finding that there was no general increase in brain size relative to body size – there have always been niches for small-brained species. However, during any given time range, carnivores have larger brains relative to body size than herbivores do.

Given the variety of taphonomic processes that can skew or alter the preservation of fossil animal remains, the estimation of body size in fossil animals can be highly problematic. A major review of body size in fossil mammals (Damuth & MacFadden 1990) illustrates the wide variety of approaches that have been used by paleontologists to reconstruct mammalian body size. An appendix in this volume lists over 900 prediction equations generated from modern species that have been used to infer body size in fossil mammals. Skull length has often been used to approximate body size in fossils (Radinsky 1972, 1979). The diameter of the foramen magnum, taken from a relatively complete fossil cranium, might be especially useful for examining the relationship between brain to body size in fossil mammals, given that the spinal cord must pass through the foramen magnum. Phillip Gingerich and his colleagues published equations allowing the prediction of body weight from the molar tooth length in fossil primates and insectivores (Gingerich *et al.* 1982, Gingerich & Smith 1985, Bloch *et al.* 1998). Gingerich (1990) also predicted mammalian body weight from the length and diameter of long-bones. The scaling relationship between limb proportions and body size was investigated for primates (Jungers 1985). Over a period of 20 years, Henry McHenry has attempted to estimate body size in fossil hominids through a variety of methods that quantify weight-bearing structures. Such structures include areas of the lumbar vertebrae, and joint surface areas of the upper and lower limbs (McHenry 1974, 1992, 1994).

Studies estimating body size in fossil hominids have been particularly abundant, and, as one might expect, these have created controversy. Different methods produce different body sizes. Fundamental criticism has even been leveled at the modes of statistical inference used to estimate fossil hominid body size (Smith 1996).

Body size and paleocommunity reconstructions

Because of the impact of body size on the evolutionary ecology of extinct species, size is routinely investigated in reconstructions of ancient community life. Peter Andrews, for example, has long used size as one of several variables separating fossil species in multivariate space (Andrews *et al.* 1979, Andrews 1996). Serge Legendre (1986) initiated the routine use of cenogram analysis by vertebrate paleontologists who were interested in questions of community structure. The cenogram is a Cartesian plot of the natural log of body size for a fossil species (Y-axis) versus its ordinal rank by decreasing body size (X-axis). If many species are available from a single fossil locality, the resulting cenogram can be compared with those generated by living species in a number of different habitats. Species diversity and size shape the cenogram, which is also thought to have paleoenvironmental implications.

I have used (Cachel 1996) Hutchinsonian ratios to examine the pattern of fossil hominoid species interactions in both early to middle Miocene sites of East Africa (20–14 mya) and in East and South African Plio-Pleistocene hominid sites (1.8–1.5 mya). In 1959, the Yale zoologist G. Evelyn Hutchinson and his student Robert MacArthur were examining species diversity in animals. They argued from empirical and theoretical evidence that competitive interactions between closely related sympatric species should generate a minimal threshold of niche similarity. This threshold was a calculated ratio between species pairs using morphological traits related to food acquisition that were important in defining niche structure (Hutchinson 1959, Hutchinson & MacArthur 1959).

Hutchinson's examples yielded a ratio of about 1.28 (rounded to 1.3) for linear structures. This was later extrapolated to a ratio of 1.69 $[(1.3)^2]$ for areas, and a ratio of 2.2 $[(1.3)^3]$ for volumes. These ratios may specify the minimal degree of separation needed to decrease competition and achieve niche separation between closely related sympatric species. They have been used to document the existence of major processes that structure animal communities.

I measured nine dental variables in the adult dentition from described specimens of African early to middle Miocene hominoid taxa in the collections of the Kenya National Museums in Nairobi. Some hominids were also measured. Dental measurements for the fossil hominids (australopithecine species and early genus *Homo*) are taken from the literature, with the exception of one index, the anterior/posterior maxillary span, which was obtained from fossil hominids in the collections of the University of the Witwatersrand Medical School, Johannesburg, and the Transvaal Museum, Pretoria. Roth (1981: 397) suggests the use of a modified Behrens–Fisher test to generate a d statistic, and test the hypothesis that a given measurement differs by a ratio of 1.3 in two species. I use a similarly modified Student's t'-test for unequal variance in the two samples, and the df is given by the integer closest to n′ in this test. Table 11.2 presents the results for single tooth measurements; values that are statistically significant at the minimal or Hutchinsonian distance are underlined. Because the t' statistics fall well below critical values at the 0.01 level with df = n′ in two-tailed tests, the null hypothesis is retained that differences between the population means are explained by the 1.3/1.69 ratios. Hence, ratios at or below these measurements demonstrate competition. The sites, site dates, species examined, and dental measurements are described in Cachel (1996b), which also presents a Hutchinsonian analysis of composite dental measurements.

What can one learn from Table 11.2? Although a vast amount of information about these ancient communities has been irretrievably lost, some understanding can still be gleaned about them. Hominoid niches were more diverse in the early to middle Miocene, where many ratios fall well above the Hutchinsonian limit. Niche differentiation appears extensive between species pairs at Kalodirr, Rusinga Island, Songhor, and the Legetet Formation. However, on all measurements, *Proconsul nyanzae* and *Proconsul heseloni* fall below the 1.3/1.69 criterion. The same is true for *Dendropithecus macinnesi* and *Limnopithecus legetet*. These species pairs may demonstrate intense competition. Body size alone is probably not the sole determinant of hominoid competition. *Rangwapithecus gordoni, Proconsul heseloni*, and *Proconsul africanus* are estimated to have the same body size, but *Rangwapithecus gordoni* may have been a linchpin species at Songhor. It demonstrates competition with *Proconsul major, Limnopithecus legetet*, and *Dendropithecus macinnesi*, because some measurements fall at or below the Hutchinsonian threshold. However, neither the similarly

Table 11.2. *Sympatric taxa which meet or exceed Hutchinson's 1.3/1.69 rule*
for minimal morphological distance (minimal distance is underlined)

Single tooth measurements		
Site/formation	Taxa	Measure and value
Swartkrans	*A. robustus,* early *Homo*	MDI1 1.30
Koobi Fora	*A. boisei,* early *Homo*	MDP4 1.32; MDM2 1.28; area M2 1.68; area M3 1.66
Olduvai Gorge	*A. boisei,* early *Homo*	MDP4 1.32; MDM2 1.28; area M2 1.68; area M3 1.66
Kalodirr	*A. turkanensis,* *T. kalakolensis*	MDP4 1.41; MDM1 1.48; area M1 2.57
Rusinga Island	*D. macinnesi,* *P. heseloni*	MDI1 1.60; MDP4 1.48; MDM1 1.34; MDM2 1.38; area M1 1.69; area M2 1.84; area M3 1.75
	D. macinnesi, *P. nyanzae*	MDI1 1.96; MDP4 1.79; MDM1 1.46; MDM2 1.70; area M1 1.91; area M2 2.49; area M3 2.45
	P. nyanzae, *P. heseloni*	All below 1.3/1.69
Songhor	*R. gordoni,* *P. major*	MDI1 1.55; MDP4 1.33; MDM1 1.34; MDM2 1.45; area M1 1.89; area M2 1.98; area M3 1.59
	R. gordoni, *L. legetet*	MDI1 1.29 MDP4 1.61; MDM1 1.34; MDM2 1.39; area M2 1.74; area M3 2.11

Table 11.2. (*cont.*)

Single tooth measurements		
Site/formation	Taxa	Measure and value
	R. gordoni,	MDI1 1.58;
	D. macinnesi	MDP4 1.44;
		MDM1 1.26;
		MDM2 1.35;
		area M3 1.96
	P. major,	MDI1 2.46;
	D. macinnesi	MDP4 1.92;
		MDM1 1.69;
		MDM2 1.96;
		area M1 2.61;
		area M2 3.19;
		area M3 3.13
	P. major,	MDI1 2.01;
	L. legetet	MDP4 2.14;
		MDM1 1.81;
		MDM2 2.01;
		area M1 2.88;
		area M2 3.44;
		area M3 3.37
	D. macinnesi,	All below 1.3/1.69
	L. legetet	
Legetet Formation	M. clarki,	MDP4 2.49;
	P. africanus	MDM1 1.64;
		MDM2 1.78;
		area M1 2.73;
		area M2 3.34;
		area M3 3.32

sized *P. heseloni* at Rusinga Island nor *P. africanus* in the Legetet Formation seem to have played an equivalent linchpin role. Because Songhor exhibits the greatest hominoid species diversity, it is possible that the greatest degree of niche overlap occurs at this site, and that *R. gordoni* plays a critical role in the competitive interplay of sympatric hominoid species.

What about hominoids much later in time? The Plio-Pleistocene hominids arise after a spectacular decline in hominoid diversity. There are many fewer species, but some sympatric species may have been experiencing competition. The comparison of the sympatric robust australopithecines *Australopithecus robustus* and *Australopithecus boisei* and early *Homo* demonstrates that species pairs are just at the 1.3/1.69 criterion when some dental measurements are examined for single teeth. Hence, although dramatic niche differences between robust australopithecines and early *Homo* have been a paleoanthropological fixture since John Robinson's work in the 1950s (which generated Robinson's classic dietary hypothesis),[1] the traditional emphasis on hominid niche differences may be over-enthusiastic. An illustration may make the point more vividly than abstract numbers (Figure 11.1). The picture of shifting tooth shape along the postcanine tooth row also shows that the Miocene hominoid taxa are an extremely diverse group. This is true even if members of the same genus are examined (*Proconsul africanus, Proconsul nyanzae, Proconsul major*). However, tooth shape among the Plio-Pleistocene hominids is remarkably similar. In summary, niche differentiation in sympatric early to middle Miocene hominoids probably exceeds that found between robust australopithecines and early *Homo*, at least as measured by dental variables. The sympatric hominids were competing with each other more closely.

Body size and behavior

The importance of controlling for body size in any biological analysis, including the analysis of behavior, is brought home by the following example. A recent primatological study appeared to demonstrate that sexual skin swellings were honest signals of female quality in wild baboons; they were affected by sexual selection as males chose mates (Domb & Pagel 2001). However, a re-analysis controlling for female body size (height is the variable used for the wild females) shows that allometry is the overriding factor, and eliminates the connection between size of the sexual skin and female quality (Domb & Pagel 2002, Zinner *et al.* 2002).

Body size has been used to examine species ranges and home range size in living mammals (Van Valen 1973b, Calder 1984). Processes revealed for animals in the present world are inferred to have some importance in determining home range size or foraging and ranging patterns in extinct species. When comparisons are made across terrestrial vertebrates, maximum body size of the largest local species appears to be clearly related to the number of home ranges that can fit within the land area encompassed by the sampled locality (Burness *et al.* 2001). The simple explanation is that food resources available within a home range determine the body size of the largest local species. Because metabolic rate affects food requirements, ectotherms and marsupial mammals, whose average metabolic rate is about 20% lower than that of placental mammals, require significantly smaller home ranges and less food. Hence, reptiles or marsupials emerge as the largest species on islands or continental areas of low productivity (Burness *et al.* 2001).

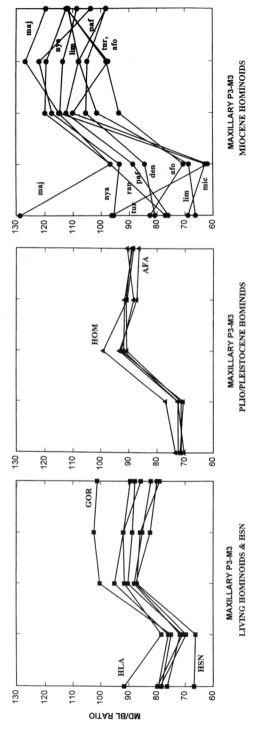

Figure 11.1. Postcanine tooth shape in three groups of hominoids. Modified from Cachel (1996b: Fig. 7). The ratios on the vertical axes are tooth length divided by width. Ticks on the horizontal axes represent P3 through M3. Left: living hominoids and Neanderthals; middle: fossil hominids of the Plio-Pleistocene (HOM = early genus *Homo*, AFA = *Australopithecus afarensis*); right: early to middle Miocene East African fossil hominoids.

However, while body mass and, broadly, metabolic rate, are known to influence home range size in living mammals, there are other factors that also impact on ranging patterns. For primates, such factors as substrate use and diet have long been recognized to be important. Crook & Gartlan (1966) determined that a fundamental dichotomy existed between terrestrial and arboreal lifestyles in living primates, and inferred that this dichotomy should be important in the evolution of primate social systems. However, it was later determined that only two variables regularly appeared to separate primates along the arboreal/terrestrial bifurcation: terrestrial primate species have larger home ranges, and juveniles of terrestrial species become behaviorally independent from their mothers at an earlier age. The importance of substrate use and diet, along with body size, in influencing home range size in living primates has recently been affirmed again (Nunn & Barton 2000). Although home range scales to body mass raised to about the 0.25 power in mammals generally, there are two exceptions: carnivores and primates (Carbone *et al.* 2005). Carnivores have larger home ranges than expected for body mass, and primates have smaller home ranges. The three-dimensional foraging of arboreal primates probably generates smaller home ranges than in other mammals.

Dietary quality is important, but can sometimes be established only by detailed analysis of the nutritional composition of foodstuffs. The distribution and abundance of critical food items and their nutritional quality can determine fundamental aspects of species ecology, such as biomass. A food may be abundant, but may be of low quality. Mature green leaves are an example. Their nutritional composition affects the biomass of leaf-eating Old World monkeys. The ratio of protein to fiber in mature leaves of the 20 most abundant tree species at a site accurately predicts the biomass of colobine monkeys at a variety of African and Asian sites (Chapman *et al.* 2002). At least seasonally, these monkeys may exist close to a critical protein threshold below which protein requirements for body maintenance and function cannot be met.

The all-too-familiar use of sexual dimorphism to infer sociality in fossil species

Adult primates, both living and fossil, may show sexual differences in body size, canine size, and skeletal traits. In living primates, soft-tissue morphology, skin color, and pelage color and pattern can be examined (Figure 11.2). A complex interplay of ecology and sexual selection affects these differences. Within each species, the final assessment of dimorphism is the product of phenotypic changes by both males and females. The limits of the fossil record generally curtail the analysis of dimorphism in fossil species. However, primate species

Figure 11.2. Dassenetch people from Ileret, northern Kenya, illustrate sexual dimorphism in living humans. Photo courtesy of Purity Kiura.

from the Paleocene through the Pleistocene are frequently scrutinized for dimorphism, and links are sought between the degree of dimorphism and ecological factors or sexual selection in fossil species (Plavcan 2001).

Inferring social behavior in fossil primate species from the degree of sexual dimorphism exhibited by the species has become a cottage industry. The species for which this has been attempted span most of the Cenozoic. The list of species extends from the first anthropoids – Eocene *Catopithecus* and Oligocene *Aegyptopithecus* – where mating systems have been inferred, to Plio-Pleistocene hominids, where different mating systems have been reconstructed, to anatomically modern humans in the terminal Pleistocene, where social interactions have been recreated.

If fossil species are dimorphic, how does one know whether individual specimens are members of the same species? Because taxonomic assignments are made on surviving morphological traits, paleontologists can usually make only limited arguments about sexual dimorphism. In many cases, if a bimodal distribution appears in an adult trait, this is enough to show dimorphism. Very infrequently, the evidence is better. For example, Andre Keyser has recently made a good case for the degree of sexual dimorphism in *Australopithecus robustus* (Keyser 2000). The South African site of Drimolen has recently yielded the first cranium and articulated mandible of a female *A. robustus* specimen. This skull (DNH 7) was found juxtaposed with a large male mandible of *A. robustus* with nearly complete dentition (DNH 8). Contrary to former expectations, the degree of variability within *A. robustus* is now seen to be comparable to that documented for the East African robust australopithecine species, *A. boisei*.

An old guideline, Rensch's rule, stated that sexual size dimorphism is positively allometric (or increases with body size) in species where males are larger, and declines with body size in species where females are larger. Rensch's rule cannot be demonstrated in sexually dimorphic body weights among platyrrhine primate species (Ford 1994). Platyrrhine primates tend to be smaller than catarrhines, and callitrichid platyrrhines are the smallest of anthropoids. However, a more recent meta-analysis powerfully supports the existence of Rensch's rule among vertebrates, and finds that primates as a whole show statistically significant positive allometry in species where males are larger (Abouheif & Fairbairn 1997). Sexual size dimorphism increases with body size.

Male-biased mortality is associated with sexual selection and sexual size dimorphism in mammals. This cost of sexual selection is very highly correlated with parasites and male-biased parasitism in wild mammal populations (Moore & Wilson 2002, Owens 2002). Primates are included in this meta-analysis, although the effect is not significant at the 0.05 level for primates. It is important to note that, when reversals of sexual size dimorphism occur in mammals, and females are the larger sex, female-biased parasitism takes place. Being male per se does not cause parasite susceptibility. Hence, body size alone is an important contributor to sex-biased parasitism in mammals, and not life-history, stress, or hormonal factors. Large animals are parasitized more because they are big – causing more exposure to ectoparasites – and consume more food – causing more exposure to endoparasites (Owens 2002). The degree of sexual size dimorphism in mammals is largely the result of sexual selection. Reversed sexual dimorphism in mammals may be associated with the degree of maternal care or female–female competition, but does not seem to involve mate competition (Ralls 1976, Moore & Wilson 2002).

There are many diverse social and mating systems in living primates. In theoretical terms, ever since Darwin wrote about sexual selection in 1871, larger male body size in animals has been related to male–male competition for access to mates. A link between the degree of sexual size dimorphism and mating system is also observed in primates (Clutton-Brock *et al.* 1977). Yet, there is no overriding link between the existence and degree of sexual dimorphism and particular mating systems in primates. Size dimorphism exists in lemuroids, along with female dominance and priority of food access. In four species of wild-trapped tamarins with large samples, adult females were 4–5% heavier than males (Garber 1990). In callitrichids, there is a near absence of size dimorphism or slightly reversed dimorphism, along with facultative polyandry, in which dominant females mate with multiple males. In hylobatids, slight size dimorphism occurs, along with social monogamy, territoriality, and territorial defense. In the Old World monkey *Cercopithecus neglectus* (De Brazza's monkey), males are approximately twice the size of females, but social monogamy exists. The Old World colobines *Presbytis potenziani* and *Nasalis concolor* (the Metawei Island langur and the simakobu monkey) are also socially monogamous. Hence, merely from this brief summary, there is reason to be suspicious about a strong link between sexual dimorphism and social system or mating system in living primates. In fact, recent examination of living primate species does not support a close relationship between measures of dimorphism and mating systems and intrasexual competition. This casts doubt upon the use of sexual dimorphism to infer social behavior in fossil species (Plavcan 2000).

Reversible body size changes in individuals

Phenotypic plasticity is a frequent occurrence in the animal world, and it illustrates how quick and widespread animal response can be to variable environments. This is in contrast to the idea of Richard Potts (1996a, 1996b, 1998) that long-term exposure to variable environments may be responsible for the origin of generalized species. Potts further argues that a novel evolutionary process known as variability selection should be recognized by theorists. However, mechanisms responsible for the origin of generalized species may be diverse.[2]

I now present some dramatic instances of phenotypic plasticity. In freshwater fish like Arctic char (*Salvelinus alpinus*) or pumpkinseed sunfish (*Lepomis gibbosus*), individuals that specialize in different diets or habitat types can vary dramatically in morphology (Robinson & Schluter 2000). The morphological differences within a species can equal or exceed that found between closely related or sibling species. In Arctic char, four sympatric morphs reflecting dietary or habitat differences can actually occur within the same lake. To a non-specialist, the differences between the morphs equal those found between different freshwater fish species (e.g., Robinson & Schluter 2000: Fig. 3.1). Dolph Schluter and his colleagues have in fact used this type of evidence to examine such major evolutionary processes as natural selection, adaptive radiation, character displacement, and parallel speciation. They have also very strongly argued that this evidence demonstrates how ecology may affect speciation. Ecological speciation is viewed as a subset of processes falling within the mode of sympatric speciation.

Phenotypic plasticity can also affect body size and weight changes in adult animals, after ontogenetic or developmental transformation is slowed or completed. Adult Galápagos marine iguanas shrink their body length by up to 20%

over 2 years in response to poor food availability during El Niño climatic fluctuations. This size change is caused principally by skeletal reduction, and is reversible after re-growth of the bone (Wikelski & Thom 2000). Individuals that shrink more survive longer, because their foraging efficiency increases while their energy expenditure decreases.

This is not the first report of shrinkage and re-growth in adult vertebrates. Reptiles continue to grow throughout their life, although at a slower rate than in their juvenile years. Unlike reptiles, mammals stop growing after they reach their adult body size. Shrinkage and re-growth in adult mammals therefore seems intrinsically unlikely. Yet, for over 50 years, it has been well documented that European shrews routinely shrink their body size and weight during the winter (Churchfield 1990, Yalden 2000). This has been called the "Dehnel effect," after the Polish mammalogist who first observed the phenomenon in 1949. Neurocranial dimensions decrease in these shrews. This is caused by resorption of the parietal and occipital bones. Body length is also shortened by resorption of the intervertebral disks in the spinal column. Photoperiod is the principal variable responsible for these changes, which are controlled by powerful parathyroid secretions during the autumn and winter. The size of the brain, liver, kidneys, spleen, testes, ovaries, adrenal, and thymus glands are also reduced. The Dehnel effect is thought to be a strategy to reduce the absolute food requirements of European shrews under a regime of low winter temperatures while the animals must still maintain a high core temperature. Re-growth occurs in the spring and summer. It is estimated that a weight loss of 30% can reduce shrew foraging time by 5 hours (Churchfield 1990). There is additional evidence that subsistence and habitat quality exercise an effect on the size of these animals. Some of the smallest shrews exist north of the Arctic circle, and shrews from poor habitats tend to be smaller than those in rich habitats.

These shrew species are not unique in the growth and shrinkage observed in the brain. Seasonal changes in the brain have been widely documented in birds. In some bird species, the hippocampus of the brain, an important anatomical region relating to memory, undergoes first seasonal growth and then regression. Seasonal behaviors that involve memory are impacted by the growth and regression of the hippocampus (Clayton 1998). Examples of such behaviors are food storage or caching or singing related to courtship. Neural centers that are only seasonally active are apparently prohibitively expensive to maintain on a year-round basis – hence their regression. The growth of new neurons in the hippocampus and olfactory bulbs has recently been documented in the adult brain of two macaque species (Kornack & Rakic 2001), although claims of constant neurogenesis in the neocortex of adult primates were not supported.

Size and shape changes: adaptation and plasticity

It has been known since the nineteenth century that body weight in widely distributed endothermic animals tends to be lower in warm environments and higher in cold environments. This negative relationship between body weight and temperature is known as Bergmann's rule, discovered in 1847. A recent comparative analysis demonstrates that, within a given species, body size is larger at higher latitudes (Ashton *et al.* 2000). Another comparative analysis shows that larger species (>0.16 kg) generally follow Bergmann's rule better than smaller species do (Freckleton *et al.* 2003). As a group, smaller species can show both statistically positive and negative correlations between body size and temperature. Ectotherms do not follow Bergmann's rule (Belk & Houston 2002). An endothermic species occurring in the coldest climate or in the poorest habitat may be smaller than expected, as in the shrew examples noted above. In this case, size reduction in poor environments reduces foraging time and conserves valuable energy in animals facing the limits of existence in marginal habitats. Another adaptation, convergently developing in endothermic animal species or subspecies exposed not only to heat but also to arid conditions, is the development of a linear body build or ectomorphy. This linearity of body build and lengthening of distal body parts in hot and arid conditions is known as Allen's rule, discovered in 1877. Bergmann's and Allen's rules affect living mammals, including humans. Studies of modern human variation routinely invoke these rules for the transformation of body size and shape that occurs in response to extremes of temperature (Coon 1965, 1982, Ruff 1994).

Body size can change quickly. Many cases of island dwarfing in large mammals like deer, elephants, and mammoths occur during the Pleistocene (Lister 1996). There may also be an example of Pleistocene insular dwarfing in hominids. *Homo floresiensis* and associated stone tools have been described from late Pleistocene deposits on the Indonesian island of Flores (Brown *et al.* 2004, Morwood *et al.* 2004). The type specimen is an adult female with a stature of 1 m, and it is suggested that this taxon is a product of insular dwarfing (Brown *et al.* 2004). Insular dwarfs, by definition, are the descendants of full-sized ancestors. There are three cases among non-human primates where the claim of dwarfing within a lineage appears fairly robust. Two cases occur among the platyrrhines (*Saimiri*, the callitrichids), and one among the catarrhines (*Miopithecus*). *Saimiri* and *Miopithecus* have high brain/body size ratios, although the callitrichids do not (Stephan 1972). This raises the question of why the type specimen of *Homo floresiensis* (LB1) has such a small cranial capacity (*c.* 380 cm^3). It is possible that this taxon is the result of long-term selection pressure for size reduction, as is apparently true for the callitrichids. *Homo floresiensis* appears to be descended from an ancestor that never had a large brain.

Because this taxon supposedly evolved from *Homo erectus*, its ancestor must have had a brain at the smallest size range for *Homo erectus*, such as D2700, the adult female from Dmanisi with a cranial capacity of 600 cm^3. The type specimen of *Homo floresiensis* is not a microcephalic human, because humans with ordinary microcephaly have large, prognathic faces relative to the neurocranium. There is a type of secondary microcephaly (Majewski's dwarfism with microcephaly)[3] where intrauterine and postnatal growth lags. Although the head is normally proportioned at birth, brain growth is retarded until true microcephaly emerges. However, humans with this condition have abnormal facial features, postcranial anomalies, and small or absent teeth (Hall *et al.* 2004), and these features are missing in LB1 – in fact, the teeth of LB1 are large. *Homo floresiensis* remains a mystery, pending further information about the skeletal biology of the taxon.

Body size reduction or increase can occur very rapidly. Body size reduction in mammals since the end of the Pleistocene is a well known and well studied phenomenon (Purdue & Reitz 1993, Seymour 1993). Intergenerational changes in body size can also be documented in living mammals. This is well known for domesticated species (e.g., size reduction in Icelandic horses and cattle), but can also be documented in humans. Genetic change need not take place, because phenotypic stunting can account for the beginning of the trend in size reduction. Body size increase is the phenomenon, famous in human biological literature, known as the "secular trend" (Frisancho 1993, Eckhardt 2000). Detailed studies documenting stature increase can be produced at the broad level of a nation (e.g., the Netherlands; Fredriks *et al.* 2000) or at the local level of a city (e.g., Belfast; Holland *et al.* 1982). It is important to note that the secular trend is now appearing in humans on a global scale, and over time spans of as little as 1–2 generations. For example, a recent study of the impact of climate on human body size and shape reveals a significant correlation between temperature and body mass and temperature and surface area/mass (Katzmarzyk & Leonard 1998). However, while adaptive responses of body weight and shape to external temperature still affect modern humans, the slopes of the regression lines between temperature and body mass and temperature and surface area/mass have declined since the last major survey of climate and human body size and shape was published (Roberts 1953). The 45 years that separate these two surveys represent perhaps only 1 or 2 generations. This short time span is enough to have changed the nutritional status of modern humans in tropical areas, which accounts for the lower recent correlation between temperature and body size and shape.

Another shift can be documented through recent historic time. Recall the discussion of resting metabolic rate early in this chapter. At the beginning of the twentieth century, two equations were developed to predict resting energy expenditure in males and females ranging in age from 6 to 70 years. These are the Harris and Benedict equations (Harris & Benedict 1919). Recent investigators have found these equations to be off by 7–24%, and new predictive equations had to be generated for healthy recent humans ranging from 19 to 78 years (Mifflin *et al.* 1990). Humans in modern populations have experienced such changes in diet, activity levels, body size, and body components like muscle and fat that they no longer conform strictly to the Harris and Benedict equations developed at the dawn of the twentieth century. The only obvious explanation for this change is modernity itself – Modern Times – a change in lifestyle that incorporates dietary and activity level shifts, and is associated with post-industrial societies and the spread of urbanization.

The normal environment of humans in post-industrial societies at the end of the twentieth century has become one in which the fatted calf is always on the table. During the last 20 years, the prevalence of obesity has increased in the U.S.A., Europe, Latin America, Japan, and China. A sharp increase in obesity occurred in the U.S.A. around 1985; England and Australia lagged behind until 1990; and there has been an increase in Brazil, with no sharp kick-point (Kopelman 2000: Fig. 2). Obesity has failed to increase only in Poland, in spite of abrupt economic alterations (Bielicki *et al.* 2001). What is happening? In post-industrial societies, there is an abundance of technologically processed food, high in both calories and fats, and this food is cheap and easily accessible (Critser 2003). This embarrassment of riches is truly excessive nourishment.[4] Another result of economic prosperity in modern America is also taking place: human activity levels have plummeted. Although a sloth-like level of activity probably occurred in the past among some nobles or people of high rank, the nearly universal spread of sedentary lifestyles and inactivity in modern cities and suburbs is unseen before in human history. In part, of course, this inactivity is supported by complex technology affecting transportation, labor-saving devices, and entertainment – what we usually applaud as "The American Way of Life." In fact, this lifestyle has resulted in American adults carrying an average of 15 kg of body fat, a number far higher than in most human populations throughout the world. For example, comparing American adults with adult Turkana pastoralists from East Africa, subcutaneous fat is far thicker in Americans than among the Turkana. As is characteristic for our species, females in both groups exceed males in the deposition of subcutaneous fat, but sexual dimorphism in the thickness of subcutaneous fat is more extreme in the U.S.A. (Little 1989: Figs 8 and 9).

The health hazards of a modern lifestyle are difficult to reverse, because the lifestyle itself is so overwhelmingly tempting. Some researchers argue that a long prehistory of hunting and gathering has given humans a genetic basis to prefer high-calorie, high-fat diets, and to prefer leisure, rather than hard labor. Ethnographic research on living hunter-gatherers shows a great deal of variability in nutrition – some may have half the nutritional intake of others. However, there is no general evidence for dietary abundance and prosperity (Foley 1999). The archeological evidence for hunter-gatherer subsistence also shows variability, but affluence is rare. There are limits to economic growth, because subsistence cannot be easily intensified the way it can be under agricultural or pastoral regimes (Rowley-Conwy 2001). What of the non-human primates? There is seasonal variation in body mass in free-ranging Japanese macaques (*Macaca fuscata*), even when they are provisioned with wheat or sweet potatoes, because the energy and protein composition varies in wild foods that they also consume (Kurita *et al.* 2002). Infant mass increases steadily after birth, but some infants take about 40% longer (400 versus 240 days) to reach a body weight of 2 kg.

The deleterious consequences of a modern lifestyle on human health have been most elegantly documented by physical anthropologists working in the years after World War II among various Amerindian communities in the American Southwest. World War II was the watershed, because Amerindian communities previously isolated by geography and culture became incorporated into the global economy during the war (Knowler *et al.* 1983, Price *et al.* 1993). Physical anthropologists routinely employ an approach that emphasizes variability within populations and an evolutionary perspective. This is different from the individual case-by-case approach employed by physicians. As a result, the physical anthropologists have revealed a link between post-World War II lifestyle changes and a catastrophic increase in obesity, diabetes (type II, non-insulin dependent diabetes mellitus), gall bladder disease, and gall bladder cancer in these Amerindian groups. This complex of health problems resulting from recent lifestyle changes has been referred to as "The New World Syndrome," because the evidence was developed from Amerindian groups, where there may be a common genetic basis for susceptibility (Weiss *et al.* 1984, Weiss 1985). In fact, individual susceptibility in the American Southwest is linked to the degree of Amerindian admixture.

"The New World Syndrome" is partly a misnomer, because it occurs in peoples of different ancestry, as well (e.g., Polynesians). It occurs in areas where changing lifestyles can also amplify the effects of underlying genetic susceptibility. Since the early 1960s, physical anthropologists have used a "thrifty gene" hypothesis to explain the increase in diabetes that is observed when traditional diets and lifestyles are abandoned in favor of modernity. In communities where seasonal food shortages are common, and periods of frank starvation are frequent, natural selection favors a "thrifty genotype." This genotype can secrete insulin quickly. Such rapid secretion removes glucose from the bloodstream, and converts it into fat depots for adaptive energy storage, anticipating the inevitable bad times (Neel 1962, 1982). Investigation has shown a genetic basis for the recently identified human "metabolic syndrome," where overeating, obesity, insulin resistance, glucose intolerance, hypertension, and dyslipidemia are associated (Masuzaki *et al.* 2001). The metabolic syndrome occurs with different frequencies in different population groups, and appears to be triggered by the build-up of a critical amount of visceral fat. The visceral fat cells overproduce glucocorticoids, which is the underlying link between elements of the syndrome. High-fat diets amplify the build-up of the visceral fat depot (Masuzaki *et al.* 2001).

The dramatic changes in human morphology that I have recounted in this section all occurred within 1–3 generations. They document the continuing evolution of living humans – in case one should be skeptical that human evolution occurs at all, or in case one is inclined to argue that evolution no longer affects living humans, although it might have occurred at a far-distant time in the Pleistocene. These recent morphological changes are so dramatic that they become an important topic for physical anthropologists who document and examine the secular trend. Given the morphological and physiological homogeneity of all catarrhine primates, one might expect that captive non-human catarrhines might also show the presence of a secular trend. Captivity translates into a sedentary existence for animals, with human caretakers provisioning zoo or colony animals with adequate food supplies that are freely available, and that require no search or foraging. However, five species of captive baboons and inter-specific hybrids show no secular trend in increasing body weight, despite relative inactivity and freely available, nourishing, constant food supplies (Jaquish *et al.* 1997). All animals were bred in captivity at the Southwest Foundation for Biomedical Research in San Antonio, Texas, and kept in large outdoor enclosures. Routine health exams required periodic weighing of each animal. The long-term colony records show no secular trend, but, despite inactivity and free access to food, nutrition is constantly monitored in captivity and diets are balanced. This may not be comparable to wild conditions where *ad libitum* food intake can sometimes access very rich food resources.

Wild African baboons, for example, freely forage at garbage dumps in villages, cities, and safari lodges. I have seen baboons calmly foraging at high noon by sitting inside garbage cans around the parking lot of the Livingstone Museum in Zambia, as they pick through discarded sandwiches and other debris from human lunches (Figure 11.3). The fat and protein-rich table scraps that baboons may acquire under these circumstances lead to weight gain, and the development of cardiovascular diseases. Weight may increase by 50% – much of this being fat – and the weight increases are more marked in females (Altmann *et al.* 1993). More important, however, is the fact that food affects reproductive success. Young wild female baboons (*Papio cynocephalus*) reach reproductive

Figure 11.3. Sitting on the rim of a garbage container, a juvenile olive baboon (*Papio anubis*) fearlessly searches for food in the parking lot of the Livingstone Museum in Zambia. The buttery fingers clutch a triangle of discarded sandwich. Note the distended cheek pouch, which is already packed with pilfered food.

maturity at an earlier age when there is a surplus of dietary protein beyond the minimal requirement, and when caloric or energy intake increases (Altmann 1991). A female baboon's lifetime fitness is established by the quality of her foraging diet even before weaning is completed, and about 3 years before the advent of puberty. Lifetime fitness is determined by reproductive span, number of infants produced, and number of infants that survive the first year of life.

A similar physiological response holds true for young human females, who reach menarche earlier when the diet is high in calories from protein and fat, and a critical level of body fat is achieved. There is a link between body fat and reproductive capacity in human females (Frisch & McArthur 1974). It is now known that the hormone leptin, produced by adipose tissue and varying with fat mass, is the cause of this link (Chehab *et al.* 1997). Leptin signals the presence of adequate energy stores in the body. Leptin influences human ovarian function, because the ovary has functional leptin receptors (Karlsson *et al.* 1997). Leptin replacement can reverse amenorrhea in leptin-deficient females with a low body weight (Welt *et al.* 2004). This treatment also increases growth hormone levels and signs of bone formation. In at least one human population (Finland), long-term differences in abundance and stability of food affects twinning (Lummaa *et al.* 1998). Dizygotic twins are more frequent in areas where food has historically been abundant and predictable.

Diet also affects growth and development in chimpanzees. Captive common chimpanzees (*Pan troglodytes*) have a much faster rate of dental maturation than wild members of the same species. The permanent teeth of wild common chimpanzees from three field sites that span the geographic range of the species erupt later than the teeth of 90% of the captive animals (Zihlman *et al.* 2004). The eruption of I2 and M2 is delayed beyond the range observed in captive animals. Wild chimpanzees also exhibit delayed growth when other indicators of maturation (e.g., body mass, limb lengths, epiphyseal fusion, testicular maturation) are considered (Zihlman *et al.* 2004). Dietary differences between wild and captive chimpanzees are the only explanation for these differences in maturation, and the differences are profound. If one were examining fossils, one might categorize these two groups as different species.

Thus, diet clearly affects the growth regimes of baboons, chimpanzees, and humans. What is the meaning of diet-mediated maturation rates and onset of puberty? It must indicate a general catarrhine predisposition, shared by humans, to respond with virtually no lag-time to an abundance of food. In a fundamental sense, what this indicates is that catarrhine populations are food-limited. Food, not climatic variables, disease, or predation, is the most important factor influencing population size. It is a density dependent factor, because the amount of available food is affected by population size. And evolutionary theory forecasts that reproductive output should be increased when resources are predictable.

The best demonstration in primates of the flexibility of reproductive output in response to food availability is found among studies of a non-catarrhine species, the common marmoset (*Callithrix jacchus*). In captivity, animals are provisioned with food. The ready availability of food and improved nutrition results in females producing triplets, quadruplets, and quintuplets, rather than the normal number of young (twins). A survey of the reproductive records of 12 common marmoset colonies shows that triplets can sometimes outnumber twins (Windle *et al.* 1999). Ultrasound investigations and hysterotomies reveal that improved nutrition allows more eggs to be released during ovulation. Litter size can be reduced during pregnancy by absorption of embryos or fetuses, rather than spontaneous abortions or miscarriages. There is a peculiar delay between implantation of the blastocyst and the beginning of the principal embryonic growth phase. The long lag-time before embryonic growth, which occurs at day 80 of a 144 day pregnancy, allows embryos/fetuses to vanish through absorption (Windle *et al.* 1999). Newborns that could not be reared in the wild are lost well before birth. The common marmoset thus possesses natural physiological mechanisms for both a very rapid increase in reproductive output when dietary conditions are good, and for a reduction in litter size throughout pregnancy when environmental conditions are poor. Because dizygotic twinning is the norm in callitrichids, it is very likely that these mechanisms also exist in other callitrichids.

The morphological changes associated with the secular trend that are observed in living humans are of the same order of magnitude as those changes that many paleoanthropologists associate with the advent of anatomically modern humans – the appearance of a new human species – during the Late Pleistocene. Data of this type cause me to be skeptical about arguments that a species-level distinction must be associated with anatomically modern humans, who succeed Neanderthal humans in Europe and coexist with them for a period of perhaps 5000–7000 years in some areas of Europe. In fact, I have developed a model in which interacting and cascading selection pressures can account for the appearance of traits associated with anatomical and behavioral modernity (Cachel 1997). Dietary shifts initiate the cascade of events. Human body size and shape changes within Ice Age Europe, but these changes reflect adaptations to ambient temperature.

Figure 11.4 shows the skeletons of three fossil humans from the late Pleistocene of Europe. All are adult males, and all are scaled to the same size. The importance of Figure 11.4 is that it illustrates the labile nature of human body shape through what – in geological terms – is a very short section of the human paleontological record. The La Ferrassie 1 specimen, which is actually more complete than shown in the figure (Fennell & Trinkaus 1997), is a Neanderthal human dating to 71–57 000 years BP. As is typical of Neanderthals, it shows biological adaptation to extremely cold climate through high body weight relative to stature (Bergmann's rule) and a body shape with short limbs and distal limb elements relative to the length of the trunk (Allen's rule). In fact, Neanderthal body shape is equivalent to that of modern Inuit (Eskimo) humans, who retain these proportions, even if they no longer pursue a traditional lifestyle. The

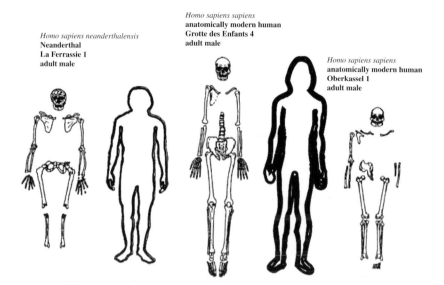

Figure 11.4. Changing human body shape in the late Pleistocene of Europe. All fossil specimens are adult males, and all are scaled to the same size. Note the differences in shape between the Middle Paleolithic Neanderthal (La Ferrassie 1) and the early Upper Paleolithic anatomically modern human (Grotte des Enfants 4). Note also the differences in shape between the earlier anatomically modern human (Grotte des Enfants 4) and the later anatomically modern human (Oberkassel 1) from the late Upper Paleolithic. The skeletons are from Pearson (2000: Fig. 5), and the body outlines are adapted from Howells (1997: Fig. 46).

Grotte des Enfants 4 specimen is an anatomically modern human dating to 28–21 000 years BP. Its low body weight relative to stature, and linear body build with long limbs and distal limb segments relative to trunk length can be used as an argument for a recent tropical African ancestry for modern humans – Grotte des Enfants 4 is then viewed as the product of a recent quick dispersion of modern humans into Europe. The Pleistocene, however, was a time of many wildly dramatic global temperature fluctuations. The last period of intense global cold (the Last Glacial Maximum) occurred at 21 000–18 000 years BP. Northern and central Europe were virtually empty of human inhabitants at this time, as people clustered in warmer and richer refuge areas to the south and west. The Oberkassel 1 specimen, dating to 16.5–11 000 years BP, is a member of a population that re-settled Northern Europe after the extreme cold ameliorated. Yet, note that the Oberkassel 1 specimen, although anatomically modern, has achieved a Neanderthal-like body shape. If one assumes that Grotte des Enfants 4 and Oberkassel 1 are sampled from two time segments of the same evolving population, this shift in body shape occurred within 4500 to 17 000 years. This should not be surprising, given the 1–3 generation transformation of human body size currently manifested in the secular trend.

Furthermore, given the nature of the fossil and archeological records, it is entirely possible that causes of the morphological shift taking place in living humans within 1–3 generations during the twentieth century would be invisible to future paleontologists and archeologists. These future researchers might be inclined to invoke the appearance of a new species (with a larger body size and reduced skeletal robusticity) to explain the obvious differences in morphology. And, surely, the relentless spread of this species was associated with technological innovations that conferred vast competitive superiority, in contrast to the smaller, leaner, laboring, and heavy burdened people that populated the globe before the coming of the new species.

In a fundamental sense, these studies illustrate the responsiveness of animal body size and shape to dietary abundance and quality, even between one generation and its successor. The broader meaning of these phenomena is the potential for significant phenotypic differences to develop even between generations within the same population. The transformative nature of dietary change can occur with virtually no lag-time, at least not at the scale that is visible in geological and paleontological time. One may use information of this sort to explore classic problems in primate and human evolutionary history. An example that will be discussed later is the increase in body size that occurred with the first specimens of African *Homo erectus* about 1.8 mya.

More food for larger individuals – hormones, health, and longevity are ultimately affected

Investigators considering the various factors involved in animal aging or senescence have discovered that one sure method of forestalling aging in mammals is to limit the intake of food. A diet restricted in calories apparently prolongs life. Mammals that are deprived of food live longer. Why is this so? Mammals whose diets are restricted in calories have low levels of insulin and insulin-like growth factor. The genetic mechanisms linking food intake and aging through hormonal signals have been studied in several laboratory model species (yeast [*Saccharomyces cerevisiae*], nematode worms [*Caenorhabditis elegans*], fruitflies [*Drosophila melanogaster*] and mice [*Mus musculus*]). Given the conservative nature of genes in animal evolution, it is probable that similar genetic mechanisms occur in animals. Three mutant strains of laboratory mice exhibit small body size and long-life. A small-bodied, long-lived mutant mouse strain, the Ames dwarf, experiences a further increase in longevity when its diet is restricted by 10, 20, and 30% increments of *ad libitum* levels of feeding (Bartke *et al.* 2001). A long-term, 6-year calorie restriction experiment in rhesus macaques (*Macaca mulatta*) was gradually instituted in 10, 20, and 30% increments. After 3 months, the animals were maintained at a 30% calorie reduction. This regimen reduced body weight, altered body composition, lowered body temperature, and delayed skeletal and sexual maturation (Lane *et al.* 1996). In another experiment, 1 month of 30% calorie restriction in young rhesus macaques was associated with a lowered body temperature, and a 24% reduction in energy expenditure (Lane *et al.* 1996). In general, although food deprivation experiments create animals that are longer-lived, these animals may not reproduce. Mating behavior itself may cease, as reproduction halts until animals can obtain an adequate food supply. Young animals treated in this fashion may not enter puberty.

Individual animals within natural populations exhibit widely varying reproductive success. Reproductive success in animals is determined by surviving to reproductive age, the length of time over which reproduction occurs, fecundity per year, and offspring surviving to reproduce. Nutrition can affect reproductive success at any of these levels. Adequate nutrition is necessary to maintain normal growth and survival of newborns. Season of birth (which affects maternal diet and lactation) and birth weight both affect neonatal survival. Maternal nutritional deprivation can cause premature birth. Experiments with sheep demonstrate that restricting maternal diet from 60 days before until 30 days after conception causes premature birth, and thus affects newborn mortality (Bloomfield *et al.* 2003). Maternal weight was reduced by just 15%, and thus the mothers were not experiencing starvation, but a relatively modest nutritional insult. Fetal size was not smaller in undernourished mothers when compared with control mothers, and thus fetal nutrient availability did not cause premature birth. Calories per se were not the triggering factor, but some minor vitamin or amino acid deficiency may have been operating. Maternal undernutrition around conception accelerated the growth of the fetal adrenal gland. This caused an early fetal surge in the two hormones, cortisol and adrenocorticotrophic hormone. Premature animals were weaker and unhealthier at birth, although they weighed the same as full-term animals. A relatively modest dietary restriction during a critical period of pregnancy can therefore profoundly affect female reproductive success. Modest religious fasting in humans can cause an increase in premature births, which has been labeled the Ramadan or Yom Kippur effect (Miller 2003). Humans in developing countries are often undernourished, and the rate of premature births in these countries can be greater than 20% (Miller 2003).

Figure 11.5. Differences in stature between Papuans from the highlands of New Guinea and a European American photographer, George Steinmetz. All are normal adult males. The European American is 188 cm in height. Photo copyright 1996 by George Steinmetz.

Population-level differences in body size

Humans: population-level differences in body size

As a species, living humans can be classified as large mammals, given that the average mammal is the size of a rat. However, on a global scale, there is a great range of variation in the body size of adult humans. In fact, among living humans, there can be a marked disparity in adult body size between members of different populations. Figure 11.5 shows how dramatic this size difference can be.

Some of the best research into population-level differences in body size has taken place in South Africa. Living humans in South Africa exhibit substantial differences in adult body size. This is especially marked when Khoisan[5] people are compared to Bantu people, or people of European ancestry. There is even some evidence that significant differences in adult human body size existed at a remote date in South Africa (Singer & Wymer 1982). The site of Klasies River Mouth contains very fragmentary human fossil material dated to about 100 000 ya that may be anatomically modern. Some of the fossil material appears to be equivalent in size to material from living Khoisan people. This raises the possibility of a great antiquity for Khoisan people, if one accepts the possibility of an unbroken lineage evolving in place in South Africa. This seems unlikely, because of the fragmentary nature of the fossils, and because of the lack of a fossil record pointing to evolution *in situ* or continuity. In the past, some researchers argued the Khoisan people were at one time more widely distributed in Africa, and were larger, suffering a reduction in size as they were pushed to marginal environments like the Kalahari Desert (Coon 1962, 1965). The South African human skeletal record is re-established much later, during the Holocene. Sealy & Pfeiffer (2000) have recently studied this record. In addition to body size, they use stable carbon and nitrogen isotopes from human bone samples to indicate diet (especially the extent of marine or terrestrial dietary components) in these Holocene individuals. Body size is small, and it remains stable through time – people are not experiencing a reduction in size. There are no major changes in diet through time, but diet does influence stature. At a time before 2000 years BP, people with the smallest stature had a diet of mainly terrestrial food items. Even more interesting is Pfeiffer's (2001) work demonstrating that Holocene Khoisan skeletons show no evidence of chronic infectious disease, and a very low incidence of skeletal trauma – much lower than one would expect given an active lifestyle in rough terrain. She suggests that small body size contributes to human strength and agility under physically hazardous circumstances. One has a vivid mental picture of people who can fall or trip, but who can bounce back up unhurt, just like a hard rubber ball. This is not unreasonable. Biomechanical analysis, as well as the practical experience of zookeepers, show that large mammals are more susceptible to trauma or death from a fall than small mammals that drop the same distance. Yet, advantages for strength and agility would accrue under any physically demanding circumstances, and so it remains to be demonstrated why Khoisan people show stability for small body size through the Holocene, if this is not caused by dietary shortfalls or the occupation of marginal environments. In any case, given the continued small adult body size of Khoisan people – in spite of gene flow and the secular trend – it is clear that Khoisan people have a genetic substrate for small body size. This may be different from the genetic underpinnings of body size in other human populations that are small on a global scale.

During the second half of the twentieth century, a long-term study of changes in human stature and body shape was conducted in Poland, and a shorter study was conducted in Mexico (Wolański & Siniarska 2001). These are genetically distinct populations, but, in both areas, stature appears to be extremely susceptible to socioeconomic changes, even within the boundaries of a single village. Shorter legs exist in individuals with poorer nutrition, especially if these episodes of poor nutrition occur during major growth periods. Males are more variable than females in responding to these changes. It is important to note that stature increase can be demonstrated in both continents, and that stature response to better nutrition is equally prompt.

Modern humans exhibit sexual dimorphism, although not to the degree that their fellow catarrhine primates do. In other catarrhines, male body weights may be twice the size of female body weights. Human dimorphism occurs in the soft tissues of adults, and the amount and patterning of subcutaneous fat is also dimorphic in modern humans. Females carry a larger percentage of subcutaneous fat than males do. In hunter-gatherer groups, where humans are physically active and lean, subcutaneous body fat as measured by skinfold thickness is 5–15% in males and 20–25% in females (Jenike 2001).

The degree of dimorphism is variable on a global basis: shoulder breadth and hip breadth varies between adults of the two sexes, but differs from one population to another. Subcutaneous fat depots that distinguish males and females also vary from population to population. The degree of dimorphism in stature is dependent on adult stature within any given population: in populations with taller adult stature, the dimorphism is greater; in populations with shorter adult stature, the dimorphism is smaller (Tobias 1975, Bielicki & Charzewski 1977, Hall 1982). Living humans therefore obey Rensch's rule. Nutrition affects differences in stature. Both males and females gain in height with improved diet, but males gain more. Females are less sensitive to poor diets, and appear to be more buffered from nutritional fluctuations (Tobias 1975, Bielicki & Charzewski 1977, Brauer 1982). Females exhibit less variation than males, even when variables other than stature are investigated (Hall 1982). Neither sexual selection (possible male–male competition reflected in marriage systems) nor male parental investment appear to affect stature dimorphism in living humans (Wolfe & Gray 1982).

A major question now arises. In terms of the broad pattern of human evolution, paleoanthropologists often argue that the reduced size dimorphism in modern humans, in comparison with other catarrhines, is the result of changes in mating patterns (e.g., Foley 1989). Living hylobatids exhibit almost no difference in adult body weight (Schultz 1968). The siamang, which is the largest species, has the most dimorphism (8%), and all hylobatids have social monogamy. The australopithecines appear to have been much more dimorphic in size than living humans, whose body weight dimorphism is 11% (Schultz 1968). It is assumed that australopithecine dimorphism must have had an impact on mating patterns. Researchers then comb through the available human fossil record to discern when size dimorphism decreases – when it does, this is thought to imply a major change in mating patterns and social structure (see Chapter 16).

However, it is also possible that the fundamental selection pressure creating reduced size dimorphism in modern humans is nutritional deprivation, and not mating patterns. The potential nutritional impact on dimorphism has been neglected as paleoanthropologists have pursued the fantastic mirage of reconstructing mating systems and detecting the origins of human pair-bonding. When reduced size dimorphism first appears, it may reflect some major change in dietary adequacy – a diminishing caloric base, and not, for example, a decline in male–male competition, or a change in factors affecting female mate choice. It is certainly the case that sexual dimorphism in stature increases in modern populations that experience improved nutrition (Tobias 1975, Bielicki & Charzewski 1977, Brauer 1982, Wolański & Siniarska 2001). The change can take place so quickly that it appears in the next generation.

Furthermore, when metabolic rate is scaled against body mass, and a regression line is drawn for mammals, human females have a slightly lower datum than human males do (MacKenzie 1999, West *et al.* 1999). Human females have a slightly lower than expected metabolic rate. They consequently need less food to support essential physiological activities. This difference appears to confirm the idea that the reduction in sexual dimorphism occurring during the course of human evolution may be the result of food stress, and not a change in ancient mating systems.

Humans: food deprivation during wartime – why large humans starve and small humans survive

Within any given species, smaller individuals require less food. Sometimes this factor is deeply embedded in the evolutionary ecology of a species. For example, among orangutans (*Pongo pygmaeus*), adult males can appear in two morphs: a large morph with the full accompaniment of the male sexually dimorphic traits of fleshy cheek pads and throat sac, and a small morph that lacks adult male phenotypic soft-tissue traits. The smaller morph appears developmentally arrested in a permanent adolescent state, but is actually a fully reproductive adult. Males distinguished by this morph pursue an alternative reproductive strategy, but also enjoy the benefit of needing less food to support their smaller body size (Maggioncalda *et al.* 1999, 2002). Large body size is a true handicap when animal species suffer food deprivation. This handicap phenomenon can affect large members of the same species, or can affect species of different sizes with similar diets – the largest individuals starve (e.g., Heinrich 2000: 212). Natural catastrophes and wars that limit food supplies yield unequivocal examples of how human body size affects dietary requirements. Warfare during the twentieth century offers many examples, both military and civilian, of how body size affects food requirements and survivorship under starvation conditions.

In World War II, documentation is especially good of widespread food shortages, food rationing and starvation, and their morphological and physiological effects on humans. Rationing and food deprivation occurred on a global scale as a result of World War II, and widespread starvation existed not only during, but also immediately after, the war in both Asia and Europe. At the same time, research was also conducted in how to reverse these effects and rehabilitate human victims. Among European countries, only Portugal, Denmark, Sweden, and Switzerland enjoyed relatively good food supplies. War-related food deprivation and even the systematic starvation of civilian populations in occupied Europe are well-documented (Keys *et al.* 1950, Wood & Jankowski 1994, Sikorski 1998). Food deprivation of imprisoned people was especially severe. For example, it is estimated that the average daily food portion allotted per person in the Potulice concentration camp under a regime of enforced heavy labor amounted to 1300 calories (Sikorski 1998).

During this time, the biology of starvation was systematically investigated under experimental conditions. At the University of Minnesota, experimental semi-starvation of healthy young American male volunteers was conducted over a 6-month span. These were conscientious objectors who became drafted civilians. More than 100 volunteers were screened down into a study group of 36. Of these, 32 completed the six months of starvation and the first half of the recovery period. Researchers studied the effects of semi-starvation and subsequent rehabilitation on the metabolism of these volunteers, and their body weight, body composition, and behavior were also quantified (Keys *et al.* 1950). Because basal metabolic rate and body temperature were reduced under this regime, the impact of these reductions would obviously be compounded by a cold environment.[6]

The siege of Stalingrad in the Soviet Union during September 1942–January 1943, yields several examples of how cold and differential body size affected human survivorship. The German Sixth Army, which initially laid siege to Stalingrad, was ultimately encircled and trapped by Russian troops. Unable to break out from the trap, starving German soldiers soon noticed a difference in survivorship based on body size. Each soldier received an identical food ration, irrespective of body size. These food rations, about 500 calories per day,[7] were enough to sustain short, lean soldiers even under conditions of extreme cold. Large, robustly built soldiers, however, were the first to die from starvation. This discrepancy in German survivorship continued after they surrendered, because their Russian captors also fed their prisoners rations that did not take human body size into account. Horses, however, were fed in proportion to their body size in the Russian labor camps, and did not experience a size-related mortality (Beevor 1999: 414).

In their survey of war-related data, the authors of the Minnesota experiment were generally skeptical about reported or estimated food intakes, because of local production of food supplies, elaborate black market interaction, and the unreliability of human memory under prison conditions (Keys *et al.* 1950). However, it is clear that, at least in Europe, protein and fat were very soon in short supply, and were sometimes entirely eliminated from human diet, with bread, potatoes, turnips, cabbage, and other vegetables accounting for most of the caloric intake in northern and central Europe. In Switzerland and in England, there was an average 10% reduction in caloric intake among civilians. This resulted in a small degree of weight loss, especially in older adults, but the weight loss reached a plateau, and caused no obvious health impairment. The severe famine during September 1944–May 1945 in the occupied western Netherlands accounted on the average for over a doubling of the previous mortality rate. However, there was a dramatic difference in mortality between men and women – the increase in male mortality was 169%; the increase in female mortality was 72% (Keys *et al.* 1950: 171). Assuming that these results reflect starvation uncomplicated by disease, this differential mortality can be accounted for by the typically greater amount of subcutaneous fat and smaller body size in adult females. Under the Minnesota experimental regime, the initial body weights of volunteers were variable. Hence, food intake was individually adjusted so that the trajectory of weight loss would be the same for everyone, and comparable to the degree of weight loss occurring under severe famine conditions. Volunteers whose initial weight put them in the heaviest group (72.30 kg) lost 17.82% of their body weight; volunteers in the lightest group (67.51 kg) lost the least amount of all groups (15.42%), and had a weight gain increment of 8.82% during rehabilitation – the greatest weight gain after starvation of all groups (Keys *et al.* 1950: Table 36).

What can be inferred from body size in fossil species?

One route to the discovery of the significance of body size in fossil species is to compare individuals of the same age and sex and different species. This eliminates the confounding factors of maturation and sexual dimorphism, but highlights the nature of body size differences between species. To illustrate this procedure, I will use the fossil hominid example of KNM-ER 1808 (female *Homo erectus*, dating to 1.7 mya) versus NME-AL 288–1 (female *Australopithecus afarensis* [the Lucy specimen], dating to 3.1 mya). See Figure 11.6. These two specimens are complete enough to allow a comparison of two fossil individuals. Surviving fossil bones or bony elements are darkened in the figure; outlines of the missing skeletal parts are also traced in. There are obvious

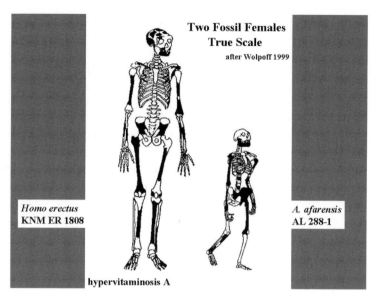

Two Fossil Females
True Scale
after Wolpoff 1999

Homo erectus
KNM ER 1808

A. afarensis
AL 288-1

hypervitaminosis A

Figure 11.6. A comparison of two adult fossil hominid females belonging to different species. The skeletons are drawn in true scale, and show marked differences in stature. Modified from Wolpoff (1999: Fig. 136).

differences in shape and body proportion between these two individuals. Nevertheless, the overwhelming first impression is one of a difference in body size. The earlier hominid, AL 288–1, is much smaller.

The paleontological evidence generally appears to show that the earliest hominids were small, although the late Miocene hominid, *Orrorin tugenensis*, dating to 6 mya, documents a body size in this species like that of a modern adult female chimpanzee – a size that is 1.5 times that of AL 288–1 (Senut *et al.* 2001). The first hominids may therefore have been larger than Lucy was, but australopithecines and early members of genus *Homo* do tend to be remarkably small. An adult female specimen of *Homo habilis* from Olduvai Gorge (OH 62) is estimated to be only a meter tall. P. V. Tobias (1991), in his definitive monograph on *Homo habilis* at Olduvai, frequently and accurately uses colorful phrases like "the Olduvai pygmoids" and the "gracile Olduvai pygmies" to describe this material. If hominid body size were initially small – and most paleontological evidence points to this – predation pressure may also have been an important factor in hominid natural selection at this time. Even if predation pressure were a major factor in the evolution of larger body size, body size could not increase without a significant augmentation of the caloric base. Dietary and other behavioral shifts must have taken place concomitantly with the advent of larger body size. Predation pressure may have been counteracted by hominid behavioral shifts, e.g., changes in vigilance or sociality.

The KNM-ER 1808 specimen is also remarkable for its pathology. KNM-ER 1808 is a diseased individual. The outer surface of all surviving skeletal elements has a peculiar, woven configuration, very unlike the smooth appearance of normal bone. This basket-weave appearance is caused by periosteal inflammation on a system-wide scale. Usually the periosteum surrounding living bone becomes infected as a result of local trauma – a climber inadvertently smashing a shinbone against a rock face, for example. This can result in a nice little patch of diseased bone which can be scrutinized by some future physical anthropologist intent on reconstructing the activity of dead humans from their surviving bones. The most reasonable explanation for the system-wide diseased state of the 1808 skeleton is periosteal inflammation caused by hypervitaminosis A – ingestion of a toxic dose of vitamin A, which affects the entire body (Walker *et al.* 1982). This explanation is reasonable. Sarcoidosis causes a similar type of inflammation, but the diseased patches of bone are very localized and discrete, and tend to occur in the hands or feet or at joint surfaces (Bell 1990). In addition, chronic myopathy or muscle wasting often ensues in sarcoidosis, resulting in loss of bone mass. There is no evidence of either well-defined, localized patches of diseased bone or of demineralization of the skeleton in the 1808 specimen. Thus, hypervitaminosis A remains the likeliest cause of the pathology in 1808.

But how does a 1.7 million year old hominid receive a toxic overdose of vitamin A? The initial suggestion was that this individual had received a poisonous dose of vitamin A through eating a carnivore liver (Walker *et al.* 1982). Skinner (1991) later proposed that the larvae of wild bees were the probable source of the vitamin A, implying that subsistence in this fossil hominid species was essentially like that of non-human primates. These animals often forage for insects and insect larvae, including the larvae of social insects, in their unending quest for food.

A significant increase in hominid body size occurs with the advent of African *Homo erectus*. The taxon is best known from the 1.53 mya specimen KNM-WT 15000, an adolescent male from Nariokotome, in the western Lake Turkana Basin, Kenya. The immaturity of this specimen is indicated by the fact that growth of the long bones is not yet complete; the epiphyses are not yet fused. Rates of enamel formation in its teeth indicate that this individual may be far younger than a comparably sized modern male. In fact, it was perhaps only about 8 years old at the time of its death (Dean *et al.* 2001). This accelerated maturation rate resembles that of earlier hominids and the great apes, and in no way compares to the delayed tempo of maturation only found in modern humans, their fossil representatives, and Neanderthals.

Growth is incomplete in this individual, but, if a growth trajectory similar to that occurring in modern human males took place, final adult stature would have exceeded 180 cm. Arboreal adaptations such as curved digits, long arms, and upwardly directed shoulder joints are lost. Unlike the case in australopithecines, the arms are short and the legs are long – in fact, body proportions are like those of later members of genus *Homo*. What causes the novel morphological changes in this taxon? The larger body size is clearly based on a higher caloric intake. However, if the arguments articulated above hold true, body size by itself may change quickly. Why does a significantly larger hominid with proportions like an australopithecine not evolve? In fact, this must have occurred before the appearance of KNM-WT 15000. Larger body size would evolve first, and might have occurred quickly. One would expect that body shape, proportions, and the loss of arboreal adaptations would change much more slowly than body size. Shape, proportion, and morphology would have changed later, as a shift to habitual terrestriality in open country took place (see Chapter 17).

So profound is the adaptive shift seen in African *Homo erectus* that this species has recently been classified as the first member of genus *Homo* in a major taxonomic revision of the hominid family (Wood & Collard 1999a). The overwhelming evidence leading to this revision in fact contradicts previous taxonomic work by Wood (1991). A principal criterion of the adaptive shift in *Homo erectus* is body size increase. This taxon was the first hominid species to emerge from Africa and to colonize far-flung portions of the Old World at an early date (Gabunia *et al.* 2000, Goren-Inbar *et al.* 2000, O'Sullivan *et al.* 2001). The greater dispersal abilities that are associated with larger body size, as well as other biological properties, ensured this initial, fundamental colonization – not advances in lithic technology or a larger relative brain size (Cachel & Harris 1995, 1996, 1998).

The sweating response, body shape, and heat adaptation

Heat adaptation is another selection pressure affecting body size and shape. In contrast to most mammals, humans have good heat tolerance, but very poor cold tolerance. The lowest ambient temperature at which unclothed, resting humans can maintain their core temperature is about 25 °C (Piantadosi 2003). However, although humans can therefore be described as tropical mammals, human body temperature must be maintained between 30 and 41 °C. Muscles generate heat when they contract, and human body temperature can reach 40 °C during prolonged heavy exercise; however, 50% mortality occurs when body temperature is elevated to 42.5 °C (Piantadosi 2003: 65–67). Thus, high external temperatures limit the activities of modern humans in tropical habitats.

Heat imposes appalling constraints on human physiology (Collins 1996). The human sweating response is the principal adaptation to ambient heat. Sweat evaporates from hairless skin over most of the body surface. This physiologically crucial adaptation is seen in all modern humans. In fact, it may be considered a species-specific trait, because it is based on the presence of eccrine sweat glands. The abundance and density of these sweat glands on the body surface is unique to humans among the mammals – they are mainly restricted to the pads of the paws and adjoining regions in other mammals. Eccrine sweat glands do not produce sebum or other fatty secretions that are associated with the scent and scent-marking that is typical for mammals. Instead, eccrine glands produce copious watery secretions spiked with salt and elements like potassium and calcium. These secretions lower the temperature of the human body surface through evaporative cooling. A normal sweating rate is 0.5–1 liter/hour, but this can be increased to 2 or sometimes 3 liters/hour in a human acclimatized to hot climate. These maximum rates usually cannot operate for more than an hour, and are based on the sweat glands being trained for increased secretion (Collins 1996). Water lost through sweating must be replenished by drinking plentiful amounts of fresh water. Sweating humans quickly develop a water deficit that can become fatal within a day. Death is caused by shock created by loss of blood volume or by heat stroke. Strangely enough, it is impossible for individuals to gauge their water shortfall directly through thirst. Thirst becomes marked only when water loss is already severe. Consequently, it is possible for humans to function in a condition of negative water balance without realizing that they are in danger until the symptoms suddenly become very severe.[8] Salts lost through sweating create much less of a survival problem.

Thus, modern humans demonstrate a number of adaptive responses to hot climate that are independent of any behavioral or technological responses that they may invent or adopt from their cultural repertoire. These biological or physiological responses include structural changes within the skin that are responsible for the unique human sweating response, as well as the general hairlessness of the skin surface. These responses indicate a fundamental and universal adaptation to heat, and point to the origin of humans within a tropical environment. Of course, one can know nothing about sweat production in fossil humans. Yet, it must be significant that human skeletal remains and archeological materials are unfailingly discovered near permanent fresh water sources until about 1.6 mya, when, for the first time, archeological traces occur near ephemeral streams.

Another adaptation to high ambient temperature relates to body weight and shape itself. Body build in fossil species can be reconstructed from a relatively complete individual specimen, or from a composite of skeletal parts from

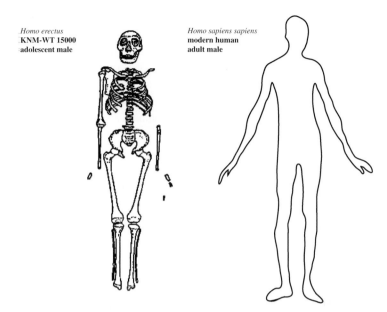

Figure 11.7. Convergent adaptation in body shape to hot, arid climate. A comparison between KNM-WT 15000, an adolescent *Homo erectus* specimen dating to 1.53 mya, and a modern adult male Dinka from the same latitude as the fossil. Growth is not complete in the fossil. In both individuals, the exaggerated ectomorphy illustrates the operation of Allen's rule. Both have a narrow bi-iliac breadth. The skeleton is from Pearson (2000: Fig. 5); the body outline is from a photo in Ruff & Walker (1993: Fig. 11.5).

individuals of the same species. The adaptive responses of human body weight or mass and body shape to extremes of temperature have been known for a long time, and are embodied in Bergmann's and Allen's rules (see Size and Shape Changes: Adaptation and Plasticity section). A small body needs less food. Because the basal metabolic rate rises when food is digested, a small body – irrespective of its shape and proportions – can function well under conditions of high temperature. Less food is eaten, less is digested, and the transient rise in metabolic rate is limited.

Analysis of the body shape of KNM-WT 15000 shows that it possesses the exaggerated ectomorphy or linearity of body build that is associated with adaptation to hot, arid climates in modern humans (Ruff & Walker 1993). Figure 11.7 illustrates this specimen, juxtaposed next to an outline of the shape of a modern human who shows an extreme adaptation in body shape to heat and aridity. This is an adult male from the Dinka community of the southern Sudan, located at the same latitude as the Nariokotome site where

Figure 11.8. Dassenetch men from Ileret, East Turkana Basin, Kenya. Note the linearity of body form (ectomorphy). Living humans with this body build are adapted to extreme heat and aridity. Photo courtesy of Purity Kiura.

KNM-WT 15000 was found. This Dinka man, who demonstrates the operation of Allen's rule, has trunk and limb proportions that nearly match those of the KNM-WT 15000 specimen (Ruff & Walker 1993: Fig. 11.7). Figure 11.8 shows this body build in two men now living in the East Turkana Basin. Ruff (1994) identified a pelvic measurement in modern humans, bi-iliac breadth, which is a proxy for body width. A narrow bi-iliac breadth is found in modern humans with climatic adaptation to high ambient temperatures. Both the Nariokotome fossil specimen and the living Dinka human have a narrow bi-iliac breadth. Because this is also the case for the two species of australopithecines with remains complete enough to yield this measurement (*Australopithecus afarensis* and *Australopithecus africanus*), Ruff (1994) infers that heat adaptation is an ancient hominid condition.

Leaving the tropics: adaptations for increasing seasonality

A significant change in Pleistocene climate occurred between 1.6 and 1.5 mya, at approximately the date of the Nariokotome specimen. The African climate became significantly more arid at about 1.6 mya (de Menocal 1995, de Menocal & Bloemendal 1995). At 1.5 mya, a major shift in global climatic variability took place – the 100 000 year cycle of continental ice-sheet oscillations began – and tropical terrestrial environments became highly perturbed (Rutherford & D'Hondt 2000). This rising climatic perturbation and mounting seasonality is probably associated with the origins of the taxon *Homo erectus*, and its dispersal out of sub-Saharan Africa (Cachel & Harris 1995, 1996, 1998). If members of this taxon had ectomorphic body proportions like the KNM-WT 15000 specimen, they would have been adapted to disperse through hot, arid open-country environments. Because body size in this taxon is also larger than in earlier hominid species, these larger hominids would also have been concomitantly adapted to colder habitats, given the operation of Bergmann's rule. A larger body retains more heat in the body core. Larger body size might facilitate the dispersal of hominids into colder, higher latitudes outside of the tropics. In fact, *Homo erectus* dispersed from sub-Saharan Africa soon after its first appearance in East Africa (Antón 2003). The Koobi Fora fossil KNM-ER 3733 is dated to 1.8 mya, and is currently the oldest firmly dated specimen of this taxon. Shortly afterwards, *Homo erectus* is found at the site of Dmanisi in the Republic of Georgia at almost 1.8 mya.

Altitude also affects ambient temperature. Higher elevations are colder, even at the equator, and hominids with a larger body size could more easily disperse into colder, higher elevations. The archeological site of Gadeb in the Ethiopian highlands, for example, was occupied at 1.4 mya. Non-human primates also demonstrate size changes with respect to the cold of high altitudes. Sexual size dimorphism in Indian Hanuman langurs is reduced at high altitudes, and much more pronounced on the hot Indian plains (Hrdy 1981). Discovered in 2003, the Arunachal macaque (*Macaca munzala*) occupies the highest altitude of any non-human primate species. This macaque species ranges between 1600 and 3500 meters above sea level. Its large body size allows it to disperse into the mountainous areas of northeastern India.

Seasonality creates problems in adaptation, even for modern humans. Modern hunter-gatherers can experience seasonal changes in energy intake of up to 50%, and many hunter-gatherer groups suffer from seasonal and long-term starvation (Foley 1999). Human biology has responded to this adaptive challenge. Not only do modern humans exhibit a species-specific sweating response, but they also exhibit an equally species-specific ability to accumulate a large amount of subcutaneous fat under the skin, that is aggregated in a number of depots throughout the body. This is not the case for other primates, who have minimal amounts of body fat in the wild. In fact, humans are unique among land mammals, because other mammals accumulate fat only before mating, migrating, or hibernating (Pond 1978). Human infants are born with a strikingly high amount of body fat. This amount is much larger than that of other mammals of equivalent body size, and nearly four times larger than the percentage of body fat in newborn baboons (Kuzawa 1998). When gestational age and body length are accounted for, there is a significant positive correlation in humans between neonatal Body Mass Index (BMI) and head circumference (Correia *et al.* 2004). Head circumference is related to brain size and cognitive function in human children and adults. Hence, a correlation between neonatal body fat and head circumference may indicate that larger fat depots in human newborns are associated with selection pressure to create fat reserves that will support the massive energy demands of an enlarged and growing brain, even if food supplied by adult caretakers is temporarily interrupted (Correia *et al.* 2004).

Not only do humans maintain large amounts of body fat, but there are also uniquely human male and female differences in the amount and distribution of this fatty tissue (Pond 1992a, 1992b). See Figure 11.2. The amount of body fat that humans maintain has apparently nothing to do with thermal insulation in cold climate or the demands of maintaining a pregnancy, although fat loss can be significant during lactation (Pond 1977). Human sexual differences in fat distribution may evolve through sexual selection (Pond 1992b). Reproductive potential in human females is linked to subcutaneous fat. Human females with large breasts and narrow waist to hip ratios have a higher probability of conceiving, as determined by daily levels of the hormone 17-β-estradiol. Human females with this body shape have levels of this hormone that are 37% higher at ovulation than females with other body shapes (Jasieńska *et al.* 2004b). Human females with low body fat either never attain menarche, or lose their menstrual periods. Normal humans of both sexes have 3–30 times more adipocytes than wild mammal species of the same body size. But polar bears and reindeer that naturally become obese and then undergo fasting only have 2–4 times more adipocytes than similarly sized species (Pond 1992b). This illustrates the peculiar human condition. How do these figures translate into human body composition? In human hunter-gatherer groups, physically active and lean individuals have subcutaneous body fat measured by skinfold thickness ranging from 5 to 15% in males and 20 to 25% in females (Jenike 2001). Note the sexual dimorphism, with the average female carrying 15% more subcutaneous fat. Among female Hadzabe hunter-gatherers, the Body Mass Index (BMI) and percentage of body fat are both good predictors of fertility (number of children born) and reproductive success (number of children surviving), but stature does not predict either of these variables (Marlowe 2004). Energy status, reflected in weight, BMI, and percentage of body fat, has a complex interaction with activity and energy expenditure. For example, high activity levels among rural Polish females suppresses ovarian function even though weight, BMI, and percentage of body fat are higher than in urban females (Jasieńska *et al.* 2004a). Moderate to heavy physical labor among rural females appears to account for this difference.

Maintaining a large amount of body fat from birth through adulthood thus seems to be a species-specific characteristic of humans. It must be an adaptation to survive fasting and near-starvation conditions. However, these conditions must have been episodic and relatively unpredictable. Feast or famine must have generally prevailed. Periods of fasting must have occurred during seasons that were very broadly or poorly defined during the annual cycle. Otherwise, like other mammals and vertebrates, humans would accumulate body fat only during a sharply defined period. Body fat depots explain how humans can easily survive periods of food reduction and "normal" starvation, such as occurred in Britain during World War II, which resulted in only an average reduction of 10% in body weight (Keys *et al.* 1950).

The evolution of body size in primates

The genetic basis of body weight is being intensively studied in modern humans as part of a broad research front exploring the genetics of human obesity (Bouchard 1994). Laboratory rodents, such as the *obese* strain of mice, have been developed to investigate the contribution of single mutant genes to obesity. Given the amazing conservatism of genetic material, such a discovery in any mammal species probably has significance throughout all living mammals. Now that the human genome has been sequenced, and found to contain only a modest 30 000 genes (The International Human Genome Mapping Consortium 2001, Venter *et al.* 2001), comparative non-human mammal research is likely to become even more significant. Human monozygotic twins are identical in their genetic makeup. If they are separated at birth and then reared apart by different families, they demonstrate that the human body mass index has a heritability of 70% (Meyer & Stunkard 1994). The body mass index (BMI), used to define obesity, is a measure of human body weight (in kilograms) for any height (in meters). This measure is given by the following equation: BMI $= kg/m^2$. Investigation of the genetic bases for human body weight has currently yielded a linkage on the long arm of chromosome 11 between the body mass index and the percentage of body fat in Pima Indians (Barsh *et al.* 2000). In human children, a high BMI may indicate either fatness or a high degree of muscularity associated with a mesomorphic body build. Only later in life does a high BMI become associated with obesity (Bielicki *et al.* 2001). Definitions of obesity are made according to sex, because the BMIs of males and females tend to differ. Aging also affects males and females differently. While both sexes decline in height after the age of 40, males begin to lose weight after 45 years of age; females of this age maintain or increase their body weight until about the age of 65 (Frisancho 1993: Figure 18.1). A recent longitudinal study of nearly 33 000 ethnically homogenous Polish adults in an urban population documented increasing BMI with age in both sexes. In males, however, a marked deceleration in BMI rise takes place after age 40; the female BMI increases with age, but is dramatically lower among females in higher social classes (Bielicki *et al.* 2001).

Like humans, well-studied non-human catarrhine primates also have a high heritability component to adult body mass. Colony members from five captive baboon species (*Papio hamadryas, Papio cynocephalus, Papio papio, Papio anubis*, and *Papio ursinus*) and inter-species hybrid individuals demonstrate a 51% genetic component for mean adult body mass; the ability to maintain body mass has a 12% heritability (Jaquish *et al.* 1997). Forty-four percent of the genetic variance in these two traits may be caused by shared genes. Significant differences in body mass and ability to maintain body mass exist between these baboon species. Females are 10.2 kg lower in body weight than males, and appear to be less shielded than males from weight perturbations. Females lose weight as they age, and have a more difficult time maintaining a stable adult body weight with increasing age. This is true even though the animals are in a colony, and are being provisioned and carefully monitored. Note that this is contrary to the human condition, as documented above. Unlike baboons, the human female BMI rises more steadily with age than it does in males, and females seem more buffered from weight loss. Again, this illustrates a species-specific human trait involving body fat. The peculiarity of this trait is heightened when humans are contrasted with close catarrhine relatives who otherwise very much resemble humans in physiology. Human subcutaneous fat is a bulwark against food scarcity and body weight fluctuations.

The ability of female baboons in colonies to maintain their body weight was independent of their individual reproductive history (Jaquish *et al.* 1997). However, females with more offspring tended to have larger body mass; this was not true for males (Jaquish *et al.* 1997). The significant positive correlation of number of offspring with female body mass demonstrates that body size in these five catarrhine primate species is an important variable affecting female reproductive success. For males in these species, body mass has no effect on reproductive success.

Results from these colony baboon species may not be a general phenomenon in mammals. The published literature on humans does not report a link between female body size and reproductive success – at least, I have not been able to document it. Increases in body fat certainly trigger menarche in humans and baboons (Altmann 1991). It is not clear whether this affects human lifetime reproductive success, although it appears to do so for baboons. But the effect of body size on reproductive success in human females has not been substantiated. The link between larger body size and reproductive success may not occur in other mammals, either. Large mothers do tend to produce large offspring, whether plants, invertebrates, or vertebrate animals are being investigated (Peters 1983, Sakai & Harada 2001). However, larger neonatal size in mammals may not translate into greater reproductive success. Neonates of large mammals must achieve more growth than neonates of smaller species. Among mammals, the growth rates of primates and ungulates rise especially slowly with respect to adult body mass (Peters 1983: Appendix VIIId). Larger neonates may be at a selective disadvantage, unless they receive greater parental investment, or are relatively immune from predators by virtue of an absolutely larger body size.[9] At weaning, the young of large mammals also must grow longer than the young of smaller species. Hence, for a variety of factors, larger neonate size may not guarantee reproductive success in mammals. For example, an increased body size in female individuals of two ungulate species creates a greater probability of surviving to old age, but has no necessary impact on an individual female's reproductive success (Gaillard *et al.* 2000).

It is possible that the baboon results linking larger female body mass with more offspring (Jaquish *et al.* 1997) demonstrate the impact of female–female food competition played out within a social group. These were colony animals, not subject to predation or intra-specific infanticide. Bigger size may be better for individual fitness in the competitive arena of a social group where females compete with other females for access to food even under captive conditions, and possibly compete for other resources that affect reproductive success. Food is a good first proxy for all of these other variables. Under these circumstances, with females nursing and caring for offspring with virtually no help from other adults, whether kin or non-kin, it is entirely possible that social groups themselves can exercise a profound disadvantage to female reproductive success. Differences in body size have notable genetic components in these catarrhine primate species – these differences have an effect on female reproductive success even under captive conditions, where diet, moving and ranging, and environmental variables such as temperature and light remain the same for all animals.

Conclusions

There are a number of consequences that are generally associated with an increase in body size across the animal world. These consequences also affect human and non-human primates. The following eight factors accompany an increase in body size. (1) A higher or more predictable caloric base is needed to support increased energy expenditure. The total energy of expenditure rises with an increase in body size. This rise may be partly offset by a decrease in basal metabolic rate or changes in the digestive system (e.g., the appearance of a ruminant stomach or symbiotic bacteria capable of digesting cellulose), and the subsequent incorporation of lower quality foods into the diet. (2) There is less danger from predation. (3) Large body size increases parasite load, from both ectoparasites and endoparasites. Large animals are parasitized more simply because they are big – which causes more exposure to ectoparasites – and they consume more food – which causes more exposure to endoparasites. (4) Inter-specific competition may be lessened, because niche boundaries are re-defined. Furthermore, among closely related species, fewer species exist at the high range of body size, in comparison to the middle range of body size. (5) There may be greater intra-specific competition, if sexual dimorphism increases with an increase in body size (i.e., Rensch's rule is operating). (6) Larger body size brings increased longevity, all else being equal. Sociality may be impacted by this – e.g., through an increase in life expectancy alone, matrilines with generational overlap can occur. (7) Dispersal abilities are greater in larger animals. Greater dispersal abilities may reduce both intra- and inter-specific competition, if novel environments are penetrated. Ecological release follows. Ecological release occurs as the result of the penetration or colonization of new environments in which no closely related competitors exist. (8) Larger animals have larger home ranges.

Note that the greater dispersal abilities and greater home range size of larger animals are advantageous in unpredictable environments. Yet conditions of unpredictable or scarce resources tend to decrease body size. In fact, larger body size necessitates a better dietary base. How may a large animal achieve a better diet in unpredictable environments? A dietary shift is probably the most parsimonious explanation for richer nutrition and a better diet. Thus, the sequence of events runs from initial dietary shift to body size increase to greater dispersal abilities. Plio-Pleistocene paleontology and archeology support this sequence in hominids, as explicated in Chapters 16 and 17.

Size has a fundamental impact on the life of animals. When the niches of living animals are examined, size is always one of the major variables determining niche structure. Size is one of the three or four niche variables that can be consistently examined for extinct species. Researchers always attempt to reconstruct size when the lifeways of fossil animals are inferred. This can be done through a variety of means. Size plays a role not only in the lifeway of a single extinct species, but can be used to examine extinct ecosystems, as well. Judgments about the competitive interaction of fossil species may also be made based on size. Furthermore, inferences about the social organization and mating systems of fossil species have been made by examining the degree of sexual dimorphism that exists.

Body size and shape and the degree of sexual dimorphism are variable in living humans. General summaries are usually made in terms of a global perspective, but differences between different human populations can be very pronounced. A long history of anthropological research supports the idea that extremes of temperature and diet influence human size and shape. The genetic bases for human size and shape are currently unknown. They must be complex and polygenic. However, like other mammals, humans quickly respond to dietary changes. Like other catarrhine primates, food appears to be the major limiting factor on human population increase. Food (or nutritional adequacy) affects human sexual dimorphism, as well. It is possible that the reduction in sexual dimorphism since the time of the australopithecines that paleoanthropologists traditionally link to changes in social structure and mating systems may have been affected by nutritional inadequacy, rather than diminishing male–male competition, for example.

With dietary change, human body weight and stature can change rapidly, from one generation to the next. Males apparently are more variable than females, whose phenotypes appear to be more safeguarded or defended from environmental change. Human size and shape also respond to climatic extremes; this is also documented in other endothermic animals that have a wide geographic distribution. Fossil hominid species also show changes in body shape, as they adapt to environmental extremes. Phenotypic plasticity and quick adaptive response is also documented in other mammals.

Body size has increased independently in many animal lineages, which implies that there are selective advantages to being large. Larger body size increases reproductive success only in females among a number of captive baboon species. Males were not affected. This phenomenon may not be true for other mammal species. Larger body size does not affect female reproductive success in two species of ungulate mammals that were investigated.

When humans are compared to other mammals, and especially their fellow catarrhine primates, two species-specific traits are highlighted. Humans possess a unique sweating response, based on the ability to generate great amounts of sweat from numerous, dense eccrine sweat glands. The evaporation of this sweat on a hairless skin surface cools the body – an adaptive response to high temperature that mandates access to abundant fresh drinking water. Humans consistently maintain a large amount of body fat under normal circumstances. In this feature, they are unique among terrestrial mammals, and very different from their fellow catarrhine primates. Sexually dimorphic patterning of body fat is also peculiar to humans. It is possible that the large depots of human body fat evolved as an adaptive response to frequent periods of starvation or near starvation in the human evolutionary past.

Endnotes

1. John Robinson argued that coexistence of robust australopithecines and genus *Homo* was possible because of pronounced dietary differences. He assumed that sympatric species could occur only with marked niche differences. This assumption has been negated by field biology evidence, which frequently discovers that minor quantitative differences separate sympatric species. Robinson argued that robust australopithecines were extremely specialized herbivores, but genus *Homo* was an omnivore, and was already eating vertebrate meat.

2. Two points should be kept in mind. First, classifying species as generalized or specialized should probably be done only within the context of a local community, because problems of species abundance, diversity, interaction, and energy flow should impact all of the species. One would expect a continuum of species along the generalized–specialized gradient, with local community variables affecting the continuum.

   Second, the theoretical expectation of evolutionary biologists is that generalized species should be out-competed by specialists in constant or increasingly more variable environments (Gilchrist 1995, Whitlock 1996). This is contrary to Potts's arguments. The fact that omnivores – animals that are generalized in diet – are significantly rarer than expected in computer simulations and even more rare in real food webs (Rosenzweig 1995: 81–82) supports the lackluster performance of generalized species in the competitive arena. This competitive failure of generalized species is the opposite of received anthropological and primatological wisdom since the nineteenth century. It was given a major impetus by the publication of Ernst Mayr's (1950) arguments about the uniquely broad and flexible nature of the hominid niche. See Chapter 17 for an extended presentation of this material.

3. The Online Mendelian Inheritance in Man (OMIM) database lists this disorder as microcephalic osteodysplastic primordial dwarfism, type II, or MOPD II. The URL of the OMIM catalogue is http://www.ncbi.nlm.nih.gov.

4. For example, common American desserts are perceived as being nauseatingly sweet, almost inedible, to Samburu pastoralists in Central Kenya who are naïve to such dietary excess.

5. The first Dutch settlers in South Africa identified three groups of indigenous people: "Bushmen" who were hunter-gatherers, "Strandlopers," who collected

shellfish and other marine resources, and "Hottentots," who were pastoralists herding sheep, goats, and cattle. These people are now collectively referred to as Khoisan people. The Strandlopers are now extinct as a cultural entity. Through the centuries, there has been substantial gene flow between the Khoisan, Bantu, and Dutch and other European populations. Nevertheless, Khoisan people remain among the smallest of African populations, although here, as elsewhere, there has been a twentieth century secular trend for stature increase.

6. Ancel Keys, who died in 2004 at the age of 99, was the chief Minnesota researcher commissioned by the U.S. government to conduct this experimental study of human starvation. Keys also advised the Department of Defense on nutritional deficiency, and invented portable nutritional meals for World War II combat soldiers – the famous emergency "K rations".

7. Dr. Hans Girgensohn, the Sixth Army pathologist, autopsied German soldiers whose deaths were unexplained. He suggested malnutrition as the cause of death, and speculated about how the effects of starvation were accelerated by cold and exhaustion (Beevor 1999: 304–306).

8. This occurs repeatedly among students enrolled in the Koobi Fora Field School. Most of the fieldwork takes place in the semi-desert areas east of Lake Turkana, in northern Kenya.

9. On the other hand, one would expect that natural selection would shorten life history stages subject to high mortality. The adaptive response is for growth to speed up, so that less time is spent in a dangerous life stage.

12   *The nature of the fossil record*

Does the fossil record faithfully record past events?

Diversity and punctuational events

How can the fossil record best be studied? Any major trends or patterns from the past can be discerned in detail only from a fossil record that is relatively complete and well-dated. This necessitates that the stratigraphic sequence incorporating fossils needs to be studied well enough to provide relative age. Many chronometric dating techniques that yield "absolute" age also exist (Figures 12.1 and 12.2). However, the terrestrial fossil record tends to be much more incomplete than the marine record, and data need to be abundant before any patterns can be discerned. This explains why paleontologists rely so heavily on marine invertebrates when limning the grand patterns of life. When the dense and well-dated record of marine invertebrates is examined through the Phanerozoic, diversity appears to increase through time, although major extinction events intervene (Sepkoski 1993). Investigation of this best of all fossil records during the Mesozoic and Cenozoic also demonstrates that higher taxa, represented by novel morphology, tend to originate in onshore environments (Jablonski & Bottjer 1991).

The actual physical determinants of the fossil record – its development through irregular bursts of sediment deposition – influence the reconstruction of ancient life and communities (Figure 12.3). The quantity of fossils discovered is directly proportional to the volume of fossil-bearing sediments. That is, the more one excavates in fossiliferous areas, the more fossils one is likely to find (Smith 2003). When marine mollusc diversity is examined over the last 60 my, diversity appears irregular. It simply correlates with the amount of surface outcrops yielding the fossils (Crampton *et al.* 2003). Peaks and troughs of diversity are artifacts, determined purely by sampling. Diversity peaks occur where the record has been well sampled, and thus diversity is over-estimated in this portion of the fossil record. The direct relationship between rock volume and numbers of fossils seriously distorts estimates of ancient biodiversity, and therefore warps ancient community reconstructions.

Figure 12.1. Badlands topography at the Reva Gap, South Dakota, U.S.A. The absence of vegetation reveals the stark sequence of vertical exposures of the Brule Formation, which contains abundant middle Oligocene mammal fossils.

Figure 12.2. Dr. J. W. K. Harris at the type site of the volcanic KBS tuff, Koobi Fora region, northern Kenya. The tuff is dated to 1.8 mya. An absolute date for the tuff can be derived through use of the potassium/argon dating technique. An ancient volcano erupting ash lay in the distant Ethiopian highlands, and the ash was blown by winds to northern Kenya. The tuff is an ancient fluvial channel fill, in which the airborne ash is mixed with silt and sand. The tuff is the light, flakey stratum directly in back of Dr. Harris.

Figure 12.3. The discontinuous nature of sedimentation. Cyclical sequences of fluvial sedimentation at Ileret, northern Kenya. The scale has five centimeter increments.

Another effect that is a signature of differential sampling has been labeled "the pull of the recent." Because the paleontological record is easier to explore in recent time ranges, and is also less likely to have been removed by erosion or the warping or compaction of strata, a greater abundance of fossils is expected in these time ranges – this is the "pull of the recent." This effect makes intuitive sense, but it has only rarely been tested. A recent examination of diversity in bivalve molluscs reveals that only 5% of the Cenozoic increase in diversity can be attributed to the "pull of the recent" (Jablonski *et al.* 2003). The increase is therefore a real biological signal, and not an artifact of over-sampling of recent time ranges.

Some organisms have a dramatically impoverished fossil record. In these cases, other lines of evidence are used to examine phylogeny or evolutionary history. Birds are an example. One of the first major cases of DNA–DNA hybridization to infer phylogeny was applied to birds, because of the paucity of their fossil record, and intense interest focused on their evolutionary history. Sometimes, however, a fossil record is known for a group, but the pattern of the record conflicts with theoretical expectations. This is the case for the primate fossil record, which shows a pattern of successive radiations and extinctions through the Cenozoic. R. D. Martin, a prominent researcher in primate evolution, is disturbed by the evidence of primate extinctions, and has produced corrections to the primate fossil record, based on the proportion of living primates whose undoubted ancestral group appears in the Plio-Pleistocene (Martin 1986b, 1990). When the proportion of expected but "missing" fossils is added to the primate fossil record, and plesiadapiform primates are removed from the order, the pattern of primate evolution then becomes an inverted cone of increasing diversity through time. There are then no major reductions in primate diversity or extinction events. Martin and his colleagues have recently used a similar diversity correction when inferring the time of primate origins (Tavaré *et al.* 2002).

Fossils are preserved only in sediment traps, areas where sediment deposition occurs, and where sediments remain intact without erosion or geological deformation. Interruptions in the fossil record are caused by the episodic nature of sediment deposition. Knowledge of earlier periods is inherently limited, and determined by the nature of the record. Breaks or pauses in deposition create a false picture of disruption in the fossil record. Only deep sea sediments preserve a relatively intact record, although this record is found only in the modern ocean basins. Fossils preserved in ancient oceans have long been destroyed by the subduction of oceanic lithosphere as ocean basins approach continental margins. The incompleteness of the fossil record and its dependence on bursts of sedimentation always creates the appearance of abrupt events. This is the natural state of the fossil record, because the geological sequence is never complete and is long only in rare and isolated cases. Furthermore, since the time of Georges Cuvier in the early nineteenth century, the fossil record has been known to show major breaks or catastrophes. Major revolutions had affected the face of the earth. Unpredictable global catastrophes had affected the history of life. These are called the "Big Five" mass extinctions, and different causes have been adduced for each. The last of these, and the most well known, was the Cretaceous/Tertiary (K/T) mass extinction. It was caused by a combination of factors, the most violent of which was the collision of an asteroid 10 km in diameter crashing into the shallow ocean waters at Chicxulub, just off the Yucatan Peninsula of North America (Chapter 4).

The question of periodicity or cyclicity in major extinction events was raised during the 1980s (Raup & Sepkoski 1984, 1986, 1988, Sepkoski & Raup 1986, Raup 1986). Examination of the relatively complete and dense marine invertebrate fossil record appeared to show a 26 my periodicity to major extinctions. An extraterrestrial source for the periodicity was hypothesized. A close pass to the solar system's remote Oort Cloud every 26 my by an unknown dark, close companion of the sun (called Nemesis) supposedly shook comets loose from the Oort Cloud. These comets then bombarded inner members of the solar system (Davis *et al.* 1984). Many paleontologists (e.g., Hoffman 1985) were convinced that the apparent cyclicity was caused by the interaction of taxonomic problems, the measurement of geological time, and the episodic creation of the geologic record. Detailed investigation of the marine invertebrate fossil record ultimately revealed that the 26 my periodicity was actually a statistical artifact (Stigler & Wagner 1987, 1988). Nevertheless, fallout from the Nemesis idea still affects the evaluation of accident or contingency in evolution (Raup 1991). At its most extreme, the contemplation of the likelihood of asteroid impacts or extraterrestrial factors affecting life on earth can lead to the idea that natural selection or adaptation are trivial factors in organic evolution, because the rules for life are constantly being rewritten by mega-catastrophes (Hsü 1986).

Punctuational events in the fossil record have been the focus of attention since Eldredge & Gould (1972) argued that virtually all evolutionary change involves disruption. They proposed that new species arise abruptly, with a sudden burst of morphological change. New species tend to emerge in small, peripheral populations, subject to accident and genetic drift. The normal mode of speciation occurs in this fashion, with morphological change limited strictly to species origins. Morphological stasis or equilibrium then characterizes the journey of a species through geological time until the inevitable point of extinction. New species then emerge in a punctuational burst of morphological novelty. Developmental constraints buffer the phenotype from change through the eons, and the phenotype is shielded from natural selection, because natural selection always tends to be a weak and ineffectual evolutionary process (Gould 1980). This theory, which is called punctuated equilibria, was proposed as an alternative to natural selection and adaptation. It was characterized as a revelation from the fossil record itself. This was evolution as documented from the fossil record, from the great procession of life, and not from mere laboratory experiments with *Drosophila* flies or abstract mathematical modeling. Eldredge & Gould (1972, Gould 1980) were contemptuous of systematists and geneticists who contributed to the New Synthesis of evolution in the 1930s and 1940s. As paleontologists, they claimed that paleontology had been slighted in the creation of the New Synthesis. If it were otherwise, punctuated equilibria would rule the day, and a new theory of evolution would arise (Gould 1980). However, paleontologists like George Gaylord Simpson, Bryan Patterson, and Glenn Jepsen had actually been present at the birth of the New Synthesis, and were not simply meek attendees. Furthermore, as detailed below, paleontologists had long been aware of different rates of evolutionary change, and did not conclude that only slow, gradual rates were possible – the "phyletic gradualism" set up as a straw man by Eldredge & Gould.

Punctuated equilibria has continued to fascinate paleoanthropologists and archeologists. Its influence has been pervasive in evolutionary anthropology (Cachel 1992). Often it is the only framework used to examine the fossil human or archeological records in major textbooks. Yet, as outlined in Chapter 5, punctuated equilibria has had a deleterious effect on the study of natural selection and adaptation. Recent work on "contemporary evolution" (Chapter 17 and below) has silenced critics who claimed that natural selection was too weak and ineffectual to be the powerhouse of evolutionary change. The impact of punctuated equilibria theory on the general study of evolutionary processes has been recently assessed by Michael Ruse, an historian of science (Ruse 1999). Rather than succumbing to rhetoric, he examined the rate of citations and the positive or negative response to punctuated equilibria theory published in the journals *Paleobiology* and *Evolution*. These are the major journals used by professionals in the field of evolutionary biology. Despite the trumpeted novelty and importance of punctuated equilibria theory in the popular press, evolutionary researchers have found it neither novel nor important. Given the dominance of punctuated equilibria theory in anthropology textbooks, it is ironic that the concept has never been used much by researchers in evolutionary theory. It has been cited by them at a diminishing pace since the middle 1980s (Ruse 1999: 151).

Divergence times; first and last appearances

The time of origin of groups of organisms has traditionally been established through the fossil record. The first appearance of a group in the fossil record is taken as its time of origin, although it is assumed that there has been some lag-time, because it is statistically improbable that the fossil record should capture the remains of the actual first members of a group. Beginning in the 1960s, the widespread use of starch gel electrophoresis began to document protein variability in living organisms, and this became a means of investigating variability at a molecular level, in addition to more traditional means of investigation.

With the documentation of an overwhelming and unsuspected amount of molecular variation in captive and in wild organisms, Kimura (1983) introduced the idea of neutral mutations. That is, the vast majority of point mutations that cause nucleotide substitutions that lead to amino acid changes and molecular variations are neutral with respect to natural selection. Allele frequencies reflect only mutation rate and random genetic drift operating on populations. Because these mutations are neutral, and are not subject to positive or negative selection pressures, they simply accumulate over time. Molecular evolution is thus, by and large, neutral. A corollary to this idea is that, since mutations are random, the number of mutations that occur in different groups should reflect the passage of time since the groups diverged. An assumption is made that mutation rates are constant. If mutation rates are constant, then the number of mutational differences between living groups of organisms establish the time since separation or a phylogeny. These phylogenies are established by a "molecular clock," that is assumed to run on time with no spurts or hesitations. The mutation rate for a particular protein is set with respect to the last common ancestor, based on the fossil record – e.g., the divergence of cartilaginous fishes from other vertebrates. However, different proteins have significantly different rates of change.

This leads to the problem of deciding which protein to select for analysis. The problem is generally resolved by using data from many proteins, because the different mutational rates average out (Feng *et al.* 1997). Direct sequencing of genes has now replaced DNA-DNA hybridization studies. However, there is controversy over whether DNA phylogenies should be constructed with the maximum number of genes or only with genes that have a robust phylogenetic signal (Benton & Ayala 2003). Phylogenies using many proteins or multiple genes are concordant with phylogenies based on multiple anatomical areas (Chapter 5).

There has been a consistent mismatch between divergence times recorded from the fossil record and divergence times inferred from molecular and genetic data. Fossils tend to yield dates that are too young, an artifact caused by missing fossils of the first, ancestral species of a group. On the other hand, molecular and genetic dates tend to be too old, which may be caused by the statistical methods used to produce a phylogeny. Recent molecular phylogenies are becoming more congruent with phylogenies produced from the fossil record (Benton & Ayala 2003). Although molecular clocks placed the divergence times of many placental mammal groups deep into the Cretaceous, analysis of preservational bias indicates that most modern placental orders arose close to their first appearance in the fossil record. Very early divergence times mandate universally very poor preservation across orders, and lower rates of speciation and extinction than those that have been actually measured (Foote *et al.* 1999).

The shape and pattern of the fossil history of any group is determined by originations and extinctions, and the differential rates between these two processes. One of the most fundamental pieces of information to be obtained about an extinct species or genus is its first and last appearance in the fossil record. This information yields two data points: the First Appearance Datum (FAD) and the Last Appearance Datum (LAD). Much paleontological research time is spent establishing these two data points in geological time. Inferences about ancient biodiversity, standing richness, community stasis, radiations, mass extinctions, and evolutionary pulses following triggering climatic events all follow from analysis of these two data points.

Paleontologists who study extinction events are aware that species with small populations are unlikely to be found in the fossil record. Even species that eventually become abundant and widespread may have miniscule populations at the beginning or end of their evolutionary span. The paleontological record consequently contains false extinctions that are created simply by sampling bias. Taxa considered extinct can suddenly reappear in the fossil record. They are termed Lazarus taxa, because they emerge from the fossil record just as the once-dead Lazarus emerged from the tomb. The two living coelacanth species are classic examples of Lazarus taxa, because coelacanths were presumed to have been extinct for 64 my until Marjorie Courtenay-Latimer unexpectedly discovered one that had been captured alive off the coast of South Africa in 1938. The Chacoan peccary (*Catagonus wagneri*) is another Lazarus taxon. It was thought to have been a victim of Pleistocene extinctions, but was nevertheless discovered alive in the Grand Chaco of Paraguay and Argentina (Wetzel *et al.* 1975).

The span of a species' existence is determined by its first and last appearance. However, the Signor-Lipps effect recognizes that the first and last appearance of a species in the fossil record may not represent the true time of its origin or extinction (Signor & Lipps 1982). The fossil record is normally very incomplete, and thus the true date of extinction will lag behind the Last Appearance Datum. Species that have been in decline have very rare members, and these members become increasingly more rare prior to the extinction point. Yet, it is possible to estimate the extinction date of species that have recently vanished or of fossil species. A probable range of extinction dates can be generated, using a series of last confirmed sightings and a statistical distribution known as the Weibull form, which makes no assumptions about the distribution of the last sightings (Roberts & Solow 2003). This method might be applied to extinct species using last appearances in the fossil record, in order to predict when true extinction occurred; and it might also be applied to living species that are listed as threatened or critically endangered, in order to predict which species might still survive undetected, and which species have already disappeared (Pimm 2003). Discussions about the incompleteness of the fossil record and methods to repair sampling bias by using probability theory are staples of modern paleontology (Jablonski 1999). Paleontological arguments about the size of an extinction event take these factors into account. Arguments about spatial or temporal gaps in the paleoanthropological record should also first consider sampling bias and human population size before invoking catastrophe or extinction.

Decimation and recovery from extinction

Viewing life from the standpoint of the present tends to generate horror about the prospects of looming extinction. Paleontology, however, is replete with extinction. Extinction has befallen over 99% of all species that ever lived on earth (Jablonski 2004). Characters that affect the probability of extinction may covary, and the rules that govern extinction probability do not apply equally to all groups. For example, large body size increases the likelihood of extinction in primates and birds, but it does not affect the likelihood of extinction in carnivores or reptiles (Jablonski 2004). The fine dating and density of the marine invertebrate record tends to establish the major perspective into deep-time. Decade-long examination of the fossil record of marine invertebrates has confirmed the existence of the Big Five mass extinctions (Jablonski 1999). At least 50% of abundant invertebrate genera are lost during each of these major extinction events (Jablonski 2001). The last of these mass extinctions, which occurred at the K/T boundary, has received the most attention. The principal cause for this mass extinction has been identified as an asteroid 10 km in diameter that crashed into the shallow ocean waters off the southeast coast of North America. However, a massive outpouring of flood basalts and global cooling have also been identified as factors to extinction. Interestingly, a global cooling event took place during the last 100 000 years of the Cretaceous (Wilf *et al.* 2003). Both terrestrial plants from North Dakota and marine foraminifera show the effects of this global cooling. Although it caused no land plant extinctions, global cooling diminished biodiversity. After the asteroid impact, the Paleocene floras demonstrate ecological collapse. All dominant Cretaceous species are lost from all of the lithofacies, species richness continues to be reduced, and specialized insect folivores disappeared almost completely (Wilf *et al.* 2003). However, palms are sometimes abundant in the earliest Paleocene – indicating no frost – and sympatric champsosaurs, crocodiles, and turtles are plentiful. From this evidence, it is inferred that only a minor degree of seasonal temperature fluctuation characterized the earliest Paleocene (Wilf *et al.* 2003). This hastened ecological recovery after the mass extinction.

David Jablonski (2001) recognizes four possible biotic responses to mass extinction: uninterrupted continuity, arrested continuity, survivorship without restoration, and explosive diversification or radiation. If a taxon persists but encounters problems, it is not clear whether these problems are caused by high background extinction intensity or selection against traits that were adaptive before the mass extinction event. Jablonski discovered that the time period directly after a mass extinction tends to have significantly more taxa that fail to persist and that go extinct before the next stage boundary. He terms this survivorship without restoration the "dead clade walking" effect (Jablonski 2001). It is not simply that survivorship does not assure future success. Surviving a mass extinction actually appears to render a taxon more vulnerable to subsequent extinction than expected by chance. Explosive diversification or radiation after extinction is not merely a matter of re-occupying ecological niches vacated by the mass extinction. This is so because the ground rules for life are completely re-written during the extinction event, and multiple lineages are simultaneously radiating to achieve the same morphology or grade of structural organization. Predicting survivorship or success of taxa is impossible under these conditions, because of complex ecological dynamics and the home field advantage of niche incumbency (Jablonski 2001). In general, because of the widespread destruction of ecosystems, extinction rates far outpace origination rates during periods of mass extinction. Even when multi-step or long-drawn out extinctions occur, there is little evidence of major phenotypic innovations taking place during these episodes (Jablonski 2001). This illustrates the universal severity of such biotic crises. Cohorts of marine invertebrate genera originate throughout the Phanerozoic. After the Paleozoic, when cohorts that recover from a mass extinction are compared with other cohorts, the survivor genera are statistically longer-lived than the other cohorts (Miller & Foote 2003). The effect is not observed before the Paleozoic. These longer-lived survivor genera did not have larger geographical ranges or ecological tolerances. However, these disaster, survivor-type taxa may have had a higher speciation rate than other taxa (Miller & Foote 2003).

A lag-time in recovery – seen in speciation – occurs after a mass extinction. For marine invertebrates, this lag-time typically appears to be around 5 my. The lag-time must involve complex feedback mechanisms after the origin of new species, because there is no exponential growth in species diversity during recovery (Erwin 2001). New ecological patterns are generated after a mass extinction, which implies that recovery necessitates rebuilding ecological space, rather than merely filling vacant ecological space. In two marine invertebrate cases (ammonoids and crinoids), morphology is affected during recovery, implying that ecological factors control recovery (Foote 1996, Saunders *et al.* 1999). Even local or regional extinctions may influence the ecology of recovering species. For example, 70% of marine invertebrate species went extinct in the West Atlantic about 3 mya, and some taxonomic groups are still rare in this region (Dietl *et al.* 2004). Modern experiments with predatory marine snails and their bivalve prey show that predation is influenced by the density of enemies and the intensity of competition. Two modes of snail predation occur, depending on whether their enemies and competitors are abundant or rare. Because these two behavioral modes leave distinctive signatures in the shells of prey, it appears that both Pleistocene and living snail species encounter less competition than their counterparts did before the late Pliocene extinction (Dietl *et al.* 2004). Hence, full recovery of this marine community has still not taken place. This confirms the idea that ecological complexity – the network of species interactions – must be rebuilt after a mass extinction (Erwin 2001).

Paleontology documents an explosive diversification or radiation of placental mammals after the Cretaceous/Tertiary mass extinction. However, molecular analyses almost uniformly place the divergence of placental mammal orders back into the Cretaceous. This implies the virtually complete absence of Cretaceous ancestors preserved in the fossil record, which leads to the question of why the record should be so biased against their preservation. If Cretaceous placental mammals were uniformly small – no larger than rodents – this might explain the absence of a record. But some Cretaceous mammals were large enough to prey upon small dinosaurs (Hu *et al.* 2005). Furthermore, small placental mammals are richly abundant in the earliest Paleocene, when they are radiating to fill an ecological vacuum. This implies that preservational bias against small species cannot explain the absence of ancestral placental orders in the Cretaceous. Does body size correlate negatively with the rate of molecular evolution? The solution might lie in a more detailed molecular phylogeny. However, even the most detailed molecular phylogeny yet produced, and analyses with a small body subset, still place the divergence between placental mammal orders within the Cretaceous (Springer *et al.* 2003). The discrepancy between paleontology and molecular phylogeny is still unresolved. Primate origins are given as between 69 and 80 mya in this analysis.

Marine molluscs demonstrate that different geographic regions can have different recovery histories. Molluscs from the U.S. Gulf Coast have a very different recovery pattern after the K/T mass extinction from molluscs in northern Europe, northern Africa, or the northern portion of the Indian tectonic plate (Jablonski 1998). The Gulf Coast recovery fauna contains more species characterized as invaders, and the bloom taxa are different. It appears intuitively obvious that this geographic variation must reflect the proximity of the Gulf Coast to the Chicxulub collision site and the angle of the asteroid impact, but this is not yet proven (Jablonski 1998).

In general, the record of fossil marine invertebrates and some other groups appears to show a link between evolutionary centers of origination, where new species are generated, and either disturbed habitats or tropical habitats (Jablonski 1999). The powerhouses of speciation thus tend to occur in these two types of habitat. Identification of the tropics as the center of species origins contradicts an old idea in vertebrate paleontology – that species tend to originate in high latitudes with marked seasonality (Matthew 1914). A tropical center of speciation, where seasonality is reduced, also contradicts a more recent idea in vertebrate paleontology – that evolutionary change is associated with climatic forcing (Vrba 1995).

Rates of evolutionary change

The study of rates of evolutionary change did not begin with Eldredge & Gould in 1972. On the contrary, evolutionary theorists began an intensive study of rates of change in the mid- to late-1940s. These theorists included G. G. Simpson – who wrote a monograph (1944) on evolutionary tempo and mode – and J. B. S. Haldane (1949). P. Gingerich (1983, 1993b) continues to study evolutionary rates. Many rates of change exist, although three divisions are usually recognized. Rates can be comparatively slow, normal, or fast – bradytelic, horotelic, and tachytelic (Simpson 1949). The topic is important, because very rapid change can create swift transformation in species morphology. The disappearance of the ancestral morphology can imitate an extinction event if the intervening morphological stages are not preserved in the fossil record. At this point, breaks or disruptions can be invoked. The important question to be considered is this: how swift does change need to be before it is labeled catastrophic or revolutionary? In 1944, Simpson coined the phrase "quantum mode" of evolution to mark change that led to an abrupt morphological transformation. Such transformation is usually invisible to paleontology, although sometimes it may be detectable where the fossil record is very dense and well-dated. Patterson (1949) used the fossil record of the taeniodonts, an extinct order of mammals, to document the occurrence of "quantum shifts" or "quantum steps" that were associated with the appearance of new adaptive complexes in the dentition, crania, and hands. However, it is important to note that details underlying such morphological revolution would usually not be detectable in the fossil record. Discussions about rates of change and the normal invisibility of details underlying morphological transformations can be valuable in paleoanthropology, although they are often missing. The result is a constant multiplication of taxa whose morphology is endlessly described, although there is little interest in the adaptive significance of the morphology itself, or the evolutionary significance of its quick transformation.

Beyond this gross perspective, however, lies a fundamental problem. The degree of change documented by paleontologists or archeologists is not necessarily the actual rate of change. This is true because of the nature of the geological record, which is composed of bursts of sedimentation. The environment of the past was as dynamic as the modern environment. Stratigraphy reflects this past dynamism, but the nature and degree of sedimentary deposition subsequently affects both the paleontological and archeological records. Fundamentally, sedimentary processes affect whether a record exists at all. Significant time intervals are missing from the record, because geomorphological processes have removed or irreversibly altered the record. Alternatively, sediments may never have been deposited at all. It is not an accident that both the paleontological and archeological records are virtually non-existent in rainforests or mountainous regions.

Textbooks on human evolution routinely describe the fossil material, and then often conclude with a series of the latest phylogenies. The pictured sequence of fossil species that irregularly appear and abruptly disappear in time contribute to the perception of breaks and rapid disturbances. In archeology, description of the material found at a single site tends to overrule broader inter-site comparisons. A myriad of single-site descriptions then create the impression of uniqueness for each site. Each site stands alone. This does not foster a study of diversity and variability between sites. Furthermore, the common archeological practice of dividing sequences of data into industries or cultural stages for purposes of comparison inevitably contributes to the implication of major cultural change or punctuational events at stage boundaries. A break in a spectrum of variability is established for purely taxonomic or classificatory reasons. This typological break subsequently appears to be an anomalous borderland. A break implies that some major novelty has appeared, and it thus demands explanation.

There is generally an inverse ratio between time-intervals and rates of evolutionary change (Gingerich 1983, 1993b). Long time-intervals are associated with slow rates of change, but short time-intervals are associated with quick rates of change. Evolutionary rates in Holocene mammals, measured over the last 10 000 years, are two orders of magnitude greater than in Tertiary mammals, measured over millions of years. The rates are 12.6 darwins versus 0.02 darwins, respectively. Pleistocene mammals show an intermediate rate of 0.5 darwins. Recent studies of "contemporary evolution" confirm the inverse relationship between time-intervals and rates of evolutionary change. "Contemporary evolution" occurs on the scale of decades, and the resulting phenotypic changes can be observed within less than several hundred generations of a species. Sometimes evolution can be observed within several generations of a species, or at the time scale of a human life. Some rates of change that are observed in the wild are equal to those obtained under laboratory conditions. This is significant, because laboratory experimentation is sometimes considered to produce unnatural rates of change that could never be sustained under field biology conditions. Yet, rates in the wild can rival those obtained in the laboratory. Furthermore, they can be four to seven orders of magnitude greater than those observed in the fossil record (Reznick *et al.* 1997: 1935). Strong selection rates and phenotypic changes in contemporary time are associated with a number of observable environmental alterations. These can be caused by climate, habitat degradation, colonization, or predation.

Many other factors also show an inverse relationship with time intervals. These factors include the level of detail in fossil morphological data, paleoecological information, and taphonomic information. The most detailed information comes from *Lagerstätten*, where instantaneous burial occurs. A splendid example is the Eocene Messel site in Germany, where sediments were deposited in an ancient lake fringed with tropical forest. Animals and plants were completely buried in the fine silt of the lake bottom. Mammals in the Messelgrube can be preserved as whole carcasses, including impressions of their soft tissues and fur, stomach contents, and bacteria (Schaal & Ziegler 1992).

Sediment accumulation rates or erosion rates are inversely related to time intervals. It has been known since the nineteenth century, for example, that it is reckless to date past geological, paleontological, or archeological events by using modern sedimentation or erosion rates. That is, one measures a geological section and estimates its antiquity from modern sedimentation or erosion rates that would yield a section of similar length. For example, in an attempt to establish the age of the earth, Darwin estimated that it would have taken about 300 my to erode material from the Weald in southeastern England. This precipitated great controversy about the age of the earth. Details of sedimentary structures also tend to be inversely related to the length of time over which they are studied. Fine points of sedimentary structure (e.g., current direction in an ancient river) tend to be lost with time, although they can sometimes be carefully extracted from the record. Time-intervals are also inversely related to the level of detail in archeological data. The most detailed information tends to come from instantaneous slices of time. The seemingly slow pace of cultural transformation during the Oldowan or Acheulean is related to the long time-interval over which it is measured, and the rapid effects of the Middle/Upper Paleolithic transition occur over a short time-interval. These inverse relationships are caused principally by time-averaging and the episodic nature of sediment deposition. Because the geological record itself is determined by bursts of deposition, it will always tend to appear irregular or erratic in nature.

Time-averaging

Through a variety of factors, time-averaging affects both paleontology and archeology. Principal confusing factors are the episodic nature of sedimentation, erosion, fluvial re-working, and wind winnowing. Soil formation, bioturbation, indeterminate causes of data mixing, and stratigraphic disorder – which distorts the contemporaneity of data and true chronologic order – also disturb the fossil and archeological records. Time-averaging is an exceptional problem in areas where habitat reconstruction and species interaction are attempted. When paleontologists examine ancient ecosystems and community structure, or when archeologists examine the nature of a site or engage in site catchment analysis, they are confounded by time-averaging.

Very rarely, paleontologists encounter instantaneous preservation caused by a volcanic eruption or mudslide. An example would be the Poison Ivy Pocket in Nebraska. Here masses of *Teleoceras*, an abundant hippo-like rhinoceros, were suddenly felled in a pond that was covered by a deep blanket of ash that was cataclysmically ejected from a volcano in the American West 10 mya. Details of diet and herd structure can be reconstructed in this extinct species. It is therefore known that this species not only resembled hippos in body size and shape, but also resembled hippos in being grazers that lived in herds, and returned from grasslands to submerge themselves in water. Yet, except under outstanding circumstances, there are ordinarily stratigraphic limits to time resolution. In paleontology, these limits make it difficult to study community structure and species relationships. Time-averaging has therefore been a topic of intense interest by paleoecologists who are studying ancient ecosystems, because they need to consider the interactions of contemporary organisms (Behrensmeyer *et al.* 1992, Damuth 1993). Given the factors that affect the formation of a fossil assemblage, which organisms are contemporary? Martin (1999) offers the contrary argument that time-averaging does not necessarily result in a loss of information, provided that one is interested in broad problems in paleoecology.

In archeology, where the signatures of human behavior are studied, broad intervals of time resolution can create a picture that is not a record of individual or group behavior, but a record of species-specific behavior. This is acceptable to a physical anthropologist, but disappointing to an archeologist, because individual human behavioral and local cultural variability are lost forever. The geological factors affecting site formation are always considered, but time-averaging usually makes it impossible even to reconstruct the environmental setting to any fine degree. Archeologists need to know about a site at a level of detail that may be impossible to reconstruct. For example, what was the degree of seasonal change? What plant species were present? Were there trees in the area where the scatter of debitage occurs? Human re-occupation and re-use of sites can further smudge the record of behavior. The behavior of multiple groups or cultures is then collapsed together, and details are impossible to resolve. If material remains are located from different cultures, debate may then occur about contact, displacement, or extermination. In general, broad archeological concerns, such as first appearance, extinction, dispersal, contact, and exchange tend to appear instantaneous, even though these events are spread over a century or more. The result is a distortion of actual events. An artificial picture of breaks, disturbances, and disruptions is created in the archeological record.

Taphonomy and experimental studies

In 1940, the Russian geologist J. A. Efremov coined the neologism "taphonomy" to describe the study of the processes that affect organisms when they die and become incorporated into the fossil record. Organic traces (e.g., molds, impressions) and vestiges of animal behavior (e.g., tracks, burrows, nests, dens) are also studied. The word "taphonomy" was formed from two Greek words that translate as "the laws of burial." What are the factors that determine whether an organism becomes fossilized? Decomposition, biological demolition, physical destruction, and dynamic reworking all operate before

Figure 12.4. Many biological factors affect fossil preservation. Here the potential for bioturbation is revealed by 1.6 my old carbonate root casts on the Karari Escarpment, northern Kenya. The scale has five centimeter increments.

final incorporation of remains into the sedimentary record (Figure 12.4). There are obvious biases that relate to the existence of hard parts. Hard parts will tend to preserve much better than soft tissues, and the relative density of hard parts is also a factor. For example, because enamel is the hardest biological substance known, teeth form the overwhelming bulk of the fossil record. Because dense bones preserve better than light bones, the fossil record of birds, which have light bones, is strikingly impoverished. Furthermore, if organisms die away from sediment traps, and are not swept into depositional basins, they will not be fossilized.

The geologist A. K. Behrensmeyer and her colleagues have continued to gather data on the factors that determine bone preservation in the fossil record (Behrensmeyer 1975, 1978, 1982, 1988, Behrensmeyer & Hill 1980). This has necessitated examining processes that affect carcass accumulation or decay (Figures 12.5 and 12.6). Other researchers have emphasized the implications of taphonomy for interpreting the paleoanthropological and archeological records (Shipman 1981, Morden 1991, Lyman 1994). Vertebrate paleontologists are rightly fascinated by the possibility of studying paleocommunities once the taphonomic filters that distort the fossil record are known (Olson 1980). At least one researcher (Martin 1999), emphasizing the invertebrate fossil record, is optimistic about the retrieval of paleoecological information, in spite of taphonomic processes. Taphonomy, he claims, does not necessarily imply the loss of

Figure 12.5. A young zebra has been killed by a lion several hours before. The carcass lies at a ford of the Ewaso Nyiro River, Central Kenya. Note that disemboweling comes first in a carnivore consumption sequence.

Figure 12.6. Remains of the carcass of a young female elephant from the Laikipia Plateau, Central Kenya, approximately one month after death. Mummified skin still preserves the integrity of much of the carcass, but note how some bones have been scattered, in spite of their large size and density.

Figure 12.7. The survival of hominid carcasses can be affected by carnivore activity. Cast of a hominid fossil (A.L. 400-1a, *Australopithecus afarensis*, Hadar, Ethiopia, 3 mya). There is extensive carnivore gnawing on both sides of the mandibular corpus, where a carnivore can easily access fatty marrow. The scale is in centimeters.

information. Yet, detailed sedimentological and geochemical analyses underlie the retrieval of such information. Martin (1999: 13–14) synthesizes 10 basic taphonomic rules from over 50 years of research, and then presents 10 additional rules of taphonomy (1999: 389–392). The last rule (number 20) states that in some, possibly many, cases, "the present may be the key to nothing at all," because past physical environmental and ecological conditions have no analog in the present (Martin 1999: 391). Nevertheless, given the tenor of the entire monograph, there is reason to be hopeful. A detailed investigation of past climates, ancient environments, and extinct species can yield some understanding of the cryptic past.

The role of animals as bone accumulators or destroyers has long been studied in an attempt to explain fossil preservation or site formation. Hughes (1954) examined hyaena habits in order to explain bone accumulations at the South African australopithecine sites (Figure 12.7). C. K. Brain (1981) dismantled Raymond Dart's evidence for the Osteodontokeratic Culture by analyzing natural agents of differential bone preservation at Swartkrans, Makapansgat, and other australopithecine sites.

Continuing long-term study of vertebrate carcass accumulation, destruction, and preservation is taking place in southern Kenya, Baja California, and elsewhere. Long-term experimental studies of the processes that affect carcass destruction, including destruction by scavengers, is taking place within the Jordan Valley of Israel (Lotan 2000). It is important to note that climate change from the 1970s through the 1990s – change discernable within a human lifetime – can affect bone preservation. Relatively minor climate change has transformed the woodlands of Amboseli Reserve in southern Kenya into open grasslands. The subsequent multiplication of hyaenas now eliminates all animal bones in the reserve that are smaller than those of a Cape buffalo, because bone-crunching hyaenas completely consume anything smaller than this (Perkins 2003). If this effect were translated into geological time, it would appear as an abrupt break in fossil preservation. The geological record of this period would be diminished in biodiversity. Naïve future paleontologists might infer that a major climatic event must have been responsible for the paucity of fossils, when, in reality, it was caused by a minor habitat shift that allowed hyaenas to proliferate in an area where they had previously been much more rare.

13   *The bipedal breakthrough*

Introduction

The origin of bipedal posture and locomotion has been a major concern in physical anthropology for over 130 years (Darwin 1871, Keith 1940, Robinson 1972, Kingdon 2003). Determination of the hominid status of equivocal 5–6 mya fossils is ultimately based on anatomical traits associated with bipedalism, and not on dental traits (Senut *et al.* 2001, Begun 2004, Haile-Selassie *et al.* 2004). Because bipedal locomotion is the principal hallmark of humans and their fossil ancestors, discussions about the mechanics of bipedalism and selection pressures associated with its inception are fundamental topics in paleoanthropology (Rose 1991, Tuttle 1994, Ward 2002). Beginning at about 4 mya, postcranial remains of early australopithecine hominids become more abundant and document unequivocal specializations for bipedalism. However, because australopithecine postcranial anatomy and body proportions differ from those of modern humans and earlier members of genus *Homo*, there is extensive debate about australopithecine lifeways. Were they habitual, obligate terrestrial bipeds (Lovejoy 1988)? Were they capable of bipedal locomotion, although still capable of arboreal climbing and seeking shelter from predators in trees (Stern & Susman 1983, Deloison 1991)? Did the retention of primitive arboreal traits compromise the efficiency of their bipedalism (Stern 2000)? Did they evolve from terrestrial knuckle-walkers like the living African great apes (Richmond & Strait 2000)? Were they actually generalized catarrhines that were quadrupedal in both the trees and on the ground (Sarmiento 1998)? The answer to these questions influences not only reconstructions of anatomy, but also reconstructions of diet, foraging and ranging behavior, sociality, division of labor, sexual dimorphism, and pair-bonding (e.g., Lovejoy 1981, Foley & Lee 1989). Thus, the analysis of bipedalism in early hominids opens into an ever-widening vista of fundamental topics in early hominid lifeways.

Ape models for bipedal origins

In 1889, Arthur Keith, a young Scottish physician, accepted a position administering to the health of gold miners working in Thailand (then Siam). Alcoholism

271

and malaria felled most of the Europeans; only two Europeans (Keith being one of them) survived a 2-year spell at the site. Being unable to stem the miners' alcoholism, Keith tried to stem the malaria. Wanting to know if the local non-human primates were subject to malaria, Keith began to dissect langurs and gibbons. He discovered a substantial series of differences between monkeys and lesser apes. To his amazement, the lesser apes resembled humans in many traits. This resemblance is the basis for the zoological superfamily category "Hominoidea," but Keith was a physician, and had never received any training in comparative anatomy. In fact, he initially thought that the gibbons were monkeys. The morphological similarity between human and gibbon was therefore a surprise to Keith (1950). Many human traits that were supposed to have appeared with the advent of erect posture and bipedal locomotion already occurred in the lesser apes, yet these species used their upper limbs as the principal mode of support and locomotion.

Keith spent some time observing gibbon posture and locomotion, which was striking. Gibbons kept their body upright or orthograde, and used their arms for propulsion during arboreal arm-swinging locomotion. That is, gibbons suspended their bodies below tree branches, and swung along below tree branches using alternate arms for propulsion. Arthur Keith coined the word "brachiation" for this arm-swinging locomotion. He argued that, when the entire weight of a suspended body is supported by a hand holding an arboreal support, the humerus becomes a lever of the third order, whose fulcrum is at the elbow (Keith 1926). Modern biomechanical analysis reveals that the gibbon ulna experiences tensile strains during arm-swinging (Swartz *et al.* 1989). Keith began publishing a series of papers documenting the differences between monkeys and hominoids (Figure 13.1). Skeletal traits unite living hominoids – they are the basis for the Superfamily Hominoidea, although the fossil record demonstrates that they appear in a mosaic fashion (Figure 13.2). Keith developed the novel idea that human erect or orthograde posture had first appeared in ancient arboreal hominoids that were using arm-swinging locomotion. Although this idea was strongly supported by William King Gregory (1927, 1930, 1934), it was rejected by many other researchers in the early twentieth century. Many, like Wood Jones (1926), argued that the ancestors of the hominids had never been quadrupeds. Others, like Adolph Schultz (1936 a, b: Fig. 21), believed that hominids evolved from a catarrhine that was a generalized arboreal quadruped, like many living cercopithecoid monkeys.

Amidst the controversy, there was a conflation of two different themes. The first concerned the ancestral group from which hominids had diverged: prosimian, generalized catarrhine, generalized hominoid, or pongid. The second concerned the type of locomotion habitually employed by this ancestor: generalized arboreal quadrupedalism, brachiation, arboreal suspension, and slow

Figure 13.1. Two figures from Arthur Keith's 1890 paper, illustrating a primate pronograde posture (left) and a primate orthograde posture (right). From Keith (1940: Fig. 1).

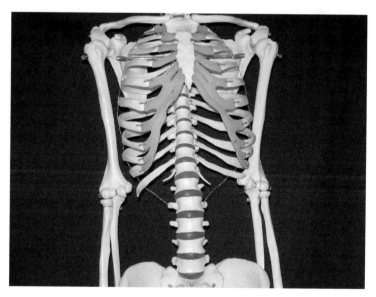

Figure 13.2. Modern human skeleton, illustrating some features characteristic of living hominoids: broad sternum with sternebrae fused in adults; broad thorax that is anteroposteriorly shallow; dorsally positioned scapula; shallow, flat glenoid fossa; long clavicle; strongly curved ribs; vertebral column protruding into the thoracic cavity; large capitulum; prominent lateral trochlear ridge; and large medial epicondyle.

arm-swinging locomotion. The problem was compounded by virtual ignorance of great ape locomotion in the wild. Hence, it was not understood that the hylobatids possessed a truly specialized form of arboreal locomotion. During the early 1960s, some researchers attempted to define brachiation more rigorously, using assessment and quantification of locomotion in captive zoo animals (Avis 1962) and skeletal proportions (Erikson 1963). Other terms appeared: e.g., "semibrachiators," "modified brachiators." After Tuttle (1969) delineated knuckle-walking as a new category of locomotion in African great apes, and established its anatomical correlates, it became clear that chimpanzees and gorillas actually possessed a specialized form of terrestrial locomotion. Their locomotion in the trees did not resemble that of gibbons and siamangs. Tuttle (1974, 1975) suggested that a new term should be coined for the specialized arboreal locomotion of gibbons and siamangs – "ricochetal brachiation." He used the term "suspensory arm locomotion" to describe suspension of the body from an arboreal support with slow arm-swinging movement. He argued that the ancestor of hominids may have been relatively small-bodied, and may have already been exhibiting some bipedal behavior along relatively horizontal arboreal substrates. He termed this a hylobatian model (Tuttle 1974, 1975). Japanese researchers (Yamazaki *et al.* 1979, Yamazaki & Ishida 1984) subsequently used biomechanical analysis and computer simulations to examine gibbon arboreal bipedal locomotion.

Using electromyography to examine the function of the shoulder muscles in the prehensile-tailed platyrrhines *Ateles* and *Lagothrix*, Stern *et al.* (1977) discovered that the electrical activity patterns of these muscles during arm-swinging resembled the patterns exhibited during vertical climbing. Shortly afterwards, it was suggested that the suite of hominoid traits evolved among animals that were climbing trees vertically, or were "vertical climbers" (Fleagle *et al.* 1981). Even more recently, it has been suggested that vertical foraging postures or bipedal foraging postures may be important (Hunt 1992, Stanford 2002). In any case, ancestral locomotion or posture is thought to pre-dispose animals to develop terrestrial bipedal locomotion.

The problem is one of equifinality. If two behaviors (suspensory posture with slow arm-swinging or vertical climbing) can lead to the same functional anatomical complex, how can one distinguish between which behavior has selected for the complex in any particular case? But is this a question of homoplasy, with the same functional anatomical complex developing independently? Hominoid traits form a unique constellation. For example, among other features, the hominoid thorax is broad but flattened anteroposteriorly, the sternum is wide, the scapula is dorsally located, the shoulder joint is cranially oriented (except in genus *Homo*), and the hominoid shoulder, elbow, and wrist are extraordinarily flexible (Tuttle 1974, Aiello & Dean 1990). Do these features evolve under conditions of vertical climbing or suspensory arm locomotion? Clearly many primates engage in vertical clinging and climbing without exhibiting hominoid musculo-skeletal specializations. What specific behaviors select for flexibility of the shoulder, elbow, and wrist? Body size may be important. Very small arboreal animals may vertically cling and climb using only claws to support their body weight with almost no lateral extension or rotation of their limbs.

However, if vertical climbing is a major locomotor category in many arboreal primate species, and if vertical climbing is thought to have been a factor in the origins of bipedalism, then the selection pressures that result in hominid bipedalism need to be resolved. Hirasaki *et al.* (2000) conducted biomechanical analyses of vertical climbing in Japanese macaques (*Macaca fuscata*) and spider monkeys (*Ateles geoffroyi*). The results are different in the two species, and there is greater extension of the hip and knee during vertical climbing in the spider monkey. Hirasaki and his colleagues conclude that different types of locomotion have been collapsed together in the category of vertical climbing. Only vertical climbing that emphasizes a vertical trunk (orthogrady) may influence the development of hominid bipedalism.

In addition to the hylobatids, other apes have served as models for the origins of bipedalism. Stern (1975) was impressed by the hindlimb musculature of the highly arboreal orangutan, and argued that the proto-hominid ancestor must have had a similarly mobile hindlimb. He also presented a biomechanical analysis showing how a highly arboreal ancestor with very long arms was destined to become bipedal once it came to the ground, because the forelimb of such a quadruped would bear too much body weight, and unbearable pressure would be placed on the hands (Stern 1975: Fig. 1). Preuschoft (2004) also endorses an orangutan model for hominid origins.

After knuckle-walking was described as a new category of locomotion for African great apes (Tuttle 1969), some researchers like S. L. Washburn quickly argued that a terrestrial knuckle-walking stage of evolution had immediately preceded bipedalism. The idea of knuckle-walking as a stage of human evolution has been revived by Richmond & Strait (2000). The anatomical evidence does not seem persuasive to me. In any case, it is not supported by Pliocene hominid fossil remains. The hands of these fossils do not possess knuckle-walking traits, and there are indications of habitual bipedalism in the foot and knee (Leakey *et al.* 1995, Ward *et al.* 1999).

Tuttle (1974, 1977) reviews the use of ape models for bipedal origins through the first three quarters of the twentieth century. Tuttle and his colleagues (Tuttle *et al.* 1979, Tuttle 1994) have also reviewed experimental studies of non-human primate posture and locomotion that affect the evolution of bipedalism.

Behavior and morphology

Some of the features associated with hominid bipedality accrue only after bipedal locomotion takes place. For example, lumbar lordosis does not occur in human neonates. Only after a youngster learns to walk does this distinguishing characteristic of bipedalism appear in the vertebral column. The typical hominid angulation of the femoral neck and the femoral bicondylar angle also depend upon the development of bipedal activity (Tardieu 1998, 1999: Figs. 1 and 4). Normal infants or humans born with congenital conditions that preclude walking exhibit a different anatomy.

How much does use determine morphology? The influence of behavior can be assessed by examining traditionally molded behavior in the Japanese macaque (*Macaca fuscata*). Beginning about AD 1000, certain Japanese macaques were trained as *sarumawashi* – performance monkeys. Traveling troops would move from village to village, and the monkeys would perform traditional routines and acrobatic displays. These performances are now viewed as entertainment, but, at one time, they enjoyed ritual status. From the age of 3, performance monkeys are trained first to stand upright, and then to walk bipedally for 2–3 km a day. Any animal can be selected for training – there is no special strain of performance monkey. Bipedal posture and locomotion takes place for about one hour per day. This is enough to cause skeletal changes to occur. Bone remodeling creates more robust hindlimb bones, and lumbar lordosis occurs in these bipedal macaques. Lordosis is maintained even when the monkeys are quadrupedal. Lordosis is greater in animals that have been trained longer, and the performance monkeys exhibit a more upright posture than untrained animals (Nakatsukasa 2004). When the trained macaques are compared with untrained control macaques walking bipedally, the effect of habitual bipedal behavior can be qualitatively and quantitatively assessed. The trained animals keep their heads up during the single-limb support phase, take longer, less frequent strides, and possess a stable head and trunk (Hirasaki *et al.* 2002). In addition, the trained animals have more extended hip and knee joints – this accounts for the longer stride length, because the hindlimb joints do not have a larger range of motion. Kinematic analysis reveals that the trajectory of the hip joint marker forms a convex curve in the trained animals, and they may have been utilizing an inverted pendulum effect to keep their heads up during single-limb support (Hirasaki *et al.* 2002). If the mass of an animal rises and falls with every step, gravitational potential energy is transformed first into forward-moving kinetic energy and then back into potential energy again, like a swaying inverted pendulum. This is a very cost-effective mode of locomotion, because it translates gravitational energy into forward movement. However, measurement of oxygen consumption shows that even highly trained performance monkeys expend 20–30% greater energy during bipedalism than during quadrupedal locomotion (Nakatsukasa 2004). Bipedalism is not energy efficient in these animals, which contrasts to the remarkable energy efficiency of bipedalism in modern humans. This is not surprising, given the fact that these macaques (as is true of all Old World monkeys) have traits that are specialized for quadrupedal locomotion, such as a long lumbar region and digitigrade feet. These traits were not present in the proto-hominid ancestor, because they do not occur in pongids. In summary, investigation of Japanese performance monkeys shows that behavioral changes precede the anatomical changes associated with bipedal locomotion. The importance of this research lies in the fact that it may limn the beginnings of bipedalism.

Bipedal efficiency

Bipedalism in modern humans walking at a normal pace requires very little muscle activity – the electromyographic (EMG) record of muscle electrical activity is very low (Basmajian 1974, Tuttle *et al.* 1979, Tuttle 1994). Analysis of living humans that are still completely dependent upon walking for transport documents the efficiency of bipedal locomotion (Musiba *et al.* 1997). However, the normal human walking pace is slow. This necessitates that whatever selection pressures were involved in the shift to habitual terrestriality in open country must have allowed for a slow traverse of the home range. Bipedalism at normal walking speeds is slow, but highly cost-effective (Steudel 1995, 1996, Steudel-Numbers 2001, Leonard & Robertson 1997, 2001). Bipedalism in humans is passive-dynamic in nature: it resembles a low-energy system of controlled falling, in which a leg swings forward in time to prevent a fall. Human children learn to walk by learning how to transfer weight smoothly without actually falling. Bipedal robots, designed after walking humans, and with anthropomorphic proportions and mass distributions, are equivalent to humans in low energy use (Collins *et al.* 2005).

In general, normal human walking expends on average only about 87% of the energy expended by a similar-sized, generalized quadrupedal mammal moving at the same rate (Steudel-Numbers 2001: Table 1). Energy saving by bipedal hominids is even more pronounced when compared with the locomotor costs observed among other primates. Primates in general and chimpanzees in particular exhibit less efficient locomotion than other quadrupeds of the same body size (Leonard & Robertson 1997, 2001). Measuring leg length as the height of the hip joint off the ground during normal stance, Alexander (2003b) generates a value known as a Froude number using the equation (velocity)^2/(gravitational acceleration)(leg length). This number allows animal movement to be compared across

Table 13.1. *Mammalian body mass and daily distance traveled. Values of body mass (kg)/daily distance traveled (km). Calculated from data in Garland (1983: Table 1)*[1]

Homo sapiens (modern San) 4.4
Non-human hominoid species (n = 5) 77.16
Non-human catarrhine primate species (n = 26) 20.96
Non-human primate species (n = 37) 17.31
Non-primate mammal species (n = 38) 133.68
All non-human mammal species (n = 75) 75.50
Mammalian carnivore species (n = 13) 8.21

a range of different sizes, because it analyzes dynamic similarity. Non-human primates are peculiar, because, at the same Froude number, they exhibit larger relative stride lengths than non-cursorial mammals do. Yet, humans exhibit about the same relative stride lengths as cursorial mammals at the same Froude number do (Alexander 2003b: 107).

Locomotion can be costly. The maximum rate of oxygen consumption during high-speed locomotion in mammals can be ten times greater than oxygen consumption at the resting metabolic rate (Garland 1983). Another method of comparing the energetic costs of moving body mass is to examine the relationship between body size and the distance over which an animal moves during the course of a day. Garland (1983) scales the cost of transport to body mass in mammals. Table 13.1 gives a measurement of body size relative to the distance traveled over the course of a day for a variety of mammalian groups. The measurements are calculated from data in Garland (1983: Table 1). Note that living humans have the lowest values of the major mammal groups listed in Table 13.1, including far lower values than other hominoids, other catarrhines, and all other primates. This reflects the greater mean daily movement of humans, in comparison to other primates, non-primate mammals, and all non-human mammals. However, mammalian carnivores also have a low value. Garland (1983) calculates that mammalian carnivores move 4.4 times as far as other mammals in terms of daily movement. He therefore believes that the cost of transport is greater in carnivores than in other mammals. The cost of transport would be equally high in humans, if they did not possess bipedalism – a very cost-effective mode of locomotion. Bipedalism allows for an economy of movement that is unexpected, given human body mass. Without bipedalism, one might expect humans to travel a daily distance that is either 4.76 or 17.54 times smaller, using other catarrhines or other hominoids, respectively, as a basis for comparison.

Terrestrial primates always have larger home ranges than arboreal primates do. Yet, even the terrestrial African great apes have very small day ranges and home ranges, in comparison to some other terrestrial mammals. For example, the modern African great apes may range through only about 400–4000 meters over the course of a day. Modern lions, leopards, and cheetahs may range through 0.5–25 kilometers over the course of a night. This is caused partly by diet, because carnivores tend to have large home ranges, and may need to locate mobile food, and patrol and defend an area. Locomotor efficiency must also affect home range size, because it affects day ranges.

A three-dimensional musculoskeletal model of an australopithecine lower body has been constructed based on the relatively complete specimen of *Australopithecus afarensis* AL 288–1 ("Lucy") from Hadar, Ethiopia. Computer simulations of bipedalism with this model demonstrate that an adult australopithecine with a body weight of 30 kg had a metabolic energy expenditure equivalent to that of modern human youngsters (Nagano *et al.* 2005). Although bipedalism in this model was more costly than in adult modern humans, a modern type of upright bipedal walking existed.

KNM-WT 15000 is a very complete specimen of *Homo erectus* from the site of Nariokotome in the West Turkana Basin of northern Kenya. This is a subadult specimen of *Homo erectus*, dating to 1.53 mya. A major feature of the Nariokotome specimen is its modern limb proportions. It does not resemble australopithecines in shape. The lower limb has increased in length, so that the relative proportions of upper and lower limbs is like that found in modern hominids. The relatively long legs of modern hominids are usually explained by presumed efficiency in bipedal locomotion. Yet recent experimental studies of locomotor efficiency in modern humans demonstrate that relatively longer legs have no substantial effect on locomotor efficiency (Steudel & Beattie 1994, Webb 1994, Steudel 1995). This is confirmed under field conditions. The Hadzabe people of northern Tanzania are efficient bipeds in spite of short legs and short stature (Musiba *et al.* 1997). Relative stride length and walking speed in the Hadzabe are greater than those of other rural and city populations that have been studied, possibly in part because they exhibit less medial and lateral rotation at the hip (Musiba *et al.* 1997). If leg length does not affect locomotor efficiency, then the evolution of relatively long lower limbs in *Homo erectus*, as exemplified by KNM-WT 15000, was driven by selection pressures either for speed or for heat adaptation. Speed is unlikely to have been a factor in increasing relative length of the legs. Heat adaptation, however, was probably a major selection factor in the increasing relative length of the Nariokotome lower limb, especially given the long distal limb segments and exaggerated ectomorphy of the whole specimen.

Paleoenvironment

Yet, a vital implication of this discussion is that a major advantage of bipedality – very little energy expenditure under normal walking conditions – would not exist if no major distances needed to be traversed in the home range. The cost-effectiveness of bipedality would disappear. It is therefore important to note that open country savannah environments appear long after the initial appearance of hominids, which dates to 6–7 mya. The earliest hominid sites contain definite indications of mosaic environments that include open woodland or forested habitats (Haile-Selassie 2001, Pickford & Senut 2001, Brunet *et al.* 2002). These earliest sites yield three hominid taxa: *Sahelanthropus tchadensis* at 6–7 mya (Brunet *et al.* 2002), *Ororrin tugenensis* at 5.7–6 mya (Senut *et al.* 2001, Sawada *et al.* 2002), and *Ardipithecus ramidus kadabba* at 5.2–5.8 mya (Haile-Selassie 2001). Unequivocal evidence of bipedality exists only in *Orrorin*, which has persuasive evidence of bipedal locomotion in the morphology of three preserved femora (Pickford *et al.* 2002). CT scans of the internal architecture of one of these proximal femora also indicate that weight-bearing stresses were associated with bipedalism (Galik *et al.* 2004). There are no postcranial remains currently associated with *Sahelanthropus*, and the argument for bipedality in *Ardipithecus ramidus kadabba* rests on a single toe bone. This is a fourth proximal foot phalanx (Haile-Selassie 2001). This is not very informative. If a first proximal foot phalanx had been found, my assessment would be different.

Paleoenvironmental research documents that widespread arid, open country conditions did not occur until about 1.6 mya in Africa (de Menocal 1995, de Menocal & Bloemendal 1995, Rutherford & D'Hondt 2000). Increasing aridity may have caused the arrival of Serengeti-like grasslands, which first appear only at 1.6 mya (Cerling 1992). However, paleosol carbonates and organic matter from sites in the Kenya Rift Valley show that a mosaic of environments appears to have persisted for the last 15.5 my. There is no evidence here of a shift to a dominant open grassland regime at any time (Kingston *et al.* 1994). Carbon isotope signals taken from herbivore teeth from the same succession of sites in Kenya show that C4 grasses are present in herbivore diets at 15.3 mya, although these grasses are not the primary food resource until 7 mya (Morgan *et al.* 1994).

Bipedal origins

Postcranial remains of *Dryopithecus laietanus* from the Miocene site of Can Llobateres in Spain demonstrate that this taxon was arboreal, with a short, orthograde trunk, and hands that were long in both the palm and fingers. The forelimb also included a long, straight humeral shaft. These features indicate suspensory arm locomotion in *Dryopithecus laietanus*. The ineluctable conclusion is that hominid bipedalism evolved directly from arboreal suspensory posture and locomotion. That is, bipedalism was not preceded by an interim stage of terrestrial quadrupedalism like the knuckle-walking of living chimpanzees and gorillas.

At 3.6 mya, hominids are undoubtedly good functional bipeds, as documented by the hominid footprint trackways at Laetoli, Tanzania (Leakey & Hay 1979, Agnew & Demas 1998). One might argue for later times that the distribution of suitable lithic raw materials for stone-tool manufacture could dictate greater hominid ranging and foraging distances. But what selected for bipedality before the widespread advent of savannah ecosystems at 1.6 mya? If hominids were bipedal in a forest, it is unlikely that they were searching for lithic raw materials. There are no stones to be discovered on the ground within a tropical forest, because the forest floor is covered with leaf litter. Kingdon (2003) argues that foraging for a variety of terrestrial food items selected for bipedality in proto-hominids. This occurred on the forest floor of a relatively open, littoral rainforest that stretched along the East African coast to South Africa. Like Jolly (1970b), who believed that the proto-hominids were grass seed-eaters with a niche similar to that of the modern gelada baboon, Kingdon envisions the proto-hominids as sitting upright on the ground while foraging for food items with their hands. Both Jolly and Kingdon argue that long periods of terrestrial sitting and squatting were necessary to ease the transition to bipedality.

These two models of hominid origins, like many others, assume that the transition to a ground-dwelling existence is difficult, and that the advent of a terrestrial lifestyle must be eased by a long period of apprenticeship. However, the transition from four legs to two legs is not difficult in catarrhine primates. In fact, one might expect bipedalism to appear at some point during catarrhine evolution. It is likely that the first appearance of bipedalism occurred in the trees. That is, hominids began engaging in facultative bipedal behavior while their locomotor repertoire was still largely dominated by arboreal climbing, suspensory foraging postures, suspensory arm locomotion, and some modest arm-swinging behavior. This has been termed a hylobatian model by Tuttle (1974, 1975), who elaborates the model, and traces its origins to Morton's (1924) study of human foot morphology. All catarrhine primates typically exhibit great amounts of orthogrady when in resting posture (Figures 13.3 and 13.4). The inevitable development of catarrhine bipedality is confirmed by the fact that approximately 10% of the locomotion of modern lesser apes (hylobatids) is bipedal. Videos reveal that arboreal bipedal locomotion by hylobatids is rapid – it is not clumsy or hesitant (Vereecke *et al.* 2004). Stanford (2002) reports an unusually high frequency of postural bipedalism in arboreal chimpanzees standing erect while foraging for fruit in *Ficus* trees. The acquisition of

Figure 13.3. An arboreal Sykes monkey (*Cercopithecus albogularis*) with an orthograde trunk in resting posture.

Figure 13.4. A terrestrial olive baboon (*Papio anubis*) with an orthograde trunk in resting posture.

arboreal bipedalism was easy, given the catarrhine substrate for this behavior. Using biomechanical analysis of living catarrhines, Preuschoft (2004: 371, 391) argues that the transition to bipedalism is only "a short step away" or "not a big one." Hence, the breakthrough to bipedalism was abrupt, although it occurred with many other types of arboreal locomotion. The transition to bipedal behavior on the ground, however, was not abrupt, given indications that the earliest hominids existed in an environmental mosaic in which open woodlands and forests were mixed with patches of grassland. This ends the centuries-long idea that severe environmental perturbations that desiccated forests and promoted the spread of grasslands were the triggering event for hominid origins. This also ends the centuries-long fascination with the idea of weak and hapless hominids suddenly confronting the catastrophic vicissitudes of life on the ground, and immediately being forced to adopt tool behavior and redesign new lifeways, because they were literally tumbled out of the trees. Because kinematic analysis of gibbon arboreal bipedalism shows different joint and foot positions than walking humans do (Vereecke *et al.* 2004), one can expect that the earliest hominids would also exhibit substantial differences in gait from modern humans.

Bipedal origins and anatomical genomics

It is now known that *Hox* genes determine the major pattern of the mammalian axial and appendicular skeleton, and organize the limbs and extremities (Muragaki *et al.* 1996, Shubin *et al.* 1997, Chiu & Hamrick 2002, Kmita *et al.* 2002, Wellik & Capecchi 2003, Papageorgiou 2004). All tetrapod limbs originated from the fleshy paddle of lobe-finned fishes like the coelacanth. In order to understand these origins, one must understand the nature and interaction of the underlying *Hox* genes. In mammals, study of the *Hox* genes is promoted by widespread use of the mouse as an experimental model in anatomical genomics. Individual genes are integrated into functional clusters, so that the deletion of a gene has less effect than a disruption of its function. The interplay of genes within these clusters makes it difficult to assign a proper function to single genes. Increased transcription of a neighboring gene can partly compensate for the deletion of a gene, and a gene may behave like another gene in the cluster when located in the same relative cluster position (Kmita *et al.* 2002). *Hox* gene expression within the tetrapod limb occurs in phases, with expression of the genes and formation of embryonic elements taking place along a vector from proximal to distal elements. The *Hox* genes are expressed in the same order as the body parts that they produce. Thus, there is a complex spatial and temporal interaction in the development of embryonic morphology. One model argues that the *Hox* genes responsible for a limb are packaged as a united linear array within a protective coating that initially isolates them from transcription factors; the limb bud is gradually extended along the proximal-distal axis like a rod as it passes through a morphogenetic gradient (Papageorgiou 2004). Genetic material that does not code for protein nevertheless has function. It has controlling, promoting or enhancing (*cis*-regulatory) functions that specify the time, cell type, and tissue type expression of proteins coded by genes on the same chromosome.

The stylopodium develops first within the developing limb bud, the zeugopodium second, and the autopodium last. The stylopodium is patterned differently in the forelimb and hindlimb (Wellik & Capecchi 2003). In the forelimb, the stylopodium (humerus) is patterned by the *Hox* 9 gene, with a smaller contribution from the *Hox* 10 gene, which also affects the zeugopodium (radius and ulna). However, in the hindlimb, the stylopodium (femur) is patterned by the *Hox* 10 gene, which also affects the zeugopodium (tibia and fibula). Comparisons of mammalian limb and extremity structure involve study of how *Hox* gene expression has changed in different taxa. *Hox* d13 is important in patterning the hands and feet (Kmita *et al.* 2002). Bones of the distal hominoid forearm are associated in a developmental module with the metacarpals and phalanges of digits II–V (Reno *et al.* 2001). Given the conservative nature of these developmental modules, this is probably generally true for mammals. Thus, in the mammalian limb, digit I (the pollex or hallux) occurs in a separate developmental module from another module that contains digits II–V and the zeugopodium (radius and ulna or tibia and fibula). Hence, digits II–V are developmentally associated with the two long bones of the distal limb, and digit I occurs in a different module. *Hox* d13 and a13 underlie the patterning of digit I, but *Hox* d11 underlies the patterning of digits II–V and the zeugopodium (Figure 13.5). The existence of these two separate modules explains major variations within the distal limbs and extremities of fossil and living mammals, including primates.

What significance does anatomical genomics have for ideas about bipedal origins? It explains the mosaic nature of traits that are evolving in the fore- and hindlimb. Because the stylopodium is patterned differently, trends affecting the morphology of humerus or femur can operate relatively independently in forelimb or hindlimb. This explains why relative length of the humerus and femur are not well correlated. Species can have a long humerus and short femur (e.g., orangutan) or a short humerus and a long femur (e.g., humans). *Hox* 10 patterns the stylopodium and zeugopodium in both fore- and hindlimb. However, the stylopodium is also patterned by *Hox* 9 in the forelimb. There is therefore a stronger link between stylopodium and zeugopodium in the

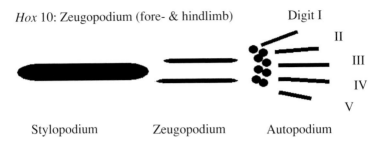

Hox 9 & 10: Stylopodium (forelimb)
Hox 10: Stylopodium (hindlimb)

Hox 10: Zeugopodium (fore- & hindlimb)   Digit I

                                                            II

                                                            III

                                                            IV

                                                            V

Stylopodium   Zeugopodium   Autopodium

Hox d11: Zeugopodium and digits II-V
Hox d13 & *Hox* a13: digit I

Figure 13.5. A schematic view of the three developmental modules of the vertebrate limb. They develop along a vector, moving from proximal to distal elements. The stylopodium contains the proximal long bone (humerus, femur), the zeugopodium contains the distal long bones (radius and ulna or tibia and fibula), and the autopodium contains the carpals/metacarpals or tarsals/metatarsals and the phalanges. The stylopodium is patterned differently in the fore- and hindlimb. *Hox* 10 and *Hox* d11 genes pattern the zeugopodium. Digits II-V of the autopodium are also patterned by *Hox* d11, but digit I is patterned by *Hox* d13 and *Hox* a13.

hindlimb, because only *Hox* 10 underlies these developmental modules. Trends for elongation or reduction of proximal or distal hindlimb segments should be more correlated. Hence, hindlimb reduction affects both stylopodium and zeugopodium (e.g., orangutan), and hindlimb elongation affects both stylopodium and zeugopodium (genus *Homo*).

Shifts in the morphology of the zeugopodium and autopodium can be examined in a similar fashion. Living pongids and hylobatids have a relatively long forearm and long digits II–V. This is also true of the australopithecines. Later hominids, members of genus *Homo*, reduce both the forearm and the lateral rays. The association of trends in the forearm and the lateral rays is explained by their origin in a single developmental module. Enlargement or reduction of the thumb separately from the other digits of the hand is explained by its position within a separate developmental module. Similarly, in the hindlimb, digit I (the hallux) can undergo change relatively independent from change occurring in digits II–V. For example, marked reduction of the hallux occurs in the living orangutan, although the lateral rays are elongated (Tuttle & Rogers 1966). The opposite trend occurred among hominids. It is likely that enlargement of the hallux was among the first anatomical changes taking place with increasingly more frequent bipedal behaviors in hominids. Elongation of the hindlimb zeugopodium (tibia and fibula) is associated with elongation of digits II–V, seen in metatarsal elongation. Reduction of the phalanges – necessary for bipedalism – occurs with a re-positioning of the inter-phalangeal joints. The genetic substrate for joint structure remains unknown, but joint position appears determined by relative degrees of cellular growth within the developing extremity. Study of digit formation in the hand of the prosimian mouse lemur (*Microcebus*) shows that programmed cell death contributes to the origin of the metacarpal-phalangeal and inter-phalangeal joints (Chiu & Hamrick 2002: Fig. 5). In contrast to other mammals, but typical for primates, the mouse lemur has short metacarpals and long phalanges.

Shoulder and pelvic morphology are also patterned by *Hox* genes, *Hox* 10 and *Hox* 11 (Wellik & Capecchi 2003). These same genes function in patterning the stylopodium and zeugopodium. The humerus is also affected by *Hox* 9, although the femur is not. This might imply a greater degree of developmental influence between the pelvis and proximal and distal segments of the hindlimb than between shoulder and forelimb, because the humerus is separately affected by *Hox* 9 function. Experimental knockout mutations in the mouse demonstrate that *Hox* 10 and 11 control the position and morphology of the pelvis to a remarkable degree (Wellik & Capecchi 2003: Fig. 2).

Is bipedalism a grade?

Bipedalism may be an adaptive grade achieved separately by different hominid lineages. This is suggested by what appear to be two types of foot anatomy in fossil hominids. There is a fossil foot (OH 8) with derived traits anteriorly and a primitive ankle anatomy, and a fossil foot (StW 573) with derived traits at the ankle and a primitive, widely abducted hallux, that is presumably capable of grasping. There is no doubt that two radically different forms of foot anatomy exist (Clarke & Tobias 1995, Harcourt-Smith & Aiello 2004). StW 573 is the foot of a virtually complete male australopithecine skeleton that is currently being excavated from the Sterkfontein breccia (Clarke 1998). The find may be substantially older than OH 8, which comes from Olduvai Gorge, and is generally presumed to be the foot of *Homo habilis*, although it may also be the foot of *Australopithecus boisei*. StW 573 may be as old as 4 mya, and thus contemporary with Lower Pliocene hominid fossils from East Africa (Partridge *et al.* 2003). This early date is controversial. A date of 3.0–3.5 mya is generally accepted. Thus, different ages or taxa may account for the differences in foot anatomy. However, the important point is the anatomical disparity. As noted by Morton (1924) in the early twentieth century, because the foot contacts the substrate and delivers the entire forward thrust of locomotion, its anatomy is crucial for the development of bipedalism. Two hominid foot morphologies therefore ineluctably indicate two hominid lineages.

The discovery of two different types of foot anatomy leads inevitably to the idea that different catarrhine lineages may have been experimenting with bipedalism during the African Pliocene. This would be a woeful realization for cladists, because it would mean that discerning the true ancestor/descendant relationships of fossil hominid taxa might be difficult. The cladistic bugbear of homoplasy might also emerge, unless different mosaics of bipedal traits could be unequivocally associated with different lineages. Yet, if different lineages were indeed experimenting with bipedalism during the late Miocene/Pliocene, and leaving descendants behind during the Plio/Pleistocene, it would be a boon for evolutionary theorists! Homoplasy in the hominid foot would make it easier to study the forces of natural selection that led to the emergence of bipedality. There would be more than only a single unique experiment in bipedal locomotion. Thus, the selection pressures that independently generate bipedalism (or different types or degrees of bipedalism) might be easier to discern.

Paleontologists who deal with other organisms do not wail and gnash their teeth when discovering the existence of grades. They seize an opportunity to examine the nature of the new adaptive zone that molded the evolution of separate, distinct lineages. For example, the transition from mammal-like reptiles to true mammals was a grade that was experimented with by a number of different lineages. Even the limited array of living mammals documents that this grade was independently achieved, because the unique middle ear bones were evolved separately in monotremes and therians (Rich *et al.* 2005). A fundamental and complex trait, often thought to be the strongest shared derived trait of living mammals, was evolved in separate lineages. A complex cranial trait, unique to living mammals, develops through homoplastic evolution. Similarly, the appearance of new fossil hominid feet reveals the diversity of the early hominid radiation and experimentation with bipedalism – the key hominid trait. If this definitive hominid trait is subject to experimentation, future analysis will almost certainly show different anatomical mosaics in different lineages experimenting with life on the ground. Multiple origins of definitive traits, and the sequence of accretion of these traits, would allow one to examine the nature of the early hominid adaptive zone. Put another way, one could examine the evolutionary biology of walking. Yet, if bipedalism does prove to be a grade, I predict there is likely to be a rush among paleoanthropologists to discover the "true" early hominid lineage – the one that leads to genus *Homo* – rather than a rush to examine adaptation.

There is another fossil hominoid taxon that contributes to ideas about the origins of bipedality. This is *Oreopithecus bambolii*.

Lessons from *Oreopithecus*

The European primate *Oreopithecus bambolii* has been known since the late nineteenth century (Gervais 1872), and has been variously described as a cercopithecid or parapithecid (specifically, like *Apidium*) on the basis of its molar crown morphology, or as a hominoid on the basis of its postcranial morphology (Delson 1986). It is generally classified in its own zoological family (Family Oreopithecidae), and is sometimes indicated to be of uncertain taxonomic status (*incertae sedis*) within the Catarrhini. Describing new material, Johannes Hürzeler (1958) argued that *Oreopithecus* was an early hominid. A hominid status was rigorously questioned by later workers (Straus 1963, Szalay & Berzi 1973).

During the late Miocene, from 9 to 6.5 mya, *Oreopithecus bambolii* existed on a large Mediterranean island composed of northern Italy, Corsica, and Sardinia. Crushed but relatively complete fossils of this endemic hominoid are found in lignite coal deposits. Paleobotanical remains indicate the existence of an ancient swamp forest (Azzaroli *et al.* 1986). In addition to *Oreopithecus*, the Tusco-Sardinian Island also contained very large endemic rodents, a relict anthracothere, and small bovids and suids. However, the only carnivores present were an omnivorous ursid and two genera of otters. There were no large predators on the island until a land connection with Europe allowed sabertooth felids to penetrate the area at 6.5 mya (Agustí & Antón 2002). Under these insular and predator-free circumstances, *Oreopithecus* developed some unique bipedal specializations. Although the body proportions of *Oreopithecus* resembled those of the modern orangutan, with long forelimbs and short hindlimbs, its lumbar, pelvic, and foot anatomy indicate a peculiar type of bipedal locomotion (Köhler & Moyà-Solà 1997). The widely abducted hallux and short lateral rays of the foot acted as a kind of tripod, enabling *Oreopithecus* to stump along on the ground. Radiographs of the ilium show a trabecular architecture that is similar to that of living humans and australopithecines (Rook *et al.* 1999). Thus, the pelvis may have supported and transmitted body weight during some form of bipedal locomotion, although one might question the frequency of bipedalism on swampy terrain. Susman (2004), in fact, argues that the long hands and curved, grasping fingers of *Oreopithecus*, as well as its orangutan-like body proportions make bipedalism unlikely.

At about 25 kg in weight (Agustí & Antón 2002), *Oreopithecus* was a primate of respectable size, but was still smaller than living pongids or known fossil hominids. Because *Oreopithecus* lived in what was then an apparently predator-free island, this seems to imply that the absence of predators may be an important factor in the development of terrestrial bipedality. Yet, Plio-Pleistocene fossil

Figure 13.6. A pride of lions in open grassland, Masai Mara Reserve, Kenya.

Figure 13.7. A cheetah in the Masai Mara Reserve, Kenya. Note how the animal has settled into the only patch of available cover. The African fossil record contains living mammalian carnivores, such as lions and cheetahs, as well as different, extinct felids, hyaenids, and canids. Some of the extinct forms, such as the saber-toothed cats and hunting hyaenas, have no living representatives.

hominids in Africa were certainly not isolated on islands, and they co-existed with many large ground-dwelling carnivores. Some of these carnivores still exist (Figures 13.6 and 13.7); many others are extinct. Fossil evidence shows the existence of a complex guild of large mammalian carnivores in Africa beginning in the late Miocene and persisting through the Pleistocene. The Plio-Pleistocene hominids were larger than *Oreopithecus*, but not large enough to discourage almost all potential predators. Both Franz Weidenreich (1946) and John Robinson (1972) argued that the first hominids must have been very large animals – otherwise they would not be safe on the ground. Yet, the record of human paleontology does not bear this out. Were Plio-Pleistocene hominids active when potential predators were sleeping? Were they careful always to keep near good refuge trees? Did they evolve vigilant behavior and division of labor with a sentinel role developing? Were they capable of putting up a good group defense? If so, how does this type of sociality evolve? These points will be discussed in Chapters 15–17.

A mixture of morphologies

One of the greatest barriers to understanding the lifeways of the earliest hominids, the australopithecines, lies in the fact that their morphology has no representation in the present world. It is difficult to understand what the mixture of arboreal and bipedal traits means. The discovery of a good data set for *Australopithecus afarensis* at Hadar, Ethiopia, initiated re-evaluation of ideas about the morphology and locomotion of the australopithecines. The first temptation is to measure the australopithecines against modern humans, and find the ancient hominids wanting. If bipedal efficiency is gauged by modern morphology, and the australopithecines lack the full complement of modern traits, then they must be relatively inefficient bipeds (Stern & Susman 1983, Stern 2000). And, of course, this also explains why they are extinct. Alternatively, arboreal traits in *A. afarensis* were considered to be primitive retentions in a fully functional biped (Lovejoy 1988).

I do not see how terrestrial bipedalism necessarily completely diminishes arboreality. If the shape and internal architecture of fossil bone demonstrates bipedal adaptation in one part of the skeleton, then bipedalism existed. Similarly, if the shape and internal architecture of fossil bone demonstrates arboreal adaptation in another part of the skeleton, then arboreality also existed.

Vertebrate paleontologists face many similar problems when they deal with a group that has a good fossil record and has modern representatives. Consider horse evolution. Unlike some paleoanthropologists, students of fossil horses do not take a progressivist view and assume that the entire course of horse evolution leads inevitably to modern horses, so that fossil three-toed species are automatically damned to extinction. Instead, they strive to understand the adaptive nature of traits that have no living exemplar, through the study of functional morphology and paleoecological analysis.

The existence of relatively long forelimbs and arboreal traits such as a vertically oriented glenoid, curved phalanges, and a relatively flexible ankle joint in australopithecines – and the apparently stable nature of this morphology – should trigger a search for the utility of these traits. Australopithecine morphology persists for a long time. In terms of chronology, it is the norm for hominids. Hence, rather than being clumsy pseudo-bipeds, the australopithecines were probably good bipeds while also being adept tree-climbers. Arboreal traits are not merely primitive retentions waiting to be winnowed out by natural selection. Arboreal traits were useful, and may have been of critical utility, given the importance of reaching shelter in trees at night.

Endnote

1. Adult weights are mean of male and female weights.

14 *The hominid radiation*

The earliest hominids

The evidence for hominid origins in the Miocene has already been reviewed (Chapter 3). Morphological, molecular, and genetic research since the 1960s has confirmed Thomas Henry Huxley's suggestion that the African pongids are the closest living relatives of the hominids. Kortlandt (1972) argues that the tectonic barriers of the African Western and Eastern Rift systems, as well as rivers, lakes, and swamps, created barriers to gene flow and initiated hominid speciation. However, the discovery of australopithecine fossils in Chad, 2500 km west of the African Rift System, indicates that hominids were a widely dispersed group of animals (Brunet *et al.* 1995). Hence, australopithecines were unlikely to be affected by the types of geographical barriers that create species in non-human primates.

The discovery of an earlier taxon from Chad, *Sahelanthropus tchadensis*, dating to 6–7 mya (Brunet *et al.* 2002), is important for two reasons. First, its existence contradicts the presumed time of hominid origin at 5–6 mya – a widely accepted date generated from the molecular clock. Clearly, if *Sahelanthropus* is a hominid, the divergence of hominids from the African great apes must have occurred before 6–7 mya. A second important feature of the *Sahelanthropus* find is its paleoenvironmental context. Sedimentological and paleontological evidence document the presence of lake deposits, but fossil aeolian dunes in the geological section provide the oldest known evidence for desert conditions, and a prevailing ancient wind regime in the Sahara region (Vignaud *et al.* 2002). Thus, although the evidence points to the existence of lacustrine, gallery forest, and savanna habitats, it also points to the proximity of a sandy desert. This is not merely a Serengeti type grassland lying nearby, but a true desert. Profoundly arid conditions must have occurred in north-central Africa at this time.

Some researchers argue that the earliest hominids emerge with little divergence. That is, late Miocene hominid material may be incorporated within a single genus, and differences reflect merely species-level or subspecies-level variability, rather than variability at a higher taxonomic level. Thus, Haile-Selassie *et al.* (2004) believe that no significant dental traits separate a newly recognized species of *Ardipithecus* (*Ardipithecus kadabba*), *Sahelanthropus tchadensis*, and *Orrorin tugenensis*. If this should prove true, then the beginning of the hominid radiation is delayed until the Pliocene, and the advent of the genus *Australopithecus*. However, the hypodigms of *Ardipithecus, Sahelanthropus*, and *Orrorin* appear different enough to warrant distinction at the genus-level.

Plio-Pleistocene hominids

The first australopithecine was described by Raymond Dart (1925) from the South African site of Taung. Beginning in the 1930s, other South African sites also yielded australopithecine fossils in breccia deposits that filled ancient fissures in the karstic bedrock (Figures 14.1 and 14.2). The Plio-Pleistocene hominids of Africa were exhaustively reviewed by Howell (1978). After this major synthesis, new specimens were recovered, in both East and South Africa. The discovery of abundant remains of a Pliocene hominid species, *Australopithecus afarensis*, including a partial skeleton (AL 288–1, "Lucy"), allows especially detailed investigation of the skeletal anatomy of this species (Johanson *et al.* 1982). Multiple dating techniques assure better chronometric control over recovered specimens. Two very complete hominid specimens have been found in South Africa (StW 573) and in East Africa (KNM-WT 15000). Noted taxonomists have reversed their opinion about the importance of adaptation. In particular, some of the foremost practitioners of the cladistic methodology have accepted the existence of adaptation, and have used adaptive complexes to define fossil hominids (Wood & Collard 1999a). Using this adaptive orientation, species are collapsed into genus *Australopithecus* if their body proportions differ from those of modern humans. *Homo habilis* and *Homo rudolfensis* thus disappear into genus *Australopithecus*. Alternatively, some researchers resurrect *Homo habilis*, and collapse *Homo rudolfensis* into this taxon (Blumenschine *et al.* 2003).

These arguments are not simply abstruse taxonomic points. The distinctions between genus *Australopithecus* and genus *Homo* may be indistinct, if postcranial remains are absent. For example, I spent 2 days in the Kenya National Museum attempting to categorize the specimen KNM-ER 1805, which is generally considered to be a specimen of *Homo habilis*. I had difficulty in assigning this specimen to either genus *Australopithecus* or genus *Homo*.

The single-species hypothesis

Shortly after the widespread acceptance of australopithecines as hominids, the systematist and evolutionary theorist Ernst Mayr (1950) argued that a taxonomic

Figure 14.1. Sterkfontein, South Africa. Sink-holes and fissures continue to form in the ground of this karstic limestone landscape. A deep chimney is forming just beneath the modern land surface. Animal remains (including hominids), stone tools, and sediments on ancient land surfaces fell or were washed into similar chimneys. With the passage of time, this material was cemented into a very hard solid matrix. The result is a composite of rocks and bones cemented together by fine-grained sediments (breccia). Current excavations at Sterkfontein conducted by the Paleo-Anthropology Research Unit of the University of the Witwatersrand are laboriously removing fossils and stone tools from the hard breccia.

revision of hominids should take place. Contrary to the general practice of vertebrate paleontology, nearly every new scrap of hominid fossil was being accorded a new taxonomic category – genus, species, or subspecies. Yet, if hominids were treated in the same fashion as other mammals, population-level variability would be recognized, and the multitude of hominid taxa would be collapsed into a far smaller number. Mayr suggested that the plentiful hominid genera and species should be collapsed into only two genera (*Homo* and

Figure 14.2. Dr. Ron Clarke stands at the base of one of the very deep Sterkfontein excavations. He is pointing to the division between Members 4 and 5 in the complex stratigraphic sequence. The angular rock fragments and animal bones within the breccia can easily be seen in the vertical wall of the excavation.

Australopithecus), each containing only a few species. Mayr presented a crucial argument: the hominid niche was so broad and flexible that only a single hominid species could exist at any given time. In fact, it was impossible for more than one hominid species to exist at any given time, given the unique properties of the hominid niche. The principle of competitive exclusion (called Gause's law), was invoked at this point. This principle states that two species cannot coexist in the same habitat if they have the same niche. It is the equivalent of the Euclidean axiom that two objects cannot occupy the same space at the same time. Hence, the course of hominid evolution became a series of hominid species progressively replacing each other. This was the single-species hypothesis, which is one of the most theoretically compelling ideas ever presented about human evolution.

The single-species hypothesis was ultimately falsified by the discovery of clearly sympatric hominid species. The sympatry of *Homo erectus* and *Australopithecus robustus* in South Africa seemed clear, but chronometric dates were not possible for the South African sites. The most compelling evidence for sympatry therefore accrued at well-dated sites in East Africa, principally in the Koobi Fora region of Kenya. The discovery of the fossil KNM-ER 1470, whose cranial capacity and morphology seemed very different from *Australopithecus boisei*, was critical in forming the consensus opinion that sympatric hominid species could coexist (Chapter 2).

Since the 1990s, fossil hominid taxa have greatly increased in number. Truly novel taxa (e.g., *Orrorin tugenensis*) have played a role in this increase (Senut *et al.* 2001). Yet, sometimes this multiplication is an artifact of an ingenuous or naïve use of cladistic methodology that discards biological meaning or adaptation. Sometimes taphonomic processes alter material enough to create the perception of novelty. For example, expanding matrix distortion may be responsible for some of the novel characteristics of *Kenyanthropus platyops*, which supposedly represents a distinctive Pliocene hominid lineage, long removed from australopithecine evolution (White 2003).

Sometimes a typological approach to fossil material multiplies hominid taxa, whereas an acknowledgment of population-level variability would reduce the number of taxa. African *Homo ergaster* can thus be collapsed into *Homo erectus*, when the African material is shown to exhibit traits seen in European or Asian material. Some researchers even collapse *Homo erectus* into *Homo sapiens* (Wolpoff *et al.* 1993). On the other hand, arguing that living primate subspecies cannot be discerned from the skeleton or dentition, but are established solely on trivial and meaningless soft tissue traits like coat color and pattern, Tattersall (1986) contends that any fossil hominid material demonstrating skeletal or dental differences should be recognized by a species distinction. He believes that this approach is more conservative than acknowledging the existence of variable fossil hominid species. Yet, dental and skeletal differences can reliably differentiate living primate subspecies and even populations. The reality of subspecies and population-level differences among living primates allows one to recognize the possibility of their existence among fossil material. Furthermore, such differences highlight the power of natural selection and ecological factors affecting speciation under field biology conditions.

In any case, amid all the taxonomic tumult, it is useful to recognize one simple fact: if every fossil species possesses a distinctive niche – which is the case for living species – then *Homo habilis, Homo rudolfensis, Homo erectus, Homo antecessor*, and *Homo heidelbergensis*, to list only some possible species of genus *Homo*, must each possess a distinctive niche. Could these niches ever be discerned from the fossil morphology? What about the behavioral component of the phenotype? Archeology should contribute to the analysis, because it is the record of hominid behavior. However, even if no difference is detected in the archeological record, sometimes unknown or unknowable psychic or mental differences are invoked to separate Neanderthal and anatomically modern humans in the Levant. And anatomically modern humans with Middle Paleolithic tools can be categorized as still another species of genus *Homo, Homo helmei*. Conroy (2003: 792–793) predicts that no fossil hominid genera will be shown to contain more than five species per genus, once remains are better known and the influences of intra-specific variability, including sexual dimorphism, are taken into account. However, this prediction is based on speciosity in living primates, and speciosity in living primates is skewed by at least two extinction events since the mid-Miocene (Chapter 4). Diversity in living primates is skewed and altered by these events. Because hominids experienced a true radiation in the late Neogene, I believe that expectations of their species richness (especially in the australopithecines) should be molded by study of primate species diversity during periods of evolutionary radiation. Hominoid diversity in the early Miocene, for example, leads one to expect more than five species in genus *Australopithecus*.

Some researchers argue that six or more species of genus *Homo* coexisted in the early Pleistocene, only to be winnowed out with the advent of *Homo sapiens*. However, it is unlikely that early genus *Homo* was so speciose. Among living mammals, the average mammalian genus contains 4.2 species, with a median of 2 species; the average primate genus contains 4.2 species, with a median of 2 species (Purvis *et al.* 2000: Footnote 17). Thus, primate speciosity precisely fits the mean and median values for mammals. Large mammals do not have as many species per genus or per family as small mammals do (Van Valen 1973b). The species richness of early *Homo* in contrast to other mammalian genera can be assessed by examining the species richness of extant mammalian genera with a similar body size. Using this method, one or two hominid species is the expected number for a mammal genus with a body size of 30–65 kg, which is the size range usually estimated for early *Homo* fossils (Conroy 2002).

Body size influences not only the number of species, but also niche structure and mode of speciation. For example, small mammals that are widespread are classified into more subspecies than widespread large mammals are. Related species of small mammal can occur in separate but neighboring distributions through a geographic region (Searle 1996). This distribution of species in contacting, but not overlapping, ranges is parapatry, and small mammal species are expected to show more parapatric speciation than large mammal species do, and perhaps more modes of speciation in general (Searle 1996). All known hominid species are large mammals. Thus, on theoretical grounds, hominids should have low numbers of species, larger home ranges, and less reliance on local habitat resources.

When body size is examined across a wide range of taxa, logarithmic transforms of size measurements are used to reveal significant patterns among animals that may be hugely different in size. Generally (e.g., Purvis & Harvey 1997) these logarithmic transformations are skewed to the right. That is, large animals are less frequent than small. There is a high peak of small animals at the left or small range of the size distribution, and a long tail out to the right of rare large animals. The skewed shape of such allometric transforms is generally thought to reflect the optimal size for certain functionally related variables shared by animals within the group. Competition between species is believed to generate this skewness. Competition within a species can also produce a skewed distribution. Theoretical modeling of optimal adult body size that assumes that digestion and respiration are affected by body size demonstrates a right skew within the same taxon, even with logarithmic transformations (Kozlowski & Weiner 1997). Computer simulations also confirm a statistically significant right skew when body size is optimized (Kindlmann *et al.* 1999). These results show how body size optimization within taxa can affect the field biology data that empirically show skews in animal size distributions within a group of related taxa or within a community. Although these investigations may appear too abstract to apply to case studies in paleoanthropology, they can affect analysis in a profound way. For example, a number of lines of evidence indicate that it is probable that more niches exist for small-bodied species than for large-bodied ones. When the fossil evidence of the hominid radiation is explored, one would expect to find many small species among sympatric taxa, and few large species.

Sympatry and multiple hominid niches

During the 1950s, John Robinson (1954, 1956) became the first researcher to articulate the idea that sympatric hominid species might have been able to coexist by dividing the available resource space. He envisioned that a difference in diet might have separated sympatric hominid species (Figure 14.3). He believed that this dietary difference must have been significant in order to allow the long coexistence of closely related species. He applied this approach first to the South African australopithecines (*Australopithecus africanus* and *Australopithecus robustus*) and later to South African genus *Homo*. In his initial development of this model, Robinson argued that robust australopithecines (*Australopithecus robustus*) were herbivorous, and gracile australopithecines (*Australopithecus africanus*) were omnivorous. *Homo erectus* was able to coexist with *Australopithecus robustus* because their niches were separated by major dietary

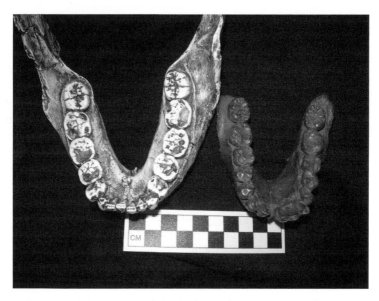

Figure 14.3. Casts of two hominid mandibles, illustrating Robinson's dietary hypothesis. On the left is *Australopithecus boisei* (Peninj 1, Lake Natron, Tanzania, 1.5 mya), representing a robust australopithecine. On the right is *Australopithecus afarensis* (A.L. 400–1a, Hadar, Ethiopia, 3 mya [missing right I1]), representing a gracile australopithecine. The robust australopithecine has large postcanine teeth relative to the anterior teeth. The gracile australopithecine has small postcanine teeth relative to the anterior teeth. The scale is in centimeters.

differences. This clever ecological approach to the hominid fossil record became known as "Robinson's dietary hypothesis." So different were members of *Australopithecus robustus* and closely related taxa ("robust australopithecines") from other hominids that Robinson endorsed the use of the genus "Paranthropus" for these taxa. In a later work, Robinson (1972) argued that not only major dietary differences, but also profound differences in habitat and locomotion separated *A. robustus* and *A. boisei* from other australopithecines ("gracile australopithecines") and genus *Homo*. In fact, he argued that *A. africanus* be sorted into genus *Homo* because of adaptive similarities – an argument also made by Mayr (1950). Robinson postulated that the robust australopithecines were more ancient than the other australopithecines, and that they may have descended from the genus *Gigantopithecus*, which represents the first hominid genus.

Yet, it is often not recognized that sympatric hominid species may exhibit only relatively minor niche differences. Contrary to Robinson's dietary hypothesis, coexisting hominid species might have been very similar in diet. Several researchers have made this point (Cachel 1975a, Winterhalder 1980, 1981).

Stable isotope analysis of the tooth enamel of South African australopithecines and genus *Homo* confirms this theoretical possibility (Sponheimer & Lee-Thorp 1999, Lee-Thorp *et al.* 2003). *Australopithecus robustus, Australopithecus africanus*, and *Homo* appear similar in diet, with all three taxa incorporating animal food items into their diet. Thus, analysis of stable carbon, nitrogen, and oxygen isotopes allows paleoanthropologists to examine the dietary differentiation between sympatric fossil species. Furthermore, the dental differences between "robust" and "gracile" australopithecines, which so impressed John Robinson, may have been caused by relatively slight genetic differences, and not a radically different genetic substrate, which would signal a profound phylogenetic distance between the two early hominid groups. Laboratory experiments on the development of lower crown morphology in mice reveal that expression levels of a single protein (ectodysplasin) can transform dental traits so much that the teeth appear to come from different species or genera. These dental traits include tooth presence, tooth size, cusp number, cusp shape, and longitudinal and transverse cresting (Kangas *et al.* 2004: Fig. 1). These changes were caused by mutations in a single gene.

There are additional ways in which the niche structures of sympatric fossil hominids can be studied. One way is to examine Hutchinsonian size ratios, which are relatively easy to retrieve from the fossil record. These might give insight into niche overlap and the degree of competition. In 1942, Julian Huxley argued that there should be limits to similarity and therefore limits to niche overlap between competing species. Character divergence producing a minimal degree of size difference should thus separate closely related sympatric species (Carothers 1986). In 1959, G. E. Hutchinson made the same argument from empirical evidence, calculating ratios between species pairs using morphological traits related to food acquisition that were important in defining niche structure (Hutchinson 1959). Hutchinson's examples yielded a ratio of about 1.28 (rounded to 1.3) for linear structures. This was later extrapolated to a ratio of 1.69 $[(1.3)^2]$ for areas, and a ratio of 2.2 $[(1.3)^3]$ for volumes. These Hutchinsonian ratios may specify the minimal degree of separation needed to decrease competition and achieve niche separation between closely related sympatric species, and they have been used as evidence for processes that structure animal communities (MacArthur & Levins 1967, Roth 1981).

There has been extensive debate about the existence and cause of the Hutchinsonian ratios (e.g., Roth 1981, Eadie *et al.* 1987). In fact, detailed studies in modern species of the relationship between size and the nature of competition are rare (but see Schoener 1984). Body weights, rather than areas or lengths, have been the principal focus of study. It has been argued (Eadie *et al.* 1987) that Hutchinsonian ratios exist only because species sizes with log normal distributions are being sampled – the natural log of 10 is 2.3, which is near the expected 2.2 volume differences of competing species pairs. Even if this is so, the cause of the log normal distribution and the appearance of the expected larger weight than length ratios must still be explained. Maiorana's (1978) explanation is that the 1.3 value is caused by variance in the size distribution of a species, which is affected by morphological variation within populations, and character displacement between species. Given the typical degree of morphological variation within traits of a species, a 1.3 ratio creates nearly complete separation between frequency distributions across species. This implies that the ratio is the minimal degree of separation necessary for species to coexist. Maiorana (1990) also argues that Hutchinsonian ratios within species clusters or guilds represent the difference in size necessary to create niche divergence between guilds, and is the result of diffuse competition within communities, rather than internal competition within a guild. Yet in local areas the body sizes of guild members show a more distinct separation, indicating that competition is occurring in local faunas and causing character displacement.

Schluter (2000) examines 61 cases of character displacement in extant species within the same genus. He defines ecological character displacement in traditional terms – as the evolution of phenotypic traits that are created or maintained by resource competition between two or more sympatric species. Resource competition is "the negative impact of one phenotype on another arising from depletion of shared resources" (Schluter 2000: S5). Only 23% (14/61) of the cases of character displacement showed evidence of resource competition between sympatric species, although this was the presumed cause of the character displacement. Even in 10 experimental cases among these 14, it was uncertain whether resource competition or some other hostile interaction was occurring between species. Species size ratios (1.4) are larger when phenotypic differences are exaggerated in sympatric species than when adjacent species are evenly distributed along a size axis (1.27). When differences in a measurement are available for species pairs under both sympatric and allopatric conditions, the character ratios are significantly greater when both species shift an equal amount (1.54) than when only one species shifts (1.29). The degree of shift may be determined by resource breadth, with a wide resource distribution causing a large shift (Schluter 2000: S8).

In experiments with sympatric stickleback fish species, Schluter (2000) documented that resource competition declines when character displacement occurs, but is pronounced when morphological or ecological differences are small. This confirms the Hutchinsonian expectation of the existence of a minimal niche separation needed to decrease competition between closely related sympatric species. Schluter's addition of another competitor to his stickleback experiment had the greatest effect on fish that most resembled the competitor in morphology and diet. A third experiment proved that the morphology and diet of the competitor caused the effect, not merely the addition of simply another competitor. Schluter (2000) summarizes his review by concluding that significant character displacement occurs in the living world, and is probably responsible for much of the diversity seen in adaptive radiations.

Hutchinsonian size ratios were discussed in Chapter 11. Table 11.2 presented the Hutchinsonian size ratios for several dental variables in sympatric *Australopithecus robustus* and *Homo* and sympatric *Australopithecus boisei* and *Homo*. Note that these sympatric hominid species fall significantly at or below the Hutchinsonian thresholds, indicating that competition is occurring. That is, the hominid niches have not separated far enough to eliminate competition and divide up the resource space without interference. This is not true for most of the sympatric Miocene hominoid taxa presented in Table 11.2. Most of these taxa lie well above the critical Hutchinsonian thresholds, indicating a wide degree of niche separation. Figure 11.1 makes the same point through the use of changing postcanine tooth shape along the dental arcade. Miocene hominoids are extremely diverse, while living hominoids are much more similar to each other. However, Plio-Pleistocene hominids are extraordinarily similar to each other. The resemblance is much closer than between living hominoids. Why is this striking? If Plio-Pleistocene hominids are all members of the same zoological family, shouldn't they resemble each other more than living hylobatids, pongids, and hominids do? However, there are molecular anthropologists who insist that the living African great apes and sometimes the orangutan are members of the same family as living humans. These researchers urge that a taxonomic change should occur, and that the great apes should be sorted with living humans into a single zoological family: Family Hominidae (Chapter 3).

Another method that might be used to examine the degree of competition among fossil hominid species is morphospace analysis (Chapter 5). One could examine the morphospace volume to see which areas are densely occupied (implying greater competition), and which areas are sparsely occupied (implying lesser competition). Because hominids were never as diverse as rodents or crinoids, there will not be the abundance of sympatric species that makes this approach most productive for examining a particular time range. However, this approach could be usefully applied to changing morphospace volume through time. For example, there is apparently a wide array of locomotor and dental anatomies, body sizes, and shapes among the earliest hominids. After the Middle Pleistocene, this is no longer the case. Species differences (if they are real) appear largely confined to craniofacial traits. The relative lack of disparity among these fossil species (*Homo erectus, Homo heidelbergensis, Homo antecessor...*), if they were real species, would imply a greater degree of competition between them than between sympatric species of the African Plio-Pleistocene.

In summary, despite nearly 50 years of paleoanthropological tradition that has presented sympatric Plio-Pleistocene hominids as being widely divergent in niche structure, it is very likely that many of these species resembled each other to a significant degree. Certainly, it is possible for closely related species to coexist without being very different in diet and locomotion. It appears that continuing research into australopithecine lifeways will reveal that they resemble each other much more than in the classical Robinson perspective (Robinson 1956, 1972). Furthermore, it is possible that the niche differences between australopithecines and members of early genus *Homo* were also not profound until a relatively late date, around 1.7–1.5 mya, when members of genus *Homo* begin generating a more complex archeological record. It is possible that behavioral differences – including tool behavior – did not become qualitatively different until this time.

Sexual dimorphism and niche structure

Given the basic catarrhine substrate, a great degree of sexual dimorphism is theoretically expected in early hominids. However, it appears that the cranium of early hominids expresses much more dimorphism than the postcranium (Plavcan 2003). Postcranial remains yield a dimorphism estimate of 35% – i.e., males are 35% larger than females on the basis of postcranial bones (McHenry 1992). But if craniofacial traits are used to estimate fossil hominid dimorphism, strikingly larger estimates of body mass dimorphism are generated: the estimates range from 46 to 222%, depending on the trait and the comparative equation developed from modern species (Plavcan 2003: 56). It is important to note that hominids run counter to other hominoids (the hylobatids and pongids) in this respect. They resemble living New and Old World monkeys, where craniofacial dimorphism increases at a faster rate than body mass dimorphism. In hominoids, craniofacial dimorphism decreases with increasing body mass. Plavcan observes (2003: 55) that one would expect spectacularly large body mass dimorphism in early hominids, if their craniofacial traits scale with body mass in a typical hominoid fashion. This overwhelmingly large body mass dimorphism is not present in the fossil postcranial remains. What is the explanation for the aberrant hominid pattern?

I believe that the explanation for the aberrant scaling pattern of early hominid craniofacial traits with body mass dimorphism with reference to other hominoids lies in the fact that they resemble New and Old World monkeys. These primates are either completely (New World monkeys) or largely (Old World monkeys) arboreal. I suggest that the early hominid pattern reflects the continuing importance of arboreality in their lifestyle. They continued to seek shelter in trees, especially at night, and probably foraged for food in trees, as well. The African great apes probably spend larger amounts of time on the ground than the early hominids did. The orangutan is arboreal, but there is a peculiar pattern of sexual size dimorphism in this taxon. Males may continue to resemble females in body size and facial traits. In spite of reaching adulthood, many males never achieve the full development of body mass dimorphism. Arboreal hylobatids are much smaller in size than the early hominids.

Another conflicting estimate of australopithecine size dimorphism has been published by Reno *et al.* (2003). Using the *Australopithecus afarensis* postcranial specimens from Hadar, they estimate a size dimorphism of 15% in this taxon. This is far lower than the 35% estimate given by McHenry (1992), and is comparable to the degree of sexual dimorphism exhibited by modern humans. From this, Reno *et al.* (2003) infer the existence of a monogamous breeding system in *A. afarensis* (Chapter 16). I compromise between these estimates, and use a figure of 25% body size dimorphism in australopithecines (Chapter 17). Using Monte Carlo simulations of composite samples from both humans and African apes, Cunningham *et al.* (2004) argue that the level of postcranial dimorphism in the Hadar AL 333 site is actually pongid-like.

Although australopithecine sexual dimorphism is thought to be reduced in genus *Homo*, there is some evidence that it remained substantial in *Homo erectus* (Vekua *et al.* 2002). Sexual dimorphism may even have increased in genus *Homo*, in contrast to contemporary robust australopithecines. When the diameter of the femoral head is examined, there is more difference between male and female *Homo* specimens at Swartkrans than male and female specimens of *Australopithecus robustus* (Susman *et al.* 2001). Because body size increases in genus *Homo*, this increase in sexual size dimorphism conforms to Rensch's Rule.

Using evolutionary game theory, Maynard Smith & Brown (1986) argue that a range of body sizes should exist in a stable population if relative size is important, and if access to resources is dependent on age, sex, or status in a hierarchy. A stable distribution of phenotypes then arises in the population, determined by competition between individuals and environmental variance. Maynard Smith and Brown believe that long-term selection for size increase occurs in many mammal lineages, and is principally focused on males, given that sexual dimorphism is greater in larger-bodied species. These trends reduce the survivorship of individual species in which they occur.

The origin of genus *Homo*

The origin of genus *Homo* has traditionally been defined by an increase in brain size relative to body size (Leakey *et al.* 1964, Tobias 1971, Tobias 1991). However, intra-specific variability in brain size exists. Some fossils classified as genus *Homo* are not dramatically different in endocranial capacity from fossils classified as genus *Australopithecus* (e.g., KNM-ER 1813, KNM-ER 1805, D 2700). They fall within the size range of 500–600 cm^3. This variability in relative brain size makes it impossible to establish the boundaries of a cerebral Rubicon, as envisioned by Arthur Keith. As noted above, even researchers who formerly eschewed adaptive complexes in classifying fossil hominids now advocate that body shape be used to separate genus *Homo* from genus *Australopithecus* (Wood & Collard 1999a, 1999b).

Brain size has recently surfaced again as a defining feature of genus *Homo*, although now brain size is thought to increase only after the removal of an external constraining factor. The discovery of a species-specific mutation in a myosin gene is responsible for reduced masticatory muscle mass in living humans (Stedman *et al.* 2004). Reduction in the temporalis muscle is viewed as releasing the neurocranium from external constraints. Brain size subsequently increases, which occurs concomitantly with a decrease in jaw size and masticatory muscle mass (Stedman *et al.* 2004). There are obvious problems with this idea about the origin of genus *Homo*. This argument runs counter to the "functional matrix" theory of craniofacial growth and form introduced during the 1960s (Moss & Young 1960, Moss 1971). This theory examines the growth of the cranium and face in terms of functional capsules. It has been thoroughly and practically tested during corrective surgery that alters and improves cranial, facial, and jaw anomalies in human infants and children. Surgical testing demonstrates that the size and shape of the human neurocranium is largely determined by the size of its contents, either by brain size or by the volume of enclosed fluids. This is seen in a number of congenital conditions. For example, in hydrocephaly, the neurocranium is expanded by the pressure of undrained cerebrospinal fluid; in congenital microcephaly, a very small brain forms a very small neurocranium; in anencephaly, absence of the cerebral hemispheres causes a complete absence of the cranial vault. Premature fusion of the cranial sutures can alter neurocranial shape, as in the case of acrocephaly or scaphocephaly. However, as long as some sutures remain unossified during the normal period of development, the vault can accommodate the growing brain through compensating growth in other regions. The size and attachment areas of the masticatory muscles can certainly affect the external morphology of the neurocranium, but its overall size is affected chiefly by the contents of the neural capsule.

In fact, the appearance of a mutation that reduced the size and strength of the masticatory muscles would be an evolutionary disaster without concomitant changes in anatomy or behavior. What animal species could tolerate a reduction in chewing muscles without accompanying anatomical or behavioral changes that would alleviate the consequences of such a reduction? However, Stedman *et al.* (2004) suggest a time of appearance for the myosin mutation (2.4 ± 0.3 mya) that overlaps with the first appearance of an archeological record. The first stone tools are dated to 2.5–2.6 mya in East Africa (Semaw *et al.* 1997, 2003), and animal bones modified by humans are associated with a small-brained australopithecine species, *Australopithecus garhi*, dating to 2.5 mya (de Heinzelin *et al.* 1999). This species had a cranial capacity of 450 cm^3 (Asfaw *et al.* 1999). Thus, I suggest that a more reasonable evolutionary scenario for the origin of genus *Homo* would run as follows. Triggered by pre-existing natural history intelligence, tool behavior allows hominids to acquire vertebrate meat and marrow using stone flakes and hammerstones (Chapters 8 and 16). The benefits of acquiring these nutrient-rich dietary resources creates new selection pressures for more efficient tool behavior. A feedback loop of selection pressure is set up that incorporates both increasing brain size and increasing tool behavior. Increasing brain size then expands the neural capsule, and alters neurocranial size and shape. At this point, a myosin mutation that reduced masticatory musculature could be tolerated, because novel dietary items and tool behavior would compensate for the lack of powerful chewing muscles.

Hominid dispersion from sub-Saharan Africa

The dispersion of hominids from Africa is timed by the occurrence of hominid fossils in the extreme Far East in Java, as well as in the Republic of Georgia in Central Asia. Javan *Homo erectus* has been radiometrically dated to 1.8–1.6 mya (Swisher *et al.* 1994). However, this antiquity is debatable. Other researchers believe that none of the Javan fossils is older than 1.1 mya (Hyodo *et al.* 2002). Multiple lines of evidence date the Dmanisi site in Georgia to 1.75 mya. The Arabian Peninsula was connected to Africa until seafloor spreading in the Afar Depression created the Red Sea. At the Strait of Bab al Mandab, which separates Ethiopia from Yemen, a seaway is present before 3.2 mya (Redfield *et al.* 2003). This seaway is presently 28 km wide at its narrowest point, and is as shallow as 137 m.

Asian ape-men: Early ideas about hominid origins in Asia

The current emphasis on hominid origins in Africa began in the 1950s. In the late nineteenth and early twentieth centuries, Asia was the focus of attention.

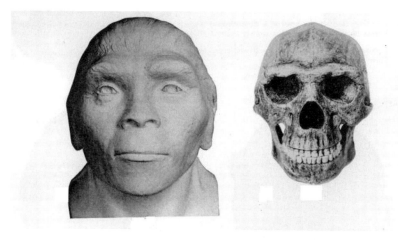

Figure 14.4. Model of the head of a female *Homo erectus* individual from Zhoukoudian (left) and the reconstructed skull upon which the model is based (right). From Weidenreich (1943: Fig. 175). Fossils from Zhoukoudian were originally identified as a novel genus and species, "Sinanthropus pekinensis." Fossil human remains from Java and China were perceived as being the earliest known hominids from the 1890s until the late 1940s. Beginning in the late 1940s, earliest hominid status shifts from Asian to African material.

This began with the 1891–1892 discovery of hominid fossils from the site of Trinil, Java, by Eugène Dubois (1895, Theunissen 1989). Attention on Asian finds culminated with the discovery of hominid fossils and archeological evidence at the Chinese site of Zhoukoudian, near Beijing. Zhoukoudian was long viewed as a Golconda of human fossils, based on initial description of the Zhoukoudian material by Davidson Black and a later magisterial description of cranial remains by Franz Weidenreich (1943). Even the inexplicable loss of most of the Zhoukoudian fossils at the outbreak of war between the United States and Japan in 1941 did not detract from their special status. Completeness of the fossils, their numbers, and the link to the much earlier material from Java (von Koenigswald & Weidenreich 1939) conferred a special status to the Zhoukoudian fossils as the best authenticated evidence for hominid evolution (Figure 14.4). Continued work at Zhoukoudian and other Chinese sites has yielded additional hominid fossils. Continued work in Java has yielded a plethora of hominid fossils from the Sangiran Dome and sites along the Solo River (Figure 14.5). Persistent volcanism at Javanese sites makes chronometric dating possible, although the association of hominid fossils with dated volcanic sediments is debatable (Hyodo *et al.* 2002). Nevertheless, it appears that the average age of *Homo erectus* fossils from Java is substantially older than those from China – perhaps as much as or more than 1 my old.

Figure 14.5. Cast of Sambungmacan 3, a *Homo erectus* calvaria from a site near the village of Poloyo on the Solo River, Central Java. Described in 2001, its age is currently unknown. The endocranial volume is 917 cc, within the range of variation for Javanese *Homo erectus* (Broadfield *et al.* 2001). The scale is in centimeters.

The origins of anatomically modern humans

Paleontological evidence points to the origins of anatomically modern humans (*Homo sapiens sapiens*) in Africa. Intriguing fragmentary fossil material from the South African site of Klasies River Mouth shows modern anatomical traits. However, the human fossil material from this site is both fragmentary and extraordinarily diverse. In addition, although the human material is dated to about 180 000 yrs BP, there is some equivocation about this date. The discovery of fossil human crania dating to 160 000–154 000 yrs BP from Herto in the Middle Awash region of Ethiopia is currently the oldest chronometrically dated material that appears to be reasonably modern (White *et al.* 2003). Photographs of the material appear to resemble most closely the South African Kabwe specimen. Stone artifacts in the Upper Herto Member are classified as "transitional Acheulean" – they show an intergradation between the Acheulean and Middle Stone Age industries, including the presence of the MSA Levallois knapping technique (Clark *et al.* 2003). Complex mortuary practices involving secondary burial of the human remains are inferred from defleshing cut-marks on the human bones (Clark *et al.* 2003). Inference of such practices is not merely an idle academic exercise deserving of a footnote. Such practices imply ritual behavior and complex cognition in ancient practitioners. Yet, mortuary practices in Neanderthals, who are not anatomically modern, are rarely used to make the case for complex cognition. If anatomical modernity is absent, the creators of a complex and intricate archeological record are always found wanting.

Nearly all of the research on anatomically modern humans has been focused on describing the fossil material and documenting its age. Virtually no attention has been paid to elucidating the forces of natural selection that might have led to the appearance of first modern anatomy, and, after a substantial lag-time, the appearance of modern behavior. Ironically, the disinterest in using natural selection to explain the origins of anatomical and behavioral modernity causes a strange resemblance between paleoanthropologists of the late nineteenth and early twenty-first century. In both cases, the appearance of modernity is attributed to an unknown and perhaps unknowable cause. Linking the origins of modernity to aspects of the human phenotype that are invisible to paleoanthropology, such as cognition, language, or an alteration of consciousness, does not explain the problem of its origins – it renders it mysterious. However, I believe that the problem is not insoluble, and I suggest that relatively subtle shifts in diet that involve an increasing abundance and predictability of dietary fat are the answer to the origins of modernity (Cachel 1997, 1999). This suggestion explains the lag-time between anatomical and behavioral modernity, and the appearance of a more abrupt transition in high latitude areas. However, this suggestion implies that the transition to modernity may have occurred in several areas of the Old World. This possibility is anathema to some researchers who emphasize the uniqueness of modern humans. Nevertheless, I feel it important to note that both anatomically modern humans and Neanderthals are associated with Middle Paleolithic lithic industries, and cannot be distinguished behaviorally by the archeological record. In addition, in areas where the archeological record is dense and well dated, the transition between the Middle and the Upper Paleolithic is not characterized by an abrupt break. Mousterian (Middle Paleolithic) and Aurignacian (Early Upper Paleolithic) people appear to have had the same lifeways (Grayson & Delpech 2003).

Emphasizing language as the sine qua non of anatomical modernity has led to a search for distinctive anatomical correlates of language (e.g., hyoid bone structure, position of the larynx) in fossil humans. These proposed anatomical correlates have not survived analysis. The search for signs of discrete language areas on fossil endocasts has also been problematic, given the faint sulcal and gyral traces usually found on endocasts. However, a strange genetic mutation has recently been implicated in human language origins. Pedigree analysis of a large extended family shows that some members with intense difficulty producing language carry an autosomal dominant mutation affecting the *FOXP2* gene (Vargha-Khadem *et al.* 2005: Fig. 1). The speech of affected individuals resembles speech that is typical of Broca's aphasia. Indeed, a special morphometric analysis of MRI scans shows that less gray matter exists within Broca's area in these individuals. The *FOXP2* gene is highly conserved among placental mammals, and shows no variability across human populations. The two amino acid changes unique to humans are thought to have been fixed about 200 000 yrs ago (Vargha-Khadem *et al.* 2005). This agrees with the origins of anatomical modernity, as indicated by the fossil evidence.

Genetic variation in modern humans

Genetic data have been used extensively to infer population history in modern humans. The most ambitious attempt to infer the history of anatomically modern humans from genetic data was mounted by Luca Cavalli-Sforza and his colleagues (Cavalli-Sforza *et al.* 1994). These researchers also attempted to integrate archeological and historical records and cranial and dental variation with the patterns of diversity revealed by multivariate analysis of the genetic data. With respect to deeper antiquity, detailed genetic analysis, using multiple lines of evidence, appears to reveal complex population histories for anatomically modern humans (Cruciani *et al.* 2002, Templeton 2002). Population history also appears to be complex when particular geographic regions are the focus of research. For example, there is an apparent ancient presence of Khoisan people in East Africa (Passarino *et al.* 1998, Chen *et al.* 2000). Although they are found today in South Africa, Khoisan people may have been widely distributed through the continent at one time. However, a recent analysis widely separates the Tanzanian Hadzabe from Khoisan people, and detects no genetic traces of Khoisan people in East Africa (Knight *et al.* 2003). From this, the authors infer a very ancient origin of click languages, spoken by both these groups. The authors further imply that the first anatomically modern humans spoke click languages.

15 *Modeling human evolution*

Baboon models

Since the early 1960s, a variety of non-human primate species have been used to model human evolution. This is especially true for studies of early hominid sociality, which leaves no residues behind until the advent of an archeological record. There is a long hiatus between hominid origins and the beginnings of an archeological record. Over time, the intellectual pressure to fill this void has resulted in a plethora of non-human primate models to reconstruct early hominid sociality (Kinzey 1987). Although most researchers now appear to favor a common chimpanzee or bonobo model, baboons served as the earliest model. They continue to focus research after over 40 years of study.

As detailed in Chapter 2, during the early 1960s, the physical anthropologist Sherwood L. Washburn first turned the attention of paleoanthropologists to the use of living non-human primate species to reconstruct the lifeways of fossil humans. Washburn convened an international symposium on "The Social Life of Early Man," and naturally invited the famous primatologist Adolph H. Schultz to present a paper. So novel was the use of non-human primates in reconstructing early hominid sociality that it bewildered Schultz. Furthermore, primatology of the day did not encompass the behavioral ecology of wild primates. Schultz began his paper with an expression of his bafflement.

> The title of our symposium seems somewhat vague to a physical
> anthropologist, used to terms that can be defined precisely. Since this is my
> first venture into social anthropology, I had to begin by consulting
> dictionaries for the exact meaning of 'social,' but found merely that it
> 'appertains to society' and is synonymous with 'companionate.' . . . The
> words 'early man' present similar puzzles. At once I was reminded of the
> proverbial 'early bird that catches the worm,' especially with my belief that
> all 'early men' did catch more parasitic worms than were good for them . . .
> With these critical remarks I merely wish to explain why at times my
> discussion may seem to go astray and not bear directly on whatever other
> specialists expect from the general title of this symposium.
>
> (Schultz 1961: 58)

Schultz considered himself not only a primatologist, but also a physical anthropologist. Note, however, that Schultz believed that non-human primate sociality falls within the purview of social anthropology, whereas today the topic is within the realm of physical anthropology. This shift was caused by Washburn's revolutionary idea that non-human primate behavioral ecology could yield insights about human evolution (Chapter 2).

No hesitation about inferring early hominid sociality characterized Washburn and his student, Irven DeVore, when they reconstructed "The Social Life of Early Man." Washburn and DeVore had a method in mind. In a summary table, they proceeded to enumerate 13 variables that compared and contrasted social behavior and ecology in early hominids and baboons (Washburn & DeVore 1961a: 102). Similarities are pervasive, but a major contrast is the absence of a home base in baboon societies, and its presence in early hominids. A focus on home bases among early hominids, their implication for complex sociality, and a search for the archeological signatures of these activity areas were to dominate paleoanthropology during the last quarter of the twentieth century (Chapter 16).

Working under the supervision of the London primatologist John Napier, Clifford Jolly wrote a 1965 doctoral dissertation on specializations of the "long-faced Cercopithecoidea." This was followed by a study of baboons, drills, and mandrills considered as an adaptive radiation (Jolly 1970a), and an extremely innovative model of hominid origins that viewed the early hominids as seed-eaters (Jolly 1970b). That is, some specialized craniofacial and dental features of the australopithecines and bipedality itself had originated in ancestors that were specialized in diet and lifeways in a manner analogous to that of the living gelada baboon (*Theropithecus gelada*). Jolly (1993, 2001) was later concerned with documenting inter-specific and inter-generic papionin hybrids and the existence of a persistent hybridization zone among living Awash baboons. These studies highlight the importance of homoplasy, the difficulty of discerning phylogeny using cladistic methodology, and the possibility of parapatric speciation as an important mode of species origin. These topics may illuminate problems of hominid origins, difficulties of discerning phylogeny amid homoplastic resemblance, and modes of speciation. Living and fossil gelada baboons continue to receive much attention (Jablonski 1993), particularly because they experienced an adaptive radiation concomitant with that of the Plio-Pleistocene hominids. Like modern humans, *Theropithecus* is now a monospecific genus. Unlike modern humans, however, gelada baboons are now restricted to a small geographic range in the highlands of Ethiopia. What evolutionary factors account for the diminishment of *Theropithecus*, and the cosmopolitan dispersal of *Homo*?

Baboons and other cercopithecoid monkeys (especially macaques and guenons) have long served as the conceptual template for ideas about early hominid sociality. They continue to generate data that contribute to ideas about human evolution. Hence, early twenty-first century studies of cercopithecoid monkeys accentuate the existence of female philopatry, matrilines, female–female cooperation (e.g., grooming associations, allomothering), female–female competition, and long-term male–female associations ("friendships"). These concepts serve as important foci for ideas about early hominid sociality. Yet, do cercopithecoids represent a derived state with respect to sociality? Di Fiore & Rendall (1994, Rendall & Di Fiore 1995) argue that cercopithecoid monkeys are highly derived with respect to primate social organization. In particular, they emphasize the unique nature of female behavior and social interactions among cercopithecoid monkeys. They argue that a complex of traits arose once in the last common ancestor of cercopithecoid monkeys. These traits include female philopatry, matrilines, female dominance hierarchies, female–female competition and cooperation, female grooming associations and allomothering (Di Fiore & Rendall 1994, Rendall & Di Fiore 1995). This similarity in cercopithecoid social organization exists in spite of the fact that cercopithecoids show great variability in habitats, locomotion and substrate use, and diet. Di Fiore and Rendall infer that a strong phylogenetic constraint generates the uniformity of cercopithecoid sociality despite marked ecological variability. They believe that some unknown features of the Miocene environment of the last common ancestor of living cercopithecoids must have molded the social organization typical of living Old World monkeys. However, because the cercopithecoid radiation is very recent – it is a Plio-Pleistocene radiation – it is possible that the uniform type of Old World monkey social organization arose during the late Miocene. Because the genus *Macaca* can be traced back to 5 mya, it is possible that the characteristic cercopithecoid sociality arose just before this time. If this is so, then the strong phylogenetic constraint inferred by Di Fiore & Rendall (1994, Rendall & Di Fiore 1995) would be an artifact of the recent nature of the cercopithecoid radiation. In any case, this social organization is adaptive in a wide variety of habitats. In fact, it is possible that the success of the cercopithecoid radiation derives as much from their social organization as from their bilophodont teeth, which confer the ability to masticate mature leaves.

Referential and conceptual models

In a highly influential paper, John Tooby & Irven DeVore (1987) analyze various models of human evolution, drawing a distinction between what they term referential and conceptual models. A referential model takes a living species, and argues that it can be used point-for-point to reconstruct the behavioral ecology of a fossil species. A conceptual model is created from various aspects of living species and then applied to the fossil species. Although this important paper was published in a volume devoted to primate models of human evolution, it is clear that no single primate species can be used as an analog for fossil hominids. Tooby and DeVore advocate the use of conceptual models, but unfortunately do not provide a framework for building such models. And it is not clear how one should go about erecting such a framework.

One of the fundamental pitfalls in exploring human evolution lies in the unconscious assumption that fossil hominid species resemble any living species, whether these be hunter-gatherers, baboons, or chimpanzees. Lewis Binford (2001) has published a monograph on the use of hunter-gatherer data to construct theoretical frameworks for interpreting the archeological record. However, his massive analysis of ethnographic data applies only to modern humans, because it attempts to explain how major shifts in human societies take place when resource limitation occurs, and "the packing threshold" is reached by human population growth.

A "composite mammal" model

By now it must be clear that I am disenchanted with any of the referential, single-species models used to reconstruct human evolution. But what is the alternative? Should it be a conceptual model using several primate species? Given the social complexity of the callitrichids, perhaps a callitrichid model should be used to explain the origins of human social complexity. However, many of the features of callitrichid sociality appear to be rooted in their mode of reproduction, which is dizygotic twinning. This mode of reproduction is unique among higher primates, and, unless one is prepared to argue that australopithecines also exhibited dizygotic twinning, a callitrichid model of sociality would not seem applicable to early hominids.

An alternative to a multi-primate model is the construction of a conceptual model based on several non-primate mammal species. Wolff (1997) presents a general conceptual model for behavioral and social evolution in placental and marsupial mammals, concentrating on the mechanisms of population regulation. Ecologists have focused on the variables that affect population size ever since the beginnings of a formal science of ecology in the early twentieth century. Wolff (1997) argues that two principal factors affect population regulation in mammals: the level of development at birth (altricial versus precocial); and offspring mobility (carried, hidden at any convenient site, independently mobile) or immobility (left in a den or burrow). Because he is concerned about self-regulatory, intrinsic mechanisms that limit populations before starvation levels are reached, Wolff (1997) classifies food, predation, and disease as extrinsic mechanisms, whereas I (Table 10.1) classify these as density-dependent

Figure 15.1. Complex social behaviors exhibited by living mongoose species. In addition to meerkats, whose behavior has been investigated with field experiments, other mongoose species also exhibit complex social behaviors. A. Sentinel behavior in the dwarf mongoose (*Helogale parvula*). B. Babysitting behavior in the dwarf mongoose. An adult cares for young while other adults forage for food. C. Group defense in the banded mongoose (*Mongos mungo*). Members of the troop rescue a companion who has been carried away by a Martial eagle. D. Mobbing behavior in the banded mongoose. The troop drives away a jackal. Troop members clump together and move toward a predator. Individual members rise up, and the whole troop moves forward. A roiling mass of animals advances upon the jackal, creating the effect of a single large and dangerous individual. Modified from Macdonald (1984: 147, 152–153).

factors that limit population size. The difference in classification is not merely rhetorical, because density-dependent factors are affected by the number of animals within a population. Wolff's approach, which concentrates on species-specific intrinsic behavior that leads to self-regulation in numbers, is one that is traditional in animal ecology (Chapter 10). This is especially the case when studying rodent populations, and Wolff is a rodent specialist.

Wolff (1997) argues that reproductive suppression of young occurs when they remain within their natal group in contact with relatives of the opposite sex, and do not encounter unrelated individuals of the opposite sex. This may explain the reproductive suppression that occurs in callitrichid groups, where older off-spring contribute to rearing of the young. A particular feature of Wolff's (1997) general discussion of mammal behavior is his conclusion that species whose populations are controlled by extrinsic factors are not vulnerable to infanticide. Because primate populations appear to be food-limited, this conclusion, if it is borne out, might signify that infanticide is unlikely to have been a major factor driving primate sociality.

A number of the traits that dominate discussions of catarrhine sociality are actually traits found among the cercopithecoid monkeys, one of the two superfamilies of living catarrhines, and the superfamily that contains the greatest diversity of catarrhine species. These cercopithecoid traits include female philopatry, female dominance hierarchies, female–female competition, the existence of matrilines within a social group, and female–female cooperation within matrilines, as documented by grooming associations, coalition formation during agonistic encounters, and allomothering (Di Fiore & Rendall 1994, Rendall & Di Fiore 1995). Because this complex of traits appears derived for cercopithecoid monkeys when compared across all of the living primates, there is no reason to expect that a proto-hominid or ancestral hominid would have possessed these traits. Why is this significant? If female philopatry, female dominance hierarchies, female–female competition, and competition between matrilines are eliminated from the suite of ancestral hominid social traits, then the stage is set for the advent of complex sociality or eusociality. This is because competitiveness or agonistic interaction between unrelated or distantly related animals within a social group eviscerates cooperation. Female philopatry, female dominance hierarchies, and matrilines are generally thought to signal stable social relationships and a persistent, unchanging core to a social group. Therefore, they are usually thought to prefigure the origins of complex social life. However, these traits actually hinder the beginnings of intricate cooperative life.

Removing female philopatry and competitive interaction aids in the construction of a "composite mammal" model for early hominid sociality, because it brings early hominids into line with the majority of extant mammal species, but the likelihood of hominids developing complex sociality is also enhanced. If one then adds some simple additional feature like sentinel behavior, observed in cooperative mammals like meerkats, then terrestrial locomotion, complex terrestrial foraging, and division of labor can readily appear among early hominids (Figure 15.1). Even communal care taking of the young may not be so difficult to evolve. It is also exhibited by meerkats. Both meerkat sentinel behavior and care taking of the young are thought to be outgrowths of "penalty-free altruism" (Chapter 9).

16 *Archeological evidence and models of human evolution*

Human antiquity

Although the archeological record, particularly the record of Plio-Pleistocene lithic artifacts, is now ineluctably associated with the idea of human evolution, this was not the case until about 1863–1865. The recognition that the archeological record extends deep into the past is unrelated to the advent of natural selection theory in 1859 (Grayson 1983). Natural selection did not cause an immediate revolution in archeology, because zoology and vertebrate paleontology were historically separated from archeology, whose origins lay in antiquarianism and conjectural history. However, antiquarians were not the only scholars skeptical about the idea of human evolution. It was also the common mindset of geologists, zoologists, and paleontologists. Major founding figures in vertebrate paleontology like Georges Cuvier and Richard Owen were of the opinion that there was no evidence for fossil humans, although fossil prosimians (*Adapis parisiensis*) and apes (*Pliopithecus antiquus, Dryopithecus fontani*) had been found in Europe before 1859. Cuvier firmly declared in 1812 that "fossil Man does not exist."[1] Even the association of human bones and the bones of extinct mammals did not automatically persuade scholars that humans and Pleistocene mammals had been contemporary. Two alternative interpretations were also held: that these mammals had survived into recent times, and had been hunted into extinction by Celts, Gauls, Teutons, and other European groups mentioned by Roman writers; or that the association of bones from humans and extinct mammals was purely accidental.

There were paleontologists, zoologists, and antiquarians in both Great Britain and the Continent who argued that humans were very ancient, based on the primitive nature of some stone tools, and the occurrence of these tools or human bones with the bones of extinct mammals. Among these scholars were Jacques Boucher de Perthes, Alfred Fontan, Édouard Lartet, John MacEnery, and Geoffroy Saint-Hilaire. Based on his excavation of stone artifacts and animal bones from Abbeville, the antiquarian Boucher de Perthes had been vigorously promoting the idea of human antiquity since the 1840s. He proclaimed in 1857 that "God is eternal, but Man is very old."[2] However, during the late eighteenth and first half of the nineteenth century, the collection and study of

317

stone artifacts was largely conducted in terms of a creationist worldview that embraced a very limited span for earth history. Neither humanity nor the earth itself was very old.

A major stratigraphic marker, found in both the Old and New Worlds, was an extensive deposit of superficial gravels and clays that the geologist William Buckland called the diluvium in 1823. The bones of extinct mammals, such as the mammoth, cave bear, and wooly rhinoceros occurred in the diluvium. The consensus opinion was that diluvial deposits had been laid down by a catastrophic worldwide flood. In England, where natural theology ruled, and the natural sciences were thought to reveal the obvious handiwork of God, diluvial deposits were considered to be evidence of the Great Flood of Noah. An alternative interpretation was advanced by Louis Agassiz, who argued in 1840 that the diluvial deposits had been laid down by glaciation, not flooding. Most geologists were unconvinced by Agassiz's interpretation. Although they could accept that glacial ice might transport gravel, rocks, and clays in mountainous areas, no one could conceive how glacial ice could appear in lowland areas, or how it could be so ubiquitous.

A limited number of scholars were convinced of human antiquity, but this did not imply that they embraced the idea of evolution. Well-known fossil mammal species that were associated with diluvial deposits were clearly extinct, and so the discovery of human remains in the diluvium did not automatically provide ancestors for living humans. Furthermore, modern environmental conditions were not present in the diluvium. Thus, the world recorded in the diluvium was not the modern world. It was the world now identified as the Pleistocene, and it seemed to be abruptly separated from the modern world by catastrophic extinction and climatic change. By the 1870s, the work of Agassiz, James Geike, and other geologists convinced even the lay public that vast continental ice sheets had covered both Europe and North America during the Pleistocene. It became clear that a Great Ice Age had assailed the world.

Because only relative dating by correlation with known geological marker beds such as "the diluvium" was possible, the absolute date of geological events, faunas, and artifacts was unknown. Even after 1859, the timeframe for evolutionary events could be very labile. For example, scholars might blithely write of an Eocene origin for hominids. This was at a time when the Eocene was recognized as the first epoch of the Tertiary Period – the Paleocene is the comparable time of hominid origins using the modern geological time scale. Many evolutionists, such as Ernst Haeckel, drew up conjectural phylogenies with hypothetical taxa (e.g., "Pithecanthropus alalus") that were completely independent of time.

Yet, more difficulties plagued scholars of evolution in addition to the indiscriminate pinning of evolutionary events to markers in the geological record. The second half of the nineteenth century witnessed one of the great crises in geological theory and interpretation. This dilemma involved key arguments about the age of the earth (Hallam 1989). Uniformitarian theory as conceived by James Hutton in 1788 and promoted by Charles Lyell during the nineteenth century implied an enormous age for the earth. The first edition of Lyell's authoritative *Principles of Geology* was published between 1830 and 1833, and the ninth edition in 1853. Central to the book is the idea that geological processes, such as erosion and sediment deposition, which can be observed today were responsible for past changes that altered the face of the earth. Geological processes in the past and in the present are uniform in nature (hence, uniformitarianism); they occur in normal, observable ways, so that catastrophes and upheavals do not need to be invoked whenever geological change occurs. A corollary to uniformitarian geological processes is that earth changes do not need to be compressed into very short periods of abrupt change. The history of the earth thus becomes very long. In Hutton's famous concluding phrase, the earth has "no vestige of a beginning, – no prospect of an end." Both Darwin and Wallace were influenced by Lyell's presentation of uniformitarianism, and argued that natural selection could bring about phenotypic alteration and speciation if enough geological time were available for natural selection to operate. A long chronology was necessary if natural processes had brought about the multifarious diversity of life, rather than one or a series of separate divine creations. However, William Thomson, Lord Kelvin, argued that the age of the earth was strictly limited by the amount of time it would take a molten body the size of the earth to cool to its present state since its time of formation. Kelvin's estimates of this time interval were continually revised downward during the course of the nineteenth century, from 400 my (1863), 100 my (1868), 50 my (1876), 50–20 my (1881), to 24 my (1897). These estimates were very short in comparison to the age of the earth projected by uniformitarian geologists. The last estimate of 24 my was equivalent to the currently recognized length of time since the beginning of the Miocene epoch, and all of earth history had to be encompassed within this span (Hallam 1989). Because Lord Kelvin was one of the chief protagonists of nineteenth century physics, geologists who argued for a long span for earth history faced a formidable foe, and one backed by the glamour of mathematics and objective physical laws. The great battle over geological time – which the geologists seemed fated to lose – was not solved until 1899–1900, when Marie Sklodowska Curie discovered radioactivity, and realized that one of the properties of radioactive decay was the emission of heat.[3] At this point, another heat source for the earth, an internal source of radioactive materials in the earth's core and mantle, finally solved the dilemma of conflicting lines of evidence over geological time.

Ultimately, the use of archeological data to establish a timeframe for human prehistory had to be delayed until geologists had worked out the principles of stratigraphy, and established a sequence of time markers for earth history.

However, a major event in the history of archeology occurred in 1859. It was completely unrelated to the advent of natural selection theory. The date is purely coincidental, because the archeological controversy had been fulminating since the late eighteenth century. Extensive debate had occurred throughout Europe about the antiquity of stone artifacts discovered in many widespread localities. Even the presence of these artifacts with the bones of extinct animals was not enough to persuade skeptics about a great age for the artifacts. The sticking point was that a great age for the artifacts necessarily meant a great age for their human makers. Perhaps the most vehement supporter of human antiquity was Boucher de Perthes, who had been publishing on his discoveries of stone artifacts since the 1840s. Following 1858 excavations at Brixham Cave in England, where artifacts were found sealed below an intact stalagmite surface indisputably in association with the bones of extinct animals, a series of prestigious English scholars finally visited Boucher de Perthes (Grayson 1983). They thoroughly examined his collections and excavations from sites near Abbeville, and were convinced by his evidence. They began delivering a series of papers on their favorable assessment of the Abbeville material. The most esteemed of these scholars was Charles Lyell, who delivered a paper on his observations at Abbeville to the 1859 meeting of the British Association for the Advancement of Science. Human antiquity, as recognized by the evidence of archeology, had been proven. The resolution of this archeological controversy was soon followed by the 1863 publication of Lyell's *Geological Evidences of the Antiquity of Man*. This magisterial book went through four editions, the last appearing in 1873. Lyell favorably reviewed the evidence for human antiquity at sites like Kent's Cavern, Brixham Cave, Hoxne, Engis, St. Acheul, and Moulin Quignon. Some of these sites had long been controversial. A spate of books on prehistoric archeology soon appeared, preeminent among them being John Lubbock's *Pre-Historic Times* in 1865. Seven editions were published, the last in 1913.

Yet, if archeology were to be used to document human presence and behavior, it was imperative that archeologists understand and incorporate geological principles needed to establish habitat type, and the context and relationship of data within the dynamic, intricate, and often confusing realm of stratigraphy. In what type of environment are finds discovered? Are finds in primary or secondary context? What is the relative date of finds? Are archeological materials contemporary, and, if so, does the spatial patterning of the material reveal anything about human behavior? Without stratigraphy and context, artifacts and archeological data remain simply objets d'art fit only to adorn an antiquarian's cabinet of curiosities, along with fossil coral, Medieval incunabulae, and the Fiji mermaid.

Up until the late seventeenth century and early eighteenth centuries, stone artifacts were routinely grouped together with fossils, unusually shaped sedimentary concretions, minerals, crystals, and meteorites. All of these things were either literally dug up from the ground (Latin: "fossils") or had fallen from the sky and been buried in the ground. Not until the arguments of Nicolaus Steno in the late seventeenth century was it clear that fossils did not coalesce within the ground like inorganic crystals, but were organic in origin (Rudwick 1972). Similarly, stone tools that were obviously artificial were ultimately recognized as being analogous to stone tools used by living humans or mentioned by Greco-Roman historians. The recognition that humans, stone tools, animals, and plants were contemporary and chronologically ancient led to the first detailed artistic representations of ancient humans as members of vibrant and dynamic ancient ecosystems (Rudwick 1992). In these representations can be seen the nucleus of modern field biology studies, because knowledge of the species composition of present habitats, vegetation cover, and the morphology and behavior of living animals are transferred to the reconstruction of the ancient world.

Recognition that the archeological record is not coeval with the human paleontological record

The oldest stone tools

The idea that tool behavior is a distinctive characteristic of humans can be traced back to Aristotle (Chapter 2). The idea held a prominent place in the conjectural history of Enlightenment scholars, because technology was associated with the beginning of humanity, and also because increasing technological complexity generated stages of increasing social complexity in schemes of conjectural history. Many models of human evolution going back to Darwin (1871) argue that there is a necessary connection between tool behavior and hominid origins. In Darwin's conception, the lack of natural defensive weapons (e.g., large canines, powerful claws) in the first hominids would make it impossible for them to become terrestrial without tool behavior. The use and manufacture of tools would make it possible for hominids to create weapons to counteract threats from ground-dwelling, contemporary large carnivores. Tool behavior was thus linked to the origin of bipedalism. Oakley (1957) presented this model in a modern format, adding food-sharing to it. Washburn (1959, 1960), in turn, argued that tool behavior had an "autocatalytic" effect. The coordination of hand and eye necessary for the knapping activities of stone tool production led to an increase in brain size.

Immediately following the general acceptance of the validity of the archeological evidence from Abbeville and elsewhere in Europe, which proved a long chronology for humans, a spate of discoveries throughout Europe and the New

Figure 16.1. An Oldowan artifact (subspheroid) newly eroded from the excavated surface of the DK site, Olduvai Gorge, Tanzania. The site was originally excavated by Mary Leakey during the early 1960s.

World appeared to demonstrate the presence of crude stone tools ("eoliths") in Miocene and Pliocene deposits (Grayson 1983). The vast majority of these eoliths have since been shown to be geofacts – produced by natural geological processes, and not human activity. However, the surge of supposedly Tertiary artifacts, which seemed to establish the presence of Tertiary humans, illustrates how close the interrelation of artifacts and humans had become in ideas about human evolution. Even the most ancient of humans must have possessed stone tools, because tools were an irreducible component of humanity.

Yet, it is now becoming abundantly clear that the archeological record, as represented by the record of stone tools, begins long after the first hominid fossils appear. The oldest stone tool industry is the Oldowan, and the earliest well-dated Oldowan material is found in East Africa (Figure 16.1). Plummer (2004) reviews African sites and Oldowan technology from 2.6 to 1.6 mya. Stone clasts or cobbles were either used as hammerstones or were flaked. Hammerstones pound and break bone to remove marrow. Hammerstones are also used to remove stone flakes from a cobble or core. These sharp stone flakes can be used to cut skin or ligament, dismember a carcass, or remove meat from bone (Figure 16.2). The oldest firmly dated material comes from the Gona region of Afar, Ethiopia, where the tools date to 2.6–2.5 mya (Semaw 1997, Semaw *et al.* 1997, 2003). Three animal bones with stone tool cut-marks and a percussion mark have been found at the Bouri locality in the Middle Awash region of Ethiopia. The archeological material dates to about 2.5 mya, and is contemporary with the australopithecine taxon *Australopithecus garhi* (de Heinzelin *et al.* 1999). Oldowan tools dating to 2.4–2.34 mya have been found in Members E and F of the

Figure 16.2. A goat being butchered with basalt flakes in a pedagogical exercise in the Koobi Fora Field School. The sharp edges of stone flakes struck from a core can be used to pierce skin, sever ligaments, and cut meat from bone.

Shungura Formation of the Lower Omo River, Ethiopia (Merrick & Merrick 1976). Lokalalei, in the western portion of the Lake Turkana Basin, Kenya, has Oldowan tools that date to 2.34 mya (Kibunjia *et al.* 1992, Kibunjia 1994, 2002, Roche *et al.* 1999). Slightly later in time, Oldowan tools dating to 2.33 mya are found in the Kada Hadar Member of the Hadar Formation, Ethiopia (Kimbel *et al.* 1996), and Oldowan tools are found in the Kanjera South sites of KS-1 and KS-2 dating to 2.2 mya in Kenya (Plummer *et al.* 1999). Archeological material from Bouri, Members E and F of the Shungura Formation, and the Kada Hadar Member of the Hadar Formation should be assessed differently from material yielded by the other sites, because the evidence is either sparse or in secondary context, having been deposited in disturbed riverine sediments.

Oldowan tools at the oldest archeological sites in the Gona region of Ethiopia occur in bedded silts or paleosols representing the banks and floodplains of the ancient Awash River (Quade *et al.* 2004). Hominids collected rounded river cobbles from a variety of volcanic rock types from gravel bars in the river that were exposed during the dry season. These cobbles are the raw material for the earliest tools. Although large volcanic cobbles occur in gravels from earlier time ranges, no stone tools were recovered in an extensive survey of a gravel dating to 3.1 mya. Despite the presence of suitable raw materials, stone tools are absent. From this, researchers infer that *Australopithecus afarensis* was incapable of making stone tools (Quade *et al.* 2004: 1537). However, the earliest stone tools appear soon after the appearance of an Awash gravel (Gm1) containing suitable cobbles. This merely represents the appearance of a new local resource for cobbles, and it therefore seems likely that stone tool technology had a longer history. Lithic technology was not simply invented just before 2.6–2.5 mya. A period of experimentation surely preceded the first known stone tools, indicating australopithecine innovation. In any case, only australopithecines are present during these time ranges. Acheulean tools at later archeological sites occur both in the axial Awash river system and along small tributary channels, indicating hominid range expansion and some transport of raw materials (Quade *et al.* 2004).

If the date of the earliest unequivocal (i.e., bipedal) hominids is taken as 6 mya (Senut *et al.* 2001, Pickford *et al.* 2002), then a temporal gap of 3.5–3.4 my separates the first undoubted hominids from the first archeological evidence. Clearly, this indicates that hominid origins are not coterminous with stone tool origins. Furthermore, pushing the boundary of the Oldowan back beyond the first known appearance of genus *Homo* ultimately signifies the existence of tool behavior in australopithecines. Is tool behavior possible in species with a pongid-like relative brain size? Yes. Researchers impressed by the tool behavior of living pongids might now summarize this behavior, implying that if pongids can make and use tools, then similar capacities can be attributed to the australopithecines. Predictably, the appearance of tool behavior in a recent inventory of orangutan cultural behavior is used to infer the presence of a substrate for material culture, including tool behavior, during the middle Miocene at about 14–15 mya in the last common ancestor of orangutans and hominids (van Schaik *et al.* 2003). Yet, no living primates, including pongids, engage in impressive amounts of tool behavior (Chapter 8). Animals that habitually use and manufacture tools, such as birds, tend to be ignored by anthropologists. It is clear, however, that animal tool behavior demonstrates no necessary connection between tool behavior and relative brain size or intelligence (Chapter 8). It is not necessary to invoke the presence of genus *Homo* – sometimes a presence invisible to the paleontological record – whenever archeological materials are discovered.

The oldest bone tools

What about bone tools? Bone tools in the archeological record are sometimes now viewed as a signature of advanced behavior that differentiates modern humans from Neanderthals and other lesser mortals. However, there is a long archeological tradition of viewing bone tools as very ancient, perhaps more ancient than stone tools. In the early twentieth century, Teilhard de Chardin and the Abbé Henri Breuil examined the possibility of bone and fire-worked antler tools occurring at the Chinese site of Zhoukoudian. Did bone tools appear first? Yet, bone tools are now frequently considered to be a species-specific signature of modern behavior, produced only by anatomically modern humans, *Homo sapiens sapiens* (Klein 1999).

Raymond Dart, describer of the first australopithecine fossil, envisioned South African *Australopithecus africanus* as a top predator that hunted significant quantities of prey. He believed that the taxon was highly carnivorous; members of this species also killed each other, and engaged in cannibalism. The evidence for this behavior could be discerned at South African sites like Makapansgat. Dart conceived of the fossil ungulates preserved at these sites as forming gigantic midden heaps of animal bones that represented species preyed upon by the carnivorous australopithecines. He came to this conclusion because of the overwhelming preponderance of bovid remains, the over-representation of cranial and lower limb bones, and what appeared to be depressed fractures on the crania. The weapons and tools used by *Australopithecus africanus* were artifacts manufactured from the bones, teeth, and horn cores of their prey. Dart termed this industry the Osteodontokeratic Culture (Dart 1949). He later compiled a monograph on the osteodontokeratic evidence from Makapansgat, publishing a formal typology of the tools, and illustrating their use (Dart 1957).

Purely scholarly objections to the Osteodontokeratic Culture and Dart's reconstructions of australopithecine behavior were made at an early period. There were two principal objections: other possible agents might explain the bone accumulations and bone damage at South African sites; and Dart had established the bone tool typology on morphology alone by intuitive resemblance (e.g., scrapers, scoops, cups), with no functional or experimental analysis. Many objections to Dart's reconstructions of australopithecine behavior, however, seem purely visceral. It is clear, for example, that the physical anthropologist Matt Cartmill (1993) is repulsed by Dart's image of the predatory "killer ape," which Cartmill envisions as a Yahoo-like figure. He sees violence and aggression as the wages of carnivory, just as the wages of sin is death. Carnivory emerges as sinful because the human capacity for reason renders humans more degraded than other animals when they succumb to violent behavior. It is true that Dart (1957) appears to revel in descriptions of australopithecine carnage. It is possible that Dart's reconstruction of australopithecines murdering and eating other australopithecines is responsible for a distasteful response, because carnivorous behavior in other mammals, such as lions, never generates such revulsion. Even though lions kill other adults and commit infanticide, they are still considered to be awesome and noble animals. However, it is likely that Cartmill's (1993) emphasis on the symbolic interactions between humans and animals, and his reflections on the pathos of dead or dying prey animals, contribute to his distaste for Dart's reconstruction of australopithecine behavior. The assumption is frequently made that there is an ineluctable connection between human hunting, malicious violence, and mindlessly aggressive behavior. The social anthropologist Colin Turnbull questions whether this is true for modern hunter-gatherers (Lee & DeVore 1968: 341), and the social anthropologist Richard Nelson (1997) questions whether this is true for modern American hunters – his book is a counterpoint to Cartmill's.

The geologist C. K. Brain mounted the most successful attack on Dart's evidence for the Osteodontokeratic Culture, by investigating the geology of site formation and taphonomic processes that affect bias in fossil preservation and the morphology of bone fracture (Brain 1981). The South African sites, which Dart envisioned as australopithecine cave lairs, had originally formed in a limestone karstic landscape as fissure fills or sinkholes that dramatically enlarged through time to become subterranean caverns. Animals either accidentally plummeted into a fissure fill, or their remains were swept into the underground cavities after their deaths. Brain (1981) also argued that leopard predation was responsible for some of the bone accumulation, if leopards were caching carcasses in trees growing up from the chimney-like sides of a fissure. He noted that the cranium of a juvenile *Australopithecus robustus* from Swartkrans had canine punctures that fitted the upper canines of an adult leopard. A remarkable diorama at the Transvaal Museum in Pretoria depicts the scene. A leopard drags off the corpse of a young *Australopithecus robustus* just before caching it in a tree. The fierce and energetic leopard strides off with the head of the dead juvenile impaled on its canines, and the nerveless limbs of the youngster trail pathetically behind. Brain (1981) ultimately concluded that the bias in bone preservation and bone fractures at the South African sites were caused by natural taphonomic processes, not by hominid activity. Hominids had been the hunted, and not the hunters. Brain appeared to have dismantled evidence for the Osteodontokeratic Culture.

Ironically, however, Brain later argued that australopithecines had used bone implements for digging up tubers, and he supervised the construction of another diorama at the Transvaal Museum showing *Australopithecus robustus* digging up tubers with such an implement. Brain's idea focused attention on tubers or underground storage organs as food items. This category of food item had previously been ignored by paleoanthropologists. Now, however, Wrangham *et al.* (1999) argue that *Homo erectus* individuals must have been eating cooked tubers. O'Connell *et al.* (1999) also emphasize the nutritional importance of plant foods, particularly cooked tubers, in the diet of *Homo erectus*. Herbivorous early hominids do not generate a visceral distaste among paleoanthropologists.

Backwell & d'Errico (2001) analyze the microwear on 85 purported bone implements from Swartkrans, Sterkfontein, and Drimolen, and test their function. They conclude that early hominids deliberately selected bone fragments with a certain size and shape to extract termites from termite mounds. Experimental studies of microwear confirm this function. Because termites yield 560 calories/100 g, and are a significant source of protein, fat, and essential amino acids (Backwell & d'Errico 2001: 1361), termite extraction might have been an important component of hominid diet. Predictably, Backwell & d'Errico (2001) draw the parallel with chimpanzee termite extraction. Yet, termites are a seasonal component of chimpanzee diet. Modern termite influx to animal diets is generally seasonal, and depends on termite migration to the external portion of a mound, unless animals (e.g., anteaters, aardvarks) are able to excavate the concrete-hard mound itself. Backwell & d'Errico (2001: 1359) observe that experimental excavation of a termite mound with a bone tool causes termites to swarm to the surface of the mound as the surface is flaked off with the point of the tool. This is entirely different from the chimpanzee situation – it allows termite consumption at any point in the annual cycle. Selection of bone as a raw material, selection of a properly shaped bone for excavation, and the activity of efficient excavation parallel to the axis of a mound sorts this behavior into the realm of natural history intelligence.

Bone tools occur at Swartkrans through a long time span between 1.8 and 1.0 mya. In addition, there is evidence for purposeful shaping and re-sharpening of bone tools at Swartkrans (d'Errico & Backwell 2003). Stone tools are also found at Swartkrans and Sterkfontein. Because *Australopithecus robustus* is sympatric with genus *Homo* at Swartkrans and Drimolen, where bone tools also occur, it is not clear which hominid was associated with the use of bone tools. Fossils of *Australopithecus robustus* far outnumber those of genus *Homo*, and so Backwell & d'Errico (2001) favor *Australopithecus robustus* as the user of the bone tools, based on probability alone, as well as the absence of *Homo* from Swartkrans Member 3, where the largest number of bone tools are found. They also note (d'Errico & Backwell 2003) that no bone tools are found after 1.0 mya, when *Australopithecus robustus* disappears from South African sites. Hence, they conclude that *Australopithecus robustus* is a better candidate for maker and user of the bone tools. I agree with Susman (1988) that the morphology of hominid hand bones from Swartkrans indicates that *Australopithecus robustus* was capable of fine manipulation. I believe that both hominid taxa were capable of tool behavior, using both bone and stone as a raw material for tools. However, as explained below, I do not think that sustained or obligate foraging for insects or other invertebrates is likely among early hominids.

An apparently complete adult male *Australopithecus* skeleton (StW 573) has been discovered at Sterkfontein, South Africa (Clarke 1998). It dates to 3.3 mya, and clearly represents the remains of an australopithecine that tumbled into a fissure-fill and died, either immediately from its injuries or from dehydration and starvation. There are no signs of carnivore damage on the bones. The associated fossil bones are being laboriously excavated from the surrounding limestone breccia. In an update on continuing excavation of this amazing find, Clarke reveals the presence of a nearly complete left cercopithecoid radius lying next to the left australopithecine femur. The cercopithecoid radius is missing only its distal joint surface, which has been gnawed off by a carnivore. Clarke (1999: 479) believes that the australopithecine may have been carrying the large cercopithecoid radius when it tumbled down the shaft of the fissure-fill. Hence, suggestions of bone tool use continue to emerge from early sites in South Africa.

A Proto-Oldowan?

Did a long period of experimentation precede the first evidence of the Oldowan at 2.6–2.5 mya? Most paleoanthropologists use non-human primates for a comparative set, and argue that tool behavior evolved in the last common ancestor of chimpanzees and humans, and independently in *Cebus* monkeys of the New World (Panger *et al.* 2002). The discovery of tool behavior in wild orangutans leads some researchers to infer that tool behavior evolved in the last common ancestor of the orangutan and humans at 14–15 mya (van Schaik *et al.* 2003). However, a broad perspective leads one to understand that tool behavior is common in the animal world. It is sometimes habitual, and can involve a degree of technical competence in manufacture (in some bird species, for example) that puts chimpanzees to shame (Chapter 8). Panda 100, a chimpanzee tool site created when the animals used unmodified stones to crack *Panda* nuts, was recently excavated at the Taï Forest in the Ivory Coast, West Africa. Hammerstones occur at the site. They were transported and curated, and show signs of battering. No intentional flaking took place, because there were no cores, or flaked pieces. Some flakes were accidentally formed by the nut-cracking activity, but 78% of the material is microshatter less than 20 mm in diameter (Mercader *et al.* 2002). Even the earliest archeological sites contain evidence of intentional flaking of cores by hominids. Characteristic bulbs of percussion and striking platforms occur on well-made flakes at the earliest archeological sites. There are obvious clear differences between the material excavated at the West African chimpanzee site and material excavated at the earliest known archeological sites in East Africa.

If one assumes that a long trail of tool behavior extends back to the middle Miocene, one could document this through the retrieval of hammerstones and hominid-modified animal bones. Analyses of trabecular orientation and cortical bones' thickness in the hand bones of early hominids might reveal hand bone function during habitual tool behavior (Panger *et al.* 2002). If one assumes that tool behavior begins during the middle Miocene, then the appearance of the first archeological sites only marks the point at which stone artifacts become concentrated into small patches, thus becoming visible to researchers (Panger *et al.* 2002). However, the assumption of a long common heritage of tool behavior rooted in the Miocene implies that tool behavior is unlikely to have evolved independently. Perusal of the widespread nature of animal tool behavior reveals that it evolves easily. It is likely, therefore, that any superficially similar tool behavior observed in hominoid primates (e.g., chimpanzee use of hammerstones to break nuts and hominid use of hammerstones to break long bones) is caused by homoplasy.

Bone modification and inferences of hominid behavior

Taphonomic investigations by Brain (1981) were underlain by actualistic studies. He experimented with goat carcasses on the modern South African veldt to determine whether natural weathering processes could eliminate parts of the skeleton and leave a biased over-representation of the remaining parts. The adjective "actualistic" is mysterious in English, until one remembers the French word *actuel*, meaning "present, at the present time." Actualistic studies supplement laboratory observations with experimentation. In this sense, they are equivalent to playback experiments and other field biology manipulations engaged in by animal behaviorists. The value of these studies to archeology becomes apparent when they identify probable agents in bone accumulation and modification.

Clear indications of hominid modification occur on animal bones at early archeological sites. Cut-marks and percussion marks on animal bones produced by stone artifacts are distinctive enough to document hominid presence, even if the stone artifacts that produced these marks are not present, and even if there are no hominid fossils. Experimental butchery marks made by flakes or cores can also be distinguished (Merritt 2000). Debate no longer occurs about these signatures of hominid behavior. Debate now centers around whether the hominid modification indicates evidence of hunting, scavenging, or both, and the degree of hominid interaction with sympatric large carnivores. Were hominids a part of the carnivore guild during the Plio-Pleistocene, or did this occur only much later in time?

Wrangham *et al.* (1999) argue that the large body size of *Homo erectus* (based on KNM-WT 15000) could not be achieved by a diet of vertebrate meat. *Homo erectus* individuals must therefore have consumed a considerable amount of other dietary items, such as invertebrates or cooked tubers. O'Connell *et al.* (1999) question whether the archeological evidence supports an interpretation of hominid big game hunting and confrontational scavenging in the Lower

Paleolithic. Alternatively, they argue for the nutritional importance of indispensable plant foods, especially tubers, which were gathered principally by post-reproductive females, and shared among matrilines. Yet, a marked increase in body size in *Homo erectus*, particularly among females, argues for the existence of a dietary shift and a significant improvement in nutrition, especially protein and fat (Chapter 11). Plant foods are unable to account for this. The nutrient quality of plant foods is too low – even if massive ingestion of cooked tubers took place, as some researchers argue (O'Connell *et al.* 1999, Wrangham *et al.* 1999).

Sustained or obligate foraging for insects or other invertebrates is also unlikely among early hominids. Diet has a major impact on body size as well as foraging behaviour. The maximum body mass of living terrestrial mammalian carnivores is highly constrained by diet (Carbone *et al.* 1999). Because of their small body size, predators that are smaller than 21.5 kg have low absolute energy requirements, and therefore can subsist on invertebrates. At a predator mass of between 21.5 and 25 kg, a transition occurs from feeding on small prey (less than half predator mass) to feeding on large prey (near the mass of the predator). Predators with a body mass above 21.5 kg cannot exist on an invertebrate diet, because ingestion rates are too low and foraging time is too costly (Carbone *et al.* 1999). The carnivore consumption rate of invertebrates is only 10% the rate of vertebrate consumption. Insectivorous mammals achieving a larger body size begin to specialize in eating social insects. Anatomical and physiological specializations follow. There is a concomitant loss of teeth, elaborate tongue and tongue muscle adaptations, and a lowered metabolic rate, because of detritus that is inescapably ingested along with the insects. Although metabolic rate may presently be difficult to infer in fossils, no fossil hominid species shows the loss of teeth, simplified molar crown patterns, or tongue and tongue muscle specializations that one would expect in organisms that preyed on social insects.

Plio-Pleistocene sites that lack tools and have only hominid-modified bones

Bunn (1994, 1997) discusses a category of Paleolithic site that was originally defined by Isaac (1971). This category encompasses sites where stone artifacts and manuports do not occur, and where the only indication of hominid presence and activity is zooarcheological evidence. Bone modification is the signature of this type of site. The evidence consists solely of cut-marks and/or percussion marks on well-preserved animal bones. For example, the archeological evidence at Bouri, Ethiopia, dating to 2.5 mya, consists of three animal bones with cut-marks and percussion marks (de Heinzelin *et al.* 1999). Hominid-modified bones are the only evidence of tool behavior, and this behavior is further considered to be "circumstantial support" that the newly recognized australopithecine species at Bouri (*Australopithecus garhi*) is the ancestor of genus *Homo* (Asfaw *et al.* 1999: 634). Consequently, a great weight is now placed on hominid-modified bones – they may be the only record of hominid behavior at some Paleolithic sites.

There are two new sites in the Koobi Fora region of the East Turkana Basin in northern Kenya that contribute significant data to this site category. These are GaJi 14 in area 103 at the Koobi Fora Ridge and FwJj 14 in area IA near Ileret (Cachel 2004, Cachel & Harris, in press). The sites occur in Okote Member deposits dating to 1.64–1.39 mya. Two sympatric hominid taxa are recorded at Koobi Fora during this time range: *Australopithecus boisei* and *Homo erectus*. *Homo erectus* was presumably the taxon responsible for the creation of the two sites. Hominid-modified bones (n = 151) have been recovered from both surface collection and excavation at these sites. However, only 1 bifacially flaked core fragment and a cobble fragment were recovered *in situ*. Two cut marked hyoid bones were excavated from FwJj 14B (Figure 16.3). These hyoids may demonstrate that hominids had primary access to a carcass (Cachel 2004). Mammalian carnivores with first access to a carcass will routinely target and fully consume the tongue, because it is a large mass of tissue that is completely unprotected by bone, and fully accessible.

Although hominids engage in focused carcass processing in these sites, they do not discard stone tools. Stone tools are therefore carefully retained or curated, in order to minimize travel to distant sources of suitable raw material for stone tools. The proximity of lithic raw material sources of a suitable clast size may have profoundly affected hominid ranging and foraging behavior. Hominids may have prospected for sources of large stone clasts suitable for flaking. The search was important, and was not simply incidental to foraging activities related to food. The extent of artifact curation can indicate the degree to which hominids had become obligate tool-users. In addition, complex foraging for dispersed lithic resources implies an intimate knowledge of the regional environment, as well as the ability to locate and predict the abundance of resources that fluctuate widely in time and space.

Climatic events and the archeological record

As reported in Chapter 3, hominid origins and turnover events (speciation and extinction) during the Plio-Pleistocene do not appear to be associated with climatic changes. However, there is a coincidence between climatic changes

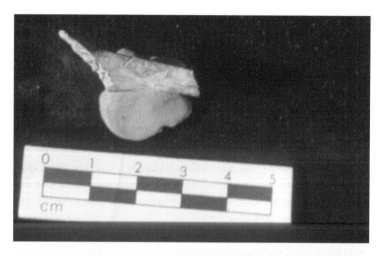

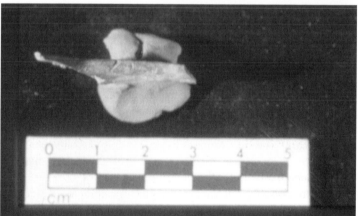

Figure 16.3. Two views of a cut-marked bovid hyoid bone (Kenya National Museum, excavation catalog no. 3124) recovered in situ from the archeological site of FwJj 14B, near Ileret, East Turkana Basin, northern Kenya. The scale is in centimeters.

and the archeological record. Specifically, paleosols record arid episodes and the spread of grasslands within the Turkana Basin at 2.52–2 and 1.81–1.58 mya (Wynn 2004). The first of these Turkana climatic episodes (2.52–2 mya) coincides with the origin of the archeological record, which appears in East Africa at 2.6–2.5 mya. The second of these Turkana climatic episodes (1.81–1.58 mya) coincides with a number of changes that occur within the archeological record in the Koobi Fora region, in the East Turkana Basin. These changes are an increase in the number of archeological sites, density of artifacts within sites, variability between sites, increasing technological sophistication

(e.g., the appearance of single-platform cores), and evidence of hominids ranging and foraging over greater distances (Cachel 2004, Cachel & Harris, in press).
Hence, if events in hominid paleontology cannot be explained by climatic
triggers, there does appear to be a coincidence between climatic and archeological events in East Africa, particularly within the Turkana Basin, where
the paleoenvironmental and archeological records are both abundant and well
dated.

What is the significance of a coincidence between climatic events and archeological changes? Because the climatic changes record episodes of aridity and an
increasing spread of grassland, both arid climate and a proliferation of grassland
and open-country habitats are associated with an increase in the density and the
complexity of the archeological record. Archeological changes appear to be triggered by paleoenvironmental changes related to intensification of arid climate
and the spread of open-country habitats. Stone artifacts and hominid-modified
bone track these changes. Thus, it is reasonable to suggest that selection pressures on hominid behavior that led to the advent of the archeological record, and
subsequent archeological change, were related to aridification and open-country
habitats. Dietary shifts may have occurred because hominids were responding
to the proliferation of available vertebrate carcasses and/or prey within the context of a more abrupt or well defined arid season, and an increase of grasslands
within habitat mosaics.

"Man the Hunter" and the new physical anthropology

S. L. Washburn established the New Physical Anthropology in the early 1950s,
and was responsible for the substitution of non-human primates for living
hunter-gatherers in scenarios about human evolution during the late 1950s/early
1960s (Chapter 2). Was Washburn also responsible for the "Man the Hunter"
model of human evolution? And was "Man the Hunter" equivalent to the New
Physical Anthropology? This is the contention of Donna Haraway (1989), who
believes that the early field-oriented studies of primates and other topics in
physical anthropology were inevitably tainted by old-fashioned positivism, the
quest for "facts," and capitalist dogma. I have argued against this historical analysis (Cachel 1990). Some of Haraway's assertions and my counterarguments
are presented in Table 16.1.

What is the "Man the Hunter" model of human evolution, and what association did Washburn and the New Physical Anthropology have with it? The
model is epitomized by a single book. In April 1965, a symposium was held at
the University of Chicago on the "Origin of Man." A year later, a symposium
on "Man the Hunter" was held at the same institution. The social anthropologist

Table 16.1. *Contentions and refutations about the "Man the Hunter" model and the "new" physical anthropology*

Contentions (Haraway 1989):

The New Physical Anthropology is equivalent to the "Man the Hunter" model of human evolution. **Refutation:** The New Physical Anthropology incorporates evolutionary theory (e.g., natural selection, adaptation, sexual selection, and other evolutionary processes) into physical anthropology. As articulated by S. L. Washburn in the early 1950s, the New Physical Anthropology also emphasized different purposes, methods, and goals from physical anthropology practiced in earlier periods (Table 2.1).

Washburn is not justly credited as the originator of the modern emphasis on field studies of primate behavior. His students lack theoretical orientation and methodological expertise. **Refutation:** Washburn is the originator of the modern emphasis on field studies of primate behavior, which is one important aspect of the integration of anatomy and behavior shaped by evolution. Early modern workers have theoretical orientation, though their methodological expertise is inchoate.

The "Man the Hunter" model stresses cooperation and affiliation in ancient humans. Ancient humans possessed cohesive social groups just as living non-human primate species exhibit stable cooperative social groups. The impetus for the "Man the Hunter" model was an anxious attempt to combat the dangers of impending global nuclear holocaust by arguing for the antiquity and consequently the inevitability of human cooperation. **Refutation:** Field-oriented studies of primate behavior were stressed in the late 1950s/early 1960s because of the theoretical problems created by the australopithecines, which the consensus opinion by then recognized as the earliest known hominids.

Sociobiology stresses individual competitiveness and not group or social cooperation. "Man the Hunter" therefore begins to resemble the insupportable idea of group selection. **Refutation:** Sociobiology is the study of how natural selection and other evolutionary processes generate social behavior and affect social life; it also examines the genetic bases for social behavior. Sociobiology studies both competition and cooperation.

The advent of sociobiology destroys the "Man the Hunter" model, and therefore the New Physical Anthropology. **Refutation:** The "Man the Hunter" model was/is one of many models of human evolution. A shift away from the model is caused when theory is tested by the paleoanthropological record (the New Archeology or processualism), not by the advent of sociobiology. The New Physical Anthropology is not extinct – it is physical anthropology as currently practiced.

Sol Tax initiated the symposium, and chaired the sessions. The published volume *Man the Hunter* was edited by Richard Lee & Irven DeVore (1968). Three reasons were given for holding the symposium. These were the accumulation of major new ethnographic data-sets on hunter-gatherers, and re-evaluation of their group structure, residence behavior, and kinship patterns; the excavation of Plio-Pleistocene "living floors" that needed hunter-gatherer data for interpretation; and the impending extinction of hunting-gathering societies (Lee & DeVore 1968: vii). The major portion of the volume consists of ethnographic case studies and vigorous discussions about subsistence and particulars of social organization. Two sections on prehistoric archeology and human evolution comprise only 26% of the text not including the references. Thus, the principal goal of the symposium was to resolve a set of issues in social anthropology, and not to provide a framework for interpreting human evolution. The five archeological papers (by Isaac, Freeman, Lewis and Sally Binford, and Clark) are circumspect and cautious. There are three papers on hunting and human evolution (by Washburn and Lancaster, Laughlin, and Steward). The Washburn and Lancaster paper was added after the symposium. Far from presenting a united front that signaled a consensus, the discussion transcripts show confusion and dissension not only on matters of social organization and the nature of inter-group relations, but also on the relative importance of hunting versus scavenging, and the defining importance of hunting as a way of life.

Of the three papers on hunting and human evolution, the most authoritative opinion that hunting molded human evolution was made by the physical anthropologist, William Laughlin, who had done research among Inuit and Aleut hunters. Laughlin's paper is so cogent and persuasive that the opening paragraph of the preprint circulated version was cited by David Hamburg in the volume's concluding discussions (Lee & DeVore 1968: 340–341). I give the final published version here:

> Hunting is the master behavior pattern of the human species. It is the organizing activity which integrated the morphological, physiological, genetic, and intellectual aspects of the individual human organisms and of the population [sic] who compose our single species. Hunting is a way of life, not simply a 'subsistence technique,' which importantly involves commitments, correlates, and consequences spanning the entire biobehavioral continuum of the individual and of the entire species of which he is a member. (Laughlin 1968: 304)

An earlier paper by Ginsberg & Laughlin (1966) also strongly argued that natural selection had molded a biobehavioral continuum for species-specific human traits.

Washburn & Lancaster (1968) emphasize that hunting necessitates tool behavior. This is not surprising, given that Washburn (1959, 1960) had long championed the autocatalytic nature of human tool behavior. Washburn & Lancaster (1968: 296) also emphasize the existence of an ancient sexual division of labor in human societies. Males hunt and butcher large game animals in cooperative, coordinated groups, and females care for the young and gather plant food. This type of society dates back to *Homo erectus*. Washburn & Lancaster (1968: 299–300) cite the idea of Konrad Lorenz (1966) that human aggression against other humans is divorced from any innate biological controls because human dependence on tools, rather than natural weapons, has not allowed these controls to develop. The pace of behavior driven by tools exceeds the pace of the evolution of biological controls.[4]

In retrospect, what is most impressive in *Man the Hunter* is the sophistication of the archeologists (Isaac, Freeman, the Binfords, Clark) and paleoanthropologists like F. Clark Howell. Howell advocated the development of new methodologies and protocols for reconstructing the behavior of fossil hominids, rather than relying largely on ethnographic evidence from living hunter-gatherers. He came to this conclusion because of the distance in time and biological remoteness of early hominids from modern humans, but he was also impressed by the diversity of opinion among social anthropologists about such critical data in living hunter-gatherers as subsistence, and the environmental variables that determined the size and mobility of social groups (Lee & DeVore 1968: 287–288).

Feminist critics later inveighed against the Man the Hunter model of human evolution, because they saw it as crediting only males with significant advances in hominization such as tool behavior and food-sharing (Tanner & Zihlman 1976, Hrdy 1981, Tanner 1981). This criticism then developed into the "Woman the Gatherer" model of human evolution, in which adult females, who collect plant foods, provide basic subsistence for a group, and are responsible for the origins of complex sociality. However, both of these models assume a sexual division of labor and degree of sociality observed in modern humans. These models were generated by ethnographic studies of living humans who hunted and gathered. Yet, the early hominids are embedded in deep time. They are taxonomically distinct from living humans, and the earliest hominids have a brain size only one-third that of living humans. It should be apparent that concerns about modern human interaction and human temperament (competitiveness, aggression, dominance) have no necessary origins in deep time among organisms that are so different from modern humans.

Food, food-sharing, and division of labor

Oakley (1957) argues that hunting mandated tool behavior, and that tool behavior subsequently molded all of human evolution. This was so because hunting meant the existence of selection pressure for intelligence, and furthermore implied the existence of food-sharing. Washburn & Lancaster (1968) also stress the importance of division of labor, but they speciously introduce the notion that only males hunt. A simple survey of carnivorous mammals would demonstrate that hunting does not imply a sexual division of labor. Washburn and Lancaster dismiss parallels to mammalian social carnivores, denigrating ". . . the old idea, recently revived, that the way of life of our ancestors was similar to that of wolves rather than that of apes or monkeys . . . Human females do not go out and hunt and then regurgitate to their young when they return. Human young do not stay in dens but are carried by their mothers. Male wolves do not kill with tools, butcher, and share with females who have been gathering" (Washburn & Lancaster 1968: 296). These arguments are largely rhetorical. No hominid is likely to evolve regurgitation at will, because this specialization is foreign to the primate order. Yet, bipedal hominids can transport food with their hands, rather than in their stomachs. If maturation rates were faster in Plio-Pleistocene hominids than in modern humans (as they seem to have been), juveniles would be independently mobile and largely free of adult handling at a much younger age (see Chapter 17). The juvenile *Australopithecus africanus* specimen from Taung may have been 3 years old, rather than 5 years; the subadult *Homo erectus* specimen from Nariokotome may have been 8 years old, rather than 12 years. Faster maturing young would not need to be carried or constantly monitored for a long period. If they did not accompany adults, they might have been left in some secure area. An area with sheltering trees would be ideal – the young could take refuge by rapidly climbing trees in the presence of danger. Perhaps the young might be under the supervision of an alert adult caretaker. This overturns the traditional assumption that hominid youngsters are helpless for years, and therefore require the constant attention of two adults (if not the entire social group) before they achieve some degree of independence. Certainly male wolves do not need tools to kill or butcher, but they do share food with all non-hunting members of the group. And it is likely that both male and female hominids gathered and shared plant foods, just as it is likely that males and females acquired and shared vertebrate meat and fat. They probably acquired meat and fat in a variety of ways – through scavenging, confrontational scavenging, and facultative hunting. Widespread hunting of large animals probably did not occur until about 1.7–1.5 mya.

Stoczkowski (2002: 113–121) analyzes the assumption of a sexual division of labor that is embedded in many models of human evolution. He discovers that old traditions of what he terms "naïve anthropology" assign distinctions to male and female roles, with males hunting and females gathering. These traditions are based on presumed female ineptitude for hard labor and the burdens of pregnancy and childcare. Stoczkowski replies to these traditions in the manner outlined above, although he lacks the most recent data on the chronological age of immature early hominid specimens. He cites six hunting-gathering societies in which female hunting occurs, and thus he questions what he terms the "universal law" that females cannot hunt. In addition, he cites work on modern hunter-gatherers by Alain Testart (Stoczkowski 2002: 116–118), who documents that taboos and social prohibitions against female hunting, as well as bans against female use of certain weapons that make hunting possible, maintain a sexual division of labor in many living hunter-gatherer groups.

A fundamental problem is that paleoanthropologists automatically adopt the perspective of modern humans, and assume that any division of labor that occurs in a complex society must be entirely or largely based on sex and behavioral sex differences. However, analysis of invertebrate or mammal societies demonstrates that division of labor can exist without a strict partitioning between the sexes (Wilson 1975). An example is seen in Cape hunting dog societies, where food sharing and complex division of labor was documented during the first field-oriented studies of this taxon (Kűhmer 1965, 1966).

The idea of an ancient sexual division of labor, with males acquiring vertebrate meat and fat and females gathering plant foods is still deeply embedded in paleoanthropological literature. It is present, for example, in O'Connell *et al.* (1999) and in Wrangham *et al.* (1999), with the recent addition of an emphasis on tuber collecting and cooking. How can this idea of a sexual division of labor be tested? Methodologies examining the chemistry of prehistoric bone and enamel may yield such tests (e.g., Bocherens *et al.* 1999, Sponheimer & Lee-Thorp 1999, Lee-Thorp *et al.* 2000, 2003, Balter *et al.* 2001). One possible test is to examine male versus female diet using both stable carbon and oxygen isotopes in tooth enamel. Another possible test is to examine male versus female ranging behavior through strontium isotopes in fossil bone. If there is a profound dichotomy between male and female activities, there should be a significant sexual distinction in diet, as well as in ranging behavior. This distinction is likely to persist in spite of any food-sharing or routine traverses of a home range, which could not hide any fundamental dichotomy in male and female activities.

Isaac (1971, 1975) emphasized tool behavior in hunting. Isaac (1978a, 1978b, 1983, Isaac & Crader 1981) later focused on the centrality of food-sharing in revolutionizing hominid sociality. The importance that Isaac attributed to hunting and food-sharing in triggering major events in hominid social evolution can be gauged from the following quotation: "Portions of a carcass are readily carried and are an important food prize when consumed at the destination. We thus favor a model in which the active delivery of some meat to fellow members of a social group developed in a reciprocal relationship with the practice of transporting and sharing some surplus plant foods. We see the model as representing a functionally integrated behavioral complex" (Isaac & Crader 1981: 93). Note that a destination exists, to which meat is transported, and at which a reciprocal exchange of food types occurs. This destination was first conceived of as a home base. It later became known as a central place, a core area in a home range to which foraging activities were tethered. Many living hunter-gatherers are central place foragers. Single individuals or foraging sub-groups spread out from and return to a central site that either contains a critical unwieldy resource like water, or is ideally positioned for acquiring multiple necessary resources (Winterhalder 2001). Foraging occurs along gradients linked to the central place.

Considerable debate took place during the Man the Hunter symposium over the degree to which either modern hunter-gatherers or Plio-Pleistocene hominids engaged in scavenging (e.g., Lee & DeVore 1968: 342). Washburn strongly rejected the idea that a scavenging phase had preceded the hunting of large vertebrate prey. Woodburn pointed out that the modern Hadza do not fear large predators and routinely engage in confrontational scavenging. DeVore maintained that modern humans might exhibit both confrontational scavenging and hunting, but that the australopithecines were unlikely confrontational scavengers, because of their smaller size and lack of sophisticated long-distance weapons. He therefore agreed with Washburn in rejecting an early scavenging phase in the hominid acquisition of vertebrate carcasses. Beginning in the 1980s, a series of researchers investigated the efficacy of scavenging in the modern Serengeti ecosystem, inferred the degree of scavenging opportunities during the African Plio-Pleistocene, and attempted to discern the archeological signatures of hunting or scavenging (Potts & Shipman 1981, Shipman & Rose 1983, Blumenschine 1986, 1987, 1995, Blumenschine & Madrigal 1993). These researchers noted that vertebrate carcasses provide not only meat, but also fat and fat-rich marrow. They recognized that fat and fat-rich marrow could be critical dietary resources, particularly during times when food was seasonally limited, and hominids were stressed for nutrients. This initiated a new emphasis on the amount, location, and temporal patterning of fat within the vertebrate body.

Alternatively, the presence of cut-marks on animal bones and a paucity of marrow processing signs might indicate hominid access to a relative glut of prime carcasses either through hunting or scavenging, seasonal avoidance of fat-depleted animals, or the availability of other food resources not visible in the archeological record. Gaudzinski (2003) analyzes 17 assemblages from the 'Ubeidiya Formation, Israel, dating to 1.4–1.0 mya. Cut-marked bones occur, but are rare. Because Gaudzinski finds no evidence of marrow processing, she infers that the 'Ubeidiya hominids probably did not engage in marrow-bone scavenging, transport, and processing. Consequently, it is possible that pronounced regional differences in hominid behavior based either on habitat distinctions or different species-specific behaviors already appear at 'Ubeidiya. Regional variation in hominid behavior certainly appears at a slightly earlier time range (1.64–1.39 mya) within different areas within the Koobi Fora region east of Lake Turkana, northern Kenya.

Survey and synthesis of modern hunter-gatherer diets reveal that environment severely constrains foraging opportunities (Cachel 1997, Jenike 2001, Kuhn & Stiner 2001). The degree to which plant foods are incorporated into the diet depends upon latitude. Equatorial regions offer the greatest opportunities for collecting plant foods. These opportunities dramatically fall off with increasing latitude. At high latitudes, with pronounced seasonality, plant foods may be unavailable for most of the year, and this may have an impact on human social organization (Cachel 1997, 1999, 2000b, 2001). Surveys of the net and gross energy yields of different food types consistently demonstrate that animal carcasses have the greatest food value per kg, and vegetation the lowest (Kuhn & Stiner 2001: Table 5.1). Roots and tubers rank significantly lower than animal carcasses – a finding that vitiates recent emphases on tuber collecting by fossil hominids (O'Connell *et al.* 1999, Wrangham *et al.* 1999). Furthermore, both tubers and grains must be cooked in order to obtain their maximum nutrient yield. Seeds and nuts have high gross energy yields, and are highly concentrated sources of energy, but the labor involved in collecting and processing them results in a low ranking in terms of net energy yields.

Pair-bonding

The establishment of stable pair-bonding has been associated with the origins of typically hominid behavior and sociality. Stable pair-bonds are often linked with a monogamous breeding system in models of human origins. This linkage began immediately after the publication of C. R. Carpenter's field study of white-handed gibbons (Carpenter 1940). Hooton (1942), for example, argued that the social system of these gibbons could well be the basis of the human nuclear family. Gibbon social groups are formed by a monogamously pair-bonded adult male and female that jointly defend the boundaries of their territory, along with sub-adult young born to the mated pair. When offspring reach reproductive maturity, they are driven out of the social group by the adult parent of the same sex. Hooton's focus on gibbon society was based largely on the fact that these animals are apes, not monkeys, and therefore have a closer genetic relationship to humans. Also, until the 1960s, gibbons remained the only ape species whose sociality had been investigated in the wild.

C. Owen Lovejoy (1981) presented a very influential version of the monogamous pair-bond idea of human social origins. Lovejoy argued that monogamy was central to human origins, because it leads to stable, permanent pair-bonds, and thus to males that are involved in rearing of the young, because they have certitude of paternity. There is a sexual division of labor. A male provides meat for his mate and his offspring. Thus, food-sharing also exists. An evolutionary ecologist updating Lovejoy's model would remark on the fact that monogamy does not necessarily lead to certitude of paternity. This is documented by genetic investigation of offspring. Hence, the term "social monogamy" is now preferred. However, the earliest hominids, the australopithecines, are known to have had a significantly greater degree of sexual size dimorphism than in modern humans. This might appear to militate against social monogamy, but Lovejoy argued that a decline in sexual dimorphism had already been occurring among the australopithecines, as seen in a major reduction in their canine size dimorphism.

Another possible argument against promiscuous mating systems in early hominids can be made on soft-tissue evidence. A number of researchers have studied the evidence for sexual selection and sperm competition in primates by examining the incidence of different sperm types within species, relative testis size, and the comparative anatomy of male and female genitalia, including the impact of the cyclical swelling of sexual skin within female catarrhine primates on the mechanics of copulation (Harcourt *et al.* 1981, Harcourt 1991, Harcourt 1997, Dixson 2002). For example, these researchers regressed testis weight against body weight in primates, and compared the results across different mating systems. Relatively large testes occur in species where multiple males mate with females, and relatively small testes occur in species with only a single breeding male in the social group. The likely explanation is that male–male breeding competition is taking place within the female reproductive tract, as sperm from different males vie to fertilize an ovum. A superabundance of sperm confers an increased likelihood of fertilization. This is known as sperm competition. The phenomenon has been well studied in insects and mammals. Modern humans exhibit variable testis sizes, depending on the geographic region or major human population group being sampled (Harcourt 1997). However, humans as a species resemble the orangutan in relative testis size – the human value is far smaller than that of the chimpanzee, but larger than that of the gorilla. Because humans are intermediate in terms of relative testis size, one might infer an ancestral mating system very unlike that of chimpanzees or gorillas, yet not resembling the social monogamy of hylobatids, either.

Reno *et al.* (2003) have recently analyzed the degree of sexual dimorphism exhibited by the sample of *Australopithecus afarensis* specimens from Hadar, including the mass death assemblage from locality AL 333. They have a sample size of 22 individuals, with an additional 7 individuals from other Hadar localities and Maka. Using femoral head diameter as a proxy for body size, they argue that skeletal size dimorphism in *A. afarensis* was about 15%, much more similar to that of modern humans than previous estimates of sexual size dimorphism in this taxon. The standard estimate of postcranial size dimorphism in early hominids is 35% (McHenry 1992), and Cunningham *et al.* (2004) argue that the AL 333 hominids had a pongid-like degree of dimorphism. If the smaller estimate of Reno and his colleagues is true, then the discrepancy between australopithecine canine size and body size dimorphism would be eliminated. From these results, Reno and his colleagues (including Lovejoy) infer the existence of monogamy in *A. afarensis*, and a concomitant reduction in aggressive male–male competition for access to mates (Reno *et al.* 2003). Thus, the earlier arguments of Lovejoy (1981) about monogamous pair-bonding in australopithecines would be supported. Yet, even if *Australopithecus afarensis* and the other australopithecines are ultimately revealed to have only a modest amount of body size dimorphism, this is no reason instantly to infer pair-bonding and a monogamous mating system. In a study of sexual dimorphism in fossil taxa, Plavcan (2000) is extremely dubious about inferring breeding systems from the degree of sexual dimorphism.

The rarity of social monogamy in the animal world – including the non-human primate world – leads one to be dubious about creating a superstructure of complex sociality based on the inference of monogamous pair-bonds in early hominids. E. O. Wilson (1975) had previously set out the three prerequisites necessary for establishing monogamy, but these were not examined by Lovejoy and his colleagues. These prerequisites are a stressful and resource-poor physical environment; a territory containing necessary scarce resources that can only be adequately defended by two animals; and an early reproductive advantage enjoyed by adults that are already mated, because they do not have to search for or compete for a reproductive partner. Only the first of these prerequisites can be examined for fossil hominids. It does not appear that the African Plio-Pleistocene was unduly stressful or poor in resources, unless one focuses solely on tropical rainforests, which were clearly severely diminished during the Pleistocene. Hence, at least one of the prerequisites for monogamy appears to have been missing for Plio-Pleistocene hominids. Furthermore, a recent comprehensive survey of social monogamy confirms its rarity in mammals, as well as the rarity of male–female partnerships in mammals (Reichard & Boesch 2003). Hence, it may be premature to focus attention on the establishment of a monogamous mating system in early hominids, and use it to explain the origins of complex social life.

In addition, the appearance of male parental investment and pair-bonding may involve factors other than mating behavior and food-sharing. For example, Palombit (1999) concludes that pair-bonding in non-human primates may often occur as a response to potential infanticide. The prolongation of male–female associations beyond estrous (i.e., "pair-bonding") may evolve as a counter-strategy to infanticide. Males and females continue to associate, and fathers may therefore actively or passively offer protection to their young from other, infanticidal males.

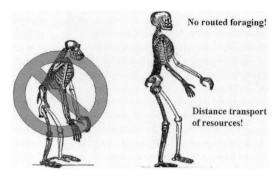

No routed foraging!

Distance transport of resources!

Figure 16.4. Contrast between the ranging and foraging behavior of pongids and *Homo erectus* individuals who were presumably responsible for creating the more complex archeological record in the Koobi Fora region between 1.64 and 1.39 mya.

Taphonomy and the nature of "sites"

Lewis Binford (1981) triggered a major revision in thinking about the nature of Plio-Pleistocene archeological sites. He was particularly critical about the idea that some Plio-Pleistocene sites represented home bases that would be the ancient equivalent of base camps among living hunter-gatherer groups, as initially argued by Isaac (1971, 1975, 1978a, 1978b). Taphonomic processes might create accidental accumulations of archeological material. Time-averaging certainly could smear any record of transient human behavior. However, Binford still advocates the use of modern hunter-gatherer evidence, albeit with extensive mathematical manipulation (Binford 2001). Glynn Isaac later revised his home base model of Plio-Pleistocene sites, and began to emphasize a model of central place foraging. There has been extensive fallout from Binford's criticism. The investigation of taphonomic processes is now de rigueur in Paleolithic archeology (e.g., Klein 1989, Lyman 1994, Stiner 1994).

But what constitutes a "site"? During the 1970s, initial surface survey in the Koobi Fora region east of Lake Turkana in northern Kenya revealed low-density scatters of stone artifacts. What did these scatters signify? Isaac & Harris (1975, Harris 1978) advocated a search for the scatter of archeological material between dense patches – the "scatter between the patches" approach. They argued that these "scatters between the patches" of dense concentrations could yield crucial information about the regional activities of early hominids. This was "landscape archeology." It emphasized the importance of studying hominid ranging, foraging, and dispersal behaviors on a broad, regional scale, rather than merely at the level of a single site, however dense its archeological evidence. In landscape archeology there is a focus on archeological variation across regional distances within constrained time units (Figure 16.4). The

Figure 16.5. Hyaena Hill, Olorgesailie, Southern Kenya. Dr. Rick Potts has randomly excavated squares in a grid pattern in order to collect data on landscape use by *Homo erectus* individuals.

protocol for investigating scatters and patches at Koobi Fora was later elaborated by Rogers and his colleagues (Rogers *et al.* 1994, Rogers 1997). A strategy was developed for random sampling of stone artifacts on transects run along an outcrop of a certain geological age, and through a certain stratigraphic interval. The Okote Member was sampled, dating from 1.64 to 1.39 mya. The general meaning of low-density artifact scatters could be inferred by extensive sampling of contemporary horizons and habitats across a broad expanse of the ancient landscape. A similar landscape approach has been undertaken for a later time range at Olorgesailie, in southern Kenya (Potts *et al.* 1999), and an earlier time range in the Olduvai Basin, in Tanzania (Blumenschine *et al.* 2003).

What has been learned from landscape archeology in the Plio-Pleistocene of East Africa? For the Koobi Fora region during Okote Member times (1.64–1.39 mya), it is clear that complex foraging for dispersed resources was occurring (Cachel & Harris 1998, Cachel 2004, Cachel & Harris in press). This implies an intimate knowledge of the regional environment, as well as the ability to locate resources that fluctuate widely in time and space and predict their abundance. This evidence contradicts Binford's (1984) routed foraging model for hominids, in which an ape-like, "feed as you go" behavior takes place. Binford argued that all fossil hominids engaged in routed foraging, until the advent of anatomically modern humans. Yet, the Koobi Fora evidence indicates a difference from pongid behavior at least as early as 1.6 mya (Figure 16.5).

The hominization process

The previous sections discussed some current important models of human origins. In the European literature, the process of acquiring typically hominid traits

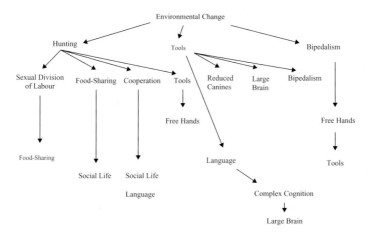

Figure 16.6. Synthesis of twenty-four classic hominization models. Twenty of these models date from the mid-1950s to the late 1980s. Explanatory factors that appear in at least three models are shown here. Note the triggering effect of environmental change and the centrality of tool behavior in these models. Modified from Stoczkowski (2002: Fig. 1).

is referred to as "hominization." Stoczkowski (2002) presents and analyzes classic hominization models, some dating to antiquity, and some to modern times. Figure 16.6 synthesizes 24 of these models, 20 of which date from the mid-1950s to the late 1980s. Explanatory factors that appear in at least three models are shown in Figure 16.6. Note that environmental changes are the initial trigger effecting change in all of these models, and that tool behavior has a central role in shaping further changes. The major conclusion of Stoczkowski's monograph is that hominization models remain static, in spite of new paleoanthropological data and methodologies, or advances in evolutionary theory. This is because these models are based on "common sense" views of human origins that are cultural constructs dating back to the Enlightenment or Greco-Roman antiquity. I note, however, that old conjectures about barbarians, savages, or hunter-gatherers are often now modernized by the introduction of non-human primate behavioral ecology, the favored species, of course, being the chimpanzee and bonobo (Chapter 2). So unthinking is "chimpocentrism" in current paleoanthropology, that it is taken as a given that chimpanzee or bonobo behavior yields detailed insights into the behavior of fossil hominids. Yet, the behavior and sociality of these two species is very different. If a chimpanzee species is chosen to reconstruct early hominid behavior, which species will serve as the model? Is it possible that some alternative model should be used? The composite mammal model might be tried in lieu of the now standard non-human primate model (Chapter 15).

A novel hominization model

In the following section, I present a novel hominization model that incorporates both anatomical and archeological factors. Unlike the classic hominization models shown in Figure 16.6, differences in sociality and cognition are not late-appearing culminations of a long series of evolutionary events. Instead, they appear early on, and are themselves factors that trigger later events in hominid evolution.

Although events are occurring against a general background of late Neogene climatic fluctuations, changes in the physical environment are not the necessary trigger for hominization. The Oligocene was an epoch of global aridity. Mosaic habitats that include some open country areas occur in the Old World since the middle Miocene. Environmental perturbation dates back to about 25 my in East Africa, since the formation of the East African rifting system. Continuing volcanism, with hyper-alkaline volcanic eruptions, characterize East Africa since the early Miocene. If environmental changes were all that was needed for hominid origins, then the first hominids should have emerged as part of the early to middle Miocene hominoid radiation. Much more important than environmental changes were changes in sociality that reduced intra-group competition. This might have occurred in either or both of two ways: through a reduction in matri-line competition, or through a reduction in dominance hierarchies. Because female philopatry, female dominance hierarchies, female–female competition, matrilines, and competition between matrilines appear to be derived features characteristic of cercopithecoid monkey societies (Di Fiore & Rendall 1994, Rendall & Di Fiore 1995), it is possible that the ancestors of the hominids never experienced the degree of intra-group competition documented among cercopithecoids. As a result of limited intra-group competition, natural history intelligence, food-sharing, and some early division of labor could evolve. If the role of vigilant sentinel was developed early – and it appears to evolve easily in non-primate mammals with complex sociality (Chapter 9) – this would facilitate exploratory foraging, including prolonged foraging on the ground that would otherwise be prohibitively hazardous for the earliest hominids. The connection between a reduction in intra-group competition and the origins of natural history intelligence is outlined in Chapter 8.

The transition to bipedalism is rapid and relatively easy, given the fundamental catarrhine substrate for bipedality. As outlined in Chapter 13, the acquisition of bipedalism does not mandate a long period of experimentation. Bipedal posture and locomotion could have first occurred in an arboreal setting, and might have been as frequent as in living hylobatids. The transition to terrestrial bipedalism was not irrevocable, given that shelter in trees was absolutely necessary until hominids developed the ability to control fire and construct artificial shelters. Hominids like the australopithecines that possessed

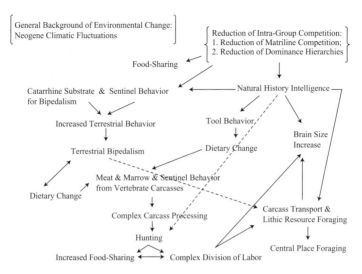

Figure 16.7. Flowchart illustrating a new model of the hominization process. It assumes a catarrhine bipedal substrate, but a different inital sociality from the catarrhine norm. Dashed lines are drawn only to enhance legibility; they are not quantitatively different from the other figure lines. Contrast this flowchart with Figure 16.6, which synthesizes traditional models of hominization.

long forelimbs, vertically oriented glenoid joints, curved phalanges, and relatively flexible ankle joints were not clumsy pseudo-bipeds. They were good bipeds and also adept tree-climbers. Arboreal behaviors were essential when shelter was needed, especially at night. At approximately the same time, natural history intelligence facilitates vigilance and tool behavior. With the addition of sentinel behavior, a relatively undemanding cooperative behavior, terrestrial habitats become easy to penetrate. With sentinel behavior, hominids could safely become terrestrial, in spite of a myriad of ravening terrestrial carnivores. Predation would not be the strong selection pressure that it is often assumed to be. This sequence of events is outlined in Figure 16.7.

How can this model be tested? To begin with, this model is congruent with the discrepancy between bipedal origins and the far later appearance of stone tools. The discovery of tool behavior in wild orangutans leads some researchers to infer a substrate for tool behavior and material culture dating back to 14–15 mya in the last common ancestor of hominids and orangutans (van Schaik *et al.* 2003). It is a generally held assumption that complex tool behavior is part of the common heritage of pongids and hominids. Yet, the model I present here views tool behavior as the outgrowth of a natural history intelligence that differentiates hominids from the pongids (Chapter 8). If this model is correct, one would predict that the archeological record begins late in time, well after the origin of hominids – certainly not with faint traces that lead back to the middle

Miocene, as implied by van Schaik *et al*. (2003). Furthermore, if natural history intelligence is involved in hominid tool behavior, there need not be a long period of experimentation with bone tools or lithic raw materials and the mechanics of knapping. The threshold to competent stone tool use and manufacture could be crossed relatively abruptly, once selection pressures begin to operate for the use of stone tools in carcass processing. Characteristic signatures of hominid modification of animal bones (cut-marks and percussion marks) might be found before the first documented lithic artifacts. The crossing of an abrupt behavioral threshold implies that a proto-Oldowan does not exist. The Oldowan itself may become more complex, as seen in the Karari Industry (Harris 1978), and the archeological record may become denser and more variable with time, but there should be no long, faint trail of bone or lithic artifact experimentation leading back to the middle Miocene.

The Karari Industry itself appears abruptly in the Koobi Fora region of the East Turkana basin at about 1.6 mya (Harris 1978). Its diagnostic artifact is the single-platform core, which allows more complete and efficient utilization of a clast (Figure 16.8). This sudden appearance of a new type of artifact supports the idea that hominid tool behavior is characteristically associated with threshold effects, and is not a marginal, desultory behavior. The stereotyped nature of the single-platform core certainly supports its creation through imitation, and not emulation. Although basalt is the major raw material used in its production, it can also be fashioned from ignimbrite, or rare lithic materials. Most of the Karari single-platform cores are very large, but some are very small, miniaturized versions of the more usual artifact form.

Furthermore, as simple as the Oldowan artifacts appear to be, if natural history intelligence is involved in their creation, study of these artifacts should demonstrate technical competency, and knowledge of the properties of different lithic raw materials. Because Oldowan artifacts can be sorted into typological categories – although technological categories are now preferred by archeologists – this implies that imitation was involved in their creation, and hence a Theory of Mind. When microwear or residue analyses of Oldowan artifacts are ultimately available, these analyses should reveal a variety of tool functions. This is expected if their makers are exploiting different subsistence behaviors, and are exploring their control of the natural world. It is even possible that specialized tool-kits appear in the Lower Paleolithic with the Acheulean Industry, rather than making their first appearance in the Middle Paleolithic. Multiple hominid species should be associated with tool behavior, although the quandary of unequivocally associating a taxon with archeological material seems currently intractable. Finally, even the earliest archeological record should reveal complex ranging and foraging behavior, unlike the routed foraging behavior of pongids.

Figure 16.8. Two views of a Karari single platform core, originally termed a Karari scraper by Harris (1978), who used it as the diagnostic tool type of the Karari Industry. The arrow points to the platform, which, in this case, is created by splitting a cobble. The single platform can also be produced on a very large flake. The lightning bolt indicates the direction of a blow delivered by a hammerstone. Flake scars can be discerned on the core. The star indicates a flake scar. The tool maker directs blows along the entire perimeter of the platform, thus removing useable flakes. This knapping procedure makes efficient and maximal use of a core. Cores can be easily transported as hominids move through a home range. Flakes can be removed whenever it is necessary to cut, scrape, or sever materials. This behavior implies that cobbles or clasts of a suitable size and raw material type are at a premium, and would consequently have been the objects of directed foraging behavior. This stone tool is a modern reproduction knapped by Dr. Christopher Monahan from a clast of red ignimbrite found in the dry bed of the ephemeral Il Eriet (Ileret) River in northern Kenya. The scale is in centimeters.

If different hominid lineages are experimenting with bipedality, as indicated by divergent foot morphology, for example, this would be another sign that a novel adaptive zone existed early on in hominid evolution. If bipedalism is a grade, this would signify that multiple lineages are scrambling to establish terrestrial lifeways. The breakthrough to a successful terrestrial niche would thus involve not merely the advent of bipedal anatomy, but also the associated cognitive and social innovations outlined above. Thus, evidence to support this new model can be found in documenting the existence of different locomotor morphology in hominid species that points to an independent acquisition of bipedalism in more than one lineage.

If predation pressure is not a strong force of natural selection among early hominids, only a relatively small number of hominid fossils should show evidence of ante-mortem carnivore damage. If sentinel behavior exists, as well as division of labor in carcass processing, then hominids did not need to be furtive while on the ground, making quick, desperate forays to acquire or process carcasses. This reconstruction is contrary to the opinion of some researchers (e.g., Potts 1988, Bunn 1994, Brantingham 1998), who argue that Plio-Pleistocene hominids needed to locate and scavenge a carcass in a cheetah-like mode of enhanced mobility, rapid processing behavior, and hurried flight before the arrival of large carnivores. Yet, this present model argues that vigilance and sentinel behavior can create a relatively safe environment in which to acquire and process carcasses. This would be true regardless of whether hominids were scavenging carcasses, engaging in confrontational scavenging, or actively hunting. If this model is true, evidence of predation on hominids should be rare. In addition, one would expect to see zooarcheological evidence of multiple animal species being processed at sites, bone modification of large species even at an early period, evidence that hominids had early access to carcasses, and evidence of leisurely, competent defleshing of carcasses. The chemistry of hominid tooth enamel should reveal omnivory, with a substantial meat component.

Endnotes

1. "*l'homme fossile n'existe pas*" (Begun 2003: 76).
2. "*Dieu est éternel, mais l'homme est bien vieux.*" (Grayson 1983: Frontispiece).
3. Marie Sklodowska Curie, Pierre Curie, and Henri Becquerel received the 1903 Nobel Prize for Physics. In 1911, Sklodowska Curie also received the Nobel Prize for Chemistry.
4. Lorenz himself was not as morose about aggression and the human condition as some other scholars – e.g., Cartmill (1993). This can be seen in the original title of *On Aggression*: *Das sogenannte Böse* [The So-Called Evil].

17 What does evolutionary anthropology reveal about human evolution?

This chapter summarizes some points discussed in a more detailed fashion elsewhere, and the appropriate chapters are cited. Other points appear here for the first time.

Phenotypic change and "contemporary evolution"

Phenotypic change obviously occurs through geological time, and it can sometimes be discerned within the framework of documented history. What is the genetic basis for rapid phenotypic change? How rapid can phenotypic change be? How rapid can natural selection be? These questions are deeply embedded in the history of anthropology. In the early twentieth century, Franz Boas argued that human cranial shape and form could be altered within a single generation, when immigrants simply entered the novel environment of America (Boas 1912). The cranium appeared endlessly malleable, and the measurement of cranial differences was therefore eschewed as meaningless. However, multivariate statistical analyses of the human cranium demonstrated a significant difference between major human population groups (Howells 1973, 1995), and forensic investigations showed the utility of cranial shape and form in inferring ancestry. A statistical re-analysis of Boas's cranial data confirms that the descendants of immigrants resembled their ancestral group, and were not homogenized in the American milieu (Sparks & Jantz 2002, 2003). The genetic underpinnings for cranial shape and form prevail over environmental plasticity. With respect to the cranium, genetics trumps the environment.

Let body size serve here as the test case, because body size and shape changes will be examined in the next section. Body size can shift quickly within the range of phenotypic plasticity documented in living animals, including humans. Yet, what is the probability of genetically based change in body size occurring? This is the type of microevolutionary change that, if it continues, yields new subspecies or species. Artificial selection for mammalian body size change is familiar. Breeding cattle, sheep, pigs, horses, and dogs for a particular size is commonplace. Laboratory studies of mice and rats serve as models for body size changes in other mammals, because they breed quickly and are easy to maintain.

Furthermore, the mouse genome has now been sequenced and compared with that of other organisms, including humans (Chinwalla *et al.* 2002). Reduced body size in mice has been achieved many times, as researchers study the physiological and genetic changes associated with body size. The heritability of body size across species is between 30 and 60%, but it is not known whether change is caused by many genes each with a small effect or a few genes each exercising a major effect (Hastings 1996: 135–136). This heritability is in line with the 51% heritability for body mass established in captive baboon species and baboon hybrids (Jaquish *et al.* 1997). In laboratory mouse strains, where some genetic variability has been lost and response to selection is presumably dampened, artificial selection can halve adult size in 50 generations, and subsequently introducing a dwarfing gene into this line creates adults that are only one-third the size of animals in the original stock (Hastings 1996). These animals are healthy. In the wild, however, thermoregulation, predation, and food would affect this response.

How fast can natural selection be in the wild? First, I must note that natural selection is abundantly documented in the wild (Endler 1986). Furthermore, contrary to some characterizations (e.g., Lewontin 1974, Gould & Eldredge 1993), natural selection in the wild can be very strong. It can occur on the scale of decades, and the resulting phenotypic changes can be observed within less than several hundred generations; sometimes it can be observed within several generations, or within a human lifetime (Reznick *et al.* 1997, Hendry & Kinnison 1999, Kinnison & Hendry 2001, Coltman *et al.* 2003, Stockwell *et al.* 2003, Olsen *et al.* 2004). This leads to the idea of "contemporary evolution" – evolution witnessed at the time scale of a human life. Some selection rates observed in the wild are equal to those obtained under laboratory conditions, and can be four to seven orders of magnitude greater (i.e., 10 thousand to 10 million greater) than those observed from the fossil record (Reznick *et al.* 1997: 1935). Strong selection rates and phenotypic changes in contemporary time are associated with environmental alteration caused by climate, habitat degradation, colonization, natural predation, or human hunting (Coltman *et al.* 2003). Some populations experiencing divergent changes show some limited reproductive isolation, or incipient speciation. Human commercial fishing for Atlantic northern cod caused a rapid evolutionary shift in life history, resulting in faster maturation and smaller body size. These shifts were caused by genetic changes, not by phenotypic responses to environmental change (Olsen *et al.* 2004).

What implication does research on "contemporary evolution" have for paleoanthropology? It establishes that phenotypic change caused by natural selection may be so quick as to be undocumented by the fossil record, because it is highly unlikely that successive members of several or several hundred generations are preserved as fossils that trace the phenotypic change through time. The appearance of a new morphology then appears inexplicable. The normal response of paleoanthropologists is then simply to describe or diagnose a new subspecies, species, or genus, without attempting to infer the evolutionary processes responsible for the new morphology. However, if the fossil record is dense enough and well-dated, phenotypic change can sometimes be observed through time (e.g., changes in brain size, body size, or body shape). The genetic basis for such change may remain unknown – although not unknowable, because the biology of living humans or other mammals can be instructive. Estimates of heritability can be made, and rates of selection gauged. Changes or alterations in the contemporary environment may include important selection factors, such as diet, habitat disruption, or predation. The degree of initial phenotypic variation must be important. Such variation would include sexual dimorphism. If one sex exhibits more phenotypic variation in a trait, that sex would have more potential to respond to selection. If the sexes differ in heritability of a trait, the sex with the higher heritability would show the faster rate of evolution. In addition, hominid population size must be an important factor, because populations that are too small might exhibit phenotypic changes that have nothing to do with natural selection. Accident or genetic drift might explain phenotypic change in small populations.

Abundant evidence has been presented in Chapter 11 that food scarcity selects against large body size. Hrdy (1981: 31) speculates that the advent of more food should reduce primate sexual size dimorphism, as females increase their body size to approach the male condition. Yet, many factors affect body size. While body size may not affect competition for mates among female mammals, it does affect female–female food competition, maternal care, and parasite load. An increase in body size should actually increase sexual size dimorphism, according to Rensch's rule. Size differences in cranial length between the Dmanisi crania D2700 and D2280 (14%) are exactly equal to cranial length size differences between KNM ER-1470 and KNM ER-1813 (14%), and these specimens are generally allocated to different species, *Australopithecus rudolfensis* and *Australopithecus habilis*, respectively (Vekua *et al.* 2002). Increased sexual dimorphism might lead to greater sexual selection.

Body size and shape changes

Chapter 11 described how a significant increase in body size occurs with the advent of early *Homo erectus*, as exemplified by the KNM-WT 15000 specimen. A higher and/or more predictable caloric base must lie behind this increase. There are both fitness costs and advantages in being large. The total energy

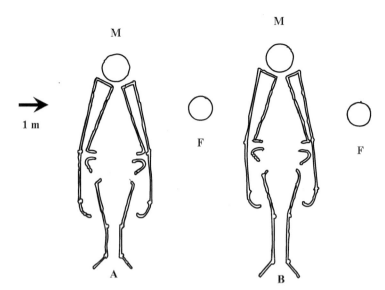

Figure 17.1. A schematic diagram illustrating changes in body proportions and sexual size dimorphism in stature when body size is increased in an australopithecine species. Female stature in A is based on NME AL-288–1 (the "Lucy" specimen). The male counterpart is a scaled-up version of the female, and there is about 25% statural dimorphism. *Australopithecus afarensis* is used only for heuristic purposes, given knowledge of the species' hypodigm. No ancestral/descendant relationship is implied with other australopithecines or genus *Homo*. A hypothetical outcome of body size increase is shown in B. The only change in male body proportions is a lengthening of the lower leg, as documented with size increase in modern humans. Note that sexual size dimorphism also increases, as expected in mammals with Rensch's rule operating.

of expenditure rises, but may be counterbalanced by the efficiency of bipedal locomotion. Heat stress increases with size, but an ectomorphic body shape dissipates heat in warm climates. In cold climates, core body temperature is easier to maintain in a larger body. Dispersal to higher altitudes in the tropics or dispersal to temperate regions is therefore affected by body size. After size increase occurs, predation pressure is reduced, inter-specific competition is lessened, and home range size is increased.

The dispersal of *Homo erectus* out of Africa was associated with all of the benefits and few of the costs of body size increase. This statement becomes clear through examining a five-stage sequence of hominid body shape and form pictured schematically in Figures 17.1 through 17.3. Three of the stages are based on known morphology (Figures 17.1A, 17.2A, 17.2C), and two are hypothetical (Figures 17.1B, 17.2B). Figure 17.3 contrasts the sharpest differences in size, shape, and sexual dimorphism occurring in this five-stage sequence.

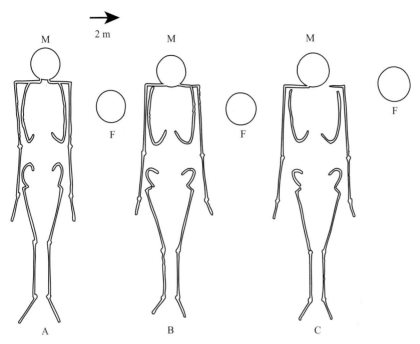

Figure 17.2. A schematic diagram illustrating changes in body proportions and sexual size dimorphism in stature in *Homo erectus* and *Homo sapiens*. The male figure in A is based on the KNM-WT 15000 subadult *Homo erectus* specimen, and shows an ectomorphic body build in the relatively long length of distal limb segments, large hands and feet, long neck, and narrow bi-iliac breadth. The male figure in B is a hypothetical specimen of *Homo erectus* without exaggerated ectomorphy. Greater body weight and body breadth would be expected in *Homo erectus* populations dispersing to higher latitudes and adapting to temperate climates. In both A and B, sexual size dimorphism in stature (14%) has decreased from the australopithecine condition seen in Figure 17.1. This estimated figure is based on cranial length differences between 1.8 mya Dmanisi specimens D2700 and D2280, which are clearly members of the same population. The male figure in C is a modern human, *Homo sapiens sapiens*. Relative brain size has increased. Sexual size dimorphism in stature is 6.25%, based on a global average.

In Figure 17.1, two australopithecines illustrate the differences in body proportions and sexual dimorphism in stature that would occur with a simple increase in body size. Female stature in Figure 17.1A is based on NME AL-288–1, the famous "Lucy" specimen of *Australopithecus afarensis*. *Australopithecus afarensis* is used only for heuristic purposes, because of the completeness of the species' hypodigm. No ancestral/descendant relationship is implied with other australopithecine species or with genus *Homo*. The male counterpart in Figure 17.1A is a scaled-up version whose statural dimorphism is 25% larger.

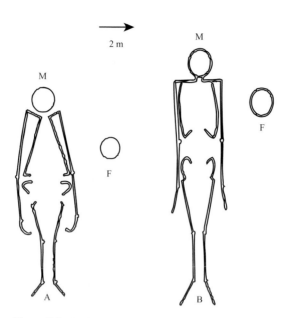

Figure 17.3. A schematic diagram illustrating the sharp contrast in body size, shape, and sexual dimorphism in stature between a hypothetical australopithecine condition with increased body size (A) and early African *Homo erectus* with heat adaptation (B). A is taken from Figure 17.1; B is taken from Figure 17.2. The greatest anatomical differences occur between these two stages. Among the most important differences in B are the following: the glenoid is horizontally oriented, the torso is reduced, the upper limbs are reduced, the femur is lengthened, the center of gravity is lowered, the femoral head is enlarged, and the phalanges are no longer curved. These changes create an obligate biped. Relative brain size has increased. In addition, B is adapted for arid open-country conditions by developing an ectomorphic body build: the distal limb segments are elongated, the neck is long, the hands and feet are large, the torso is antero-posteriorly constricted, and the bi-iliac breadth is narrow.

This number is a compromise. Traditional paleoanthropological inference is that sexual dimorphism in australopithecines was far more extreme than in modern humans, and may have been as extreme as in other catarrhines, where it can reach 100%. The standard estimate for body size dimorphism is 35% (McHenry 1992), and Cunningham *et al.* (2004) support this. On the other hand, Reno *et al.* (2003) argue that sexual dimorphism was reduced in *Australopithecus afarensis*. They believe that it was 15%, approximating the degree of dimorphism in modern humans, from which they infer ancient pair-bonding and other behavioral consequences (Chapter 16). I compromise between these two estimates, and use 25%. However, the 25% degree of dimorphism illustrated in Figure 17.1A is not merely an abstract compromise. It is in accordance with the degree of sexual dimorphism unequivocally present in the Plio-Pleistocene South African Dri-molen fossils of *Australopithecus robustus* (Keyser 2000, Keyser *et al.* 2000).

The lower postcanine tooth row of DNH8, an adult male, is 24% longer than the lower postcanine tooth row of DNH7, an adult female (Keyser 2000: Table 1). Lest one should deny that this figure has significance for *Australopithecus afarensis*, Keyser (2000) notes that this degree of dimorphism is comparable to that noted for another australopithecine species, *Australopithecus boisei*.

The diagram in Figure 17.1B is hypothetical. The increase in relative length of the lower leg occurs in modern human populations where male children experience better socioeconomic conditions and grow larger. Obviously ontogenetic conditions in the australopithecines are unknown. Yet, it is reasonable to assume that body size increase in australopithecines would be associated with an increase in lower limb length, given the fact that long bone length is significantly correlated with body size in living pongids (Schultz 1933). Sexual size dimorphism has increased in part B from the condition illustrated in part A. This is expected in mammals such as catarrhine primates in which Rensch's rule is operating (Chapter 11).

In light of Figure 17.1B, it is interesting that the 2.5 mya taxon *Australopithecus garhi* from Bouri, Ethiopia, has increased lower limb length, while retaining a relatively long forearm, typical of the australopithecine condition (Asfaw *et al*. 1999). A femur and partial fibula exist – the femur is elongated, but it is not clear whether the lower leg is also elongated. The relative size of the postcanine teeth and relative brain size are also australopithecine-like. The describers of this taxon argued that it might be a possible ancestor for genus *Homo*. Irrespective of the final taxonomic judgment about the *A. garhi* material, it is clear that body size increase can occur in early hominids without a complete overhaul of body shape and form. That is, body size is the first factor to change – in fact, it changes easily. Also note that lower limb length increases first, as indicated in Figure 17.1B. Furthermore, three fossil ungulate bones showing hominid modification are in spatial association with the hominid material, indicating the existence of lithic technology and hominid reliance on vertebrate meat and marrow at Bouri (de Heinzelin *et al*. 1999).

Figure 17.2 illustrates three members of genus *Homo*: the ectomorphic KNM-WT 15000 *Homo erectus* specimen, a *Homo erectus* individual with no ectomorphic specializations, and a modern human. Sexual dimorphism in stature is also indicated.

Figure 17.3 contrasts the greatest differences in body size, shape, and sexual dimorphism occurring in this sequence.

Predictions associated with body size and shape changes

Figures 17.1 and 17.2 illustrate five stages in hominid body size and shape transformation. These are the following: stage 1, early australopithecine

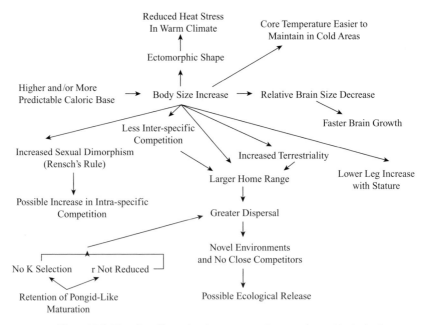

Figure 17.4. Flowchart illustrating the sequence of events triggered by body size increase.

(Figure 17.1A); stage 2, australopithecine with body size increase (Figure 17.1B); stage 3, generalized *Homo erectus* (Figure 17.2B); stage 4, ectomorphic *Homo erectus* (Figure 17.2A); and stage 5, modern *Homo sapiens* (Figure 17.2C). Figure 17.4 is a flowchart illustrating factors associated with body size and shape differences in this sequence. A number of predictions can be generated from the sequence.

The body size increase in stage 2 is associated with australopithecine-like proportions. Because a dietary shift (presumably habitually acquiring vertebrate meat and marrow) allowed body size increase, archeological evidence of this dietary shift (flake production, bone modification) should first occur at this stage, or prior to this stage. That is, archeological evidence should predate the appearance of genus *Homo*. Leg length (especially lower leg length) should increase with the first increase of body size in stage 2, because lower leg length is the first trait to respond to better dietary quality in modern human males. The greatest degree of sexual dimorphism should be exhibited in stage 2, because of Rensch's rule. Stage 3 is associated with increased bipedal efficiency, as the center of gravity is lowered through reduction of upper body mass. Because increased bipedal efficiency in stage 3 lowers the Total Energy of Expenditure, the first members of genus *Homo* should be taller in stature and larger in body mass than australopithecines. The reduction in the Total Energy of Expenditure in stage 3 increases dispersal abilities and home range size. These factors potentially reduce male–male competition, and thus allow sexual size dimorphism to decline. Hence, the first members of genus *Homo* exhibit decreased sexual size dimorphism in spite of increased adult stature and body mass. A further decrease in sexual size dimorphism in stage 5 modern humans may be explained by selection for severe periodic food limitation, rather than by decreasing sexual selection. Male stature and body mass is more affected than female stature and body mass. Greater dispersal abilities and greater latitudinal range distributions should be exhibited onwards from stage 2. Strontium isotope analysis of fossil bone and enamel should yield geographic differences between individuals, based on bedrock distinctions established by ranging variations.

General discussion of body size changes

Terrestriality is associated with larger home ranges. Open country foraging and foraging over increased distances are documented for hominids at 1.6–1.7 mya in the Koobi Fora region of the East Turkana Basin (Cachel & Harris 1996, 1998). One can speculate what resources are responsible for implementing the shift to terrestriality. As detailed in Chapter 13, bipedalism requires very little muscle activity in modern humans walking at a normal pace (Basmajian 1974), and the normal human walking pace is slow. What is on the ground? – What can be foraged for, and what is worth foraging for? The small expenditure of energy is offset by the hazards conveyed by prolonged terrestrial movement. Furthermore, even at a later period 1.7–1.6 mya, whatever selection pressures were involved in the shift to habitual terrestriality in open country must have allowed for a slow traverse of the home range.

Bipedalism at normal walking speeds may be slow, but it is highly cost-effective (Steudel 1995, 1996, Steudel-Numbers 2001, Leonard & Robertson 1997, 2001). Normal human walking expends on average only about 87% of the energy expended by a similar-sized, generalized quadrupedal mammal moving at the same rate (Steudel-Numbers 2001: Table 1). Energy saving by bipedal hominids is even more pronounced when compared to the locomotor costs documented among other primates. Primates in general, and chimpanzees in particular, exhibit less efficient locomotion than other quadrupeds of the same body size (Leonard & Robertson 1997, 2001). Locomotor costs are high in primates. They have no economy of movement, and the total energy cost of locomotion is great. This is probably a reflection of the ancient, embedded blueprint for arboreal locomotion in primates, which allows for movement in a discontinuous canopy substrate, and affects relative limb length, gait patterns, and other features (Chapter 1). Even if arboreal bipedalism occurs among hylobatids at a rate of 10%, it would confer very little benefit in the trees if the substrate were uneven, and bipedalism was interspersed with scrambling and climbing.

Yet, a fundamental implication of this discussion is that a major advantage of bipedality – very little energy expenditure under normal walking conditions – would not exist if no major distances needed to be traversed in the home range. The cost-effectiveness of bipedality would disappear. Paleoenvironmental research documents that the spread of arid, open country conditions, and the appearance of Serengeti-like grasslands occurred only at about 1.6 mya in Africa (de Menocal 1995, de Menocal & Bloemendal 1995, Rutherford & D'Hondt 2000). This was long after the initial appearance of hominids. Yet, at 3.6 mya hominids are undoubtedly good functional bipeds, as documented by the hominid footprint trackways at Laetoli, Tanzania. During later time ranges, one might argue that the distribution of suitable lithic raw materials for stone-tool manufacture could dictate greater hominid ranging and foraging distances. But what selected for bipedality before 1.6 mya? This remains a major mystery in paleoanthropology.

In addition, some of the costs associated with larger body size (e.g., a long period of development, late reproduction) did not occur in the taxon *Homo erectus*. Somatic growth was completed faster. If the physiological processes triggering puberty were the same in *Homo erectus* as they are in modern humans, other catarrhines, and laboratory mammal species, a dietary shift could affect female reproductive maturation. A dietary change that increases body fat can accomplish this. Subcutaneous body fat releases the hormone leptin, which initiates menarche. Body size increase with pongid-like maturation contributes to greater dispersal abilities. A relatively higher intrinsic rate of increase (r), characteristic of good colonizing species, would contribute to the seemingly rapid geographic spread of this taxon through major areas of the Old World. Natural selection tends to promote and favor r-selection in environments with few or no competitors, or environments with pronounced seasonality. These conditions would occur more often as *Homo erectus* dispersed to novel Eurasian localities. Hence, the dispersal of *Homo erectus* was characterized by many benefits and few costs of body size increase.

The greater dispersal abilities, greater home range size, and faster female reproductive maturation noted above are advantageous in unpredictable environments. Body size increase implies the existence of a better dietary base. How can a better dietary base be achieved in unpredictable environments? A dietary shift must have occurred. Otherwise, one would expect a decrease in body size to take place when resources are unpredictable or scarce. Chapter 16 summarized evidence indicating that this shift involved incorporating vertebrate meat, fat, and marrow into the diet, first through scavenging and confrontational scavenging, and then through hunting.

Figure 17.4 summarizes a number of the arguments made above. The major cost incurred with a larger body size is the increase in the total energy of expenditure. However, an increase in the total energy of expenditure is offset by the dietary change that underlies and supports larger body size. Larger body size alone may have augmented terrestrial behavior. The biggest change in the total energy of expenditure takes place between stage 1 and stage 2 – that is, between Figures 17.1 A and 17.1 B. Another possible cost might be increased sexual selection resulting from an increase in sexual size dimorphism. Stage 3 (Figure 17.2 B) reduces the total energy of expenditure by increasing bipedal efficiency. Increased bipedal efficiency (achieved chiefly through a lowering of the center of gravity as the thorax and upper limbs are reduced in size) counterbalances an increased total energy of expenditure. The femur and the lower limb lengthens. The head of the femur is enlarged. The glenoid joint becomes horizontally oriented, and phalangeal curvature is lost. These hominids are obligate bipeds. In stage 4 (Figure 17.2A) an alteration of body shape through exaggerated ectomorphy reduces heat stress. A larger body size even without ectomorphy would enable African hominids to range and forage in higher African elevations and disperse to higher latitudes. Until fossil postcranial material from Dmanisi is published, the body shape of hominids initially dispersing from Africa is unknown. It is possible that hominids dispersing during the Plio-Pleistocene were ectomorphic, but this is not necessary.

What factors are responsible for the origin of generalized species?

Phenotypic variation can be accounted for by four principal factors: morphological integration or its disruption, developmental constraint (canalization) or change, adaptive response to selection, the pleiotropic effects of single genes, or the additive properties of several genes. Phenotypic plasticity (including behavioral plasticity) underlies the ability of organisms to be fit in different environments. Such organisms, which are adapted to a variety of habitats, are classified as generalized species. Ernst Mayr (1950) was the first modern researcher to argue that hominids were preeminent among generalized species, because they possessed not only a broad ecological niche, but a uniquely flexible one. The dimensions of this niche, especially during later time ranges, were affected by culture and tool behavior.

Examining the evidence of evolution and climate change at Olorgesailie, in central Kenya, Richard Potts (1996a, 1996b, 1998) argued that long-term exposure to variable environments might be responsible for the origin of generalized species. He suggested that a novel evolutionary process known as variability selection should be recognized by theorists. Potts viewed hominids, especially members of genus *Homo*, as the epitome of generalized species, and his arguments about the flexibility of the hominid niche echo those of Mayr (1950). However, mechanisms responsible for the origin of generalized species may be diverse.

Classifying species as generalized or specialized should probably be done only within the context of a local community, because problems of species abundance, diversity, interaction, and energy flow should influence all of the species. One would expect a continuum of species along a gradient from generalized to specialized. Local community variables can affect the continuum.

The theoretical expectation of evolutionary biologists is that generalized species should be out-competed by specialists in constant or increasingly more variable environments (Gilchrist 1995, Whitlock 1996). Increased phenotypic trait plasticity in one environment reduces fitness across all environments. When selection pressure is strong, and there are costs to phenotypic plasticity, theory predicts that variable environments should lead to the evolution of specialists that have more extreme phenotypes than generalists. This is contrary to the argument about variability selection (Potts 1996a, 1996b, 1998). Generalized species tend to deliver a lackluster performance in the competitive arena. They usually cannot deliver a high level of performance over a wide range of habitats and resources. This assessment is supported by the fact that omnivores – animals that are generalized in diet – are significantly rarer than herbivores or carnivores in real food webs (Van Valen 1973b), and also significantly rarer than expected in computer simulations (Rosenzweig 1995: 81–82). Theoretical studies also support the problematic status of generalists. The Lotka–Volterra competition equations, which were developed in the 1920s, are routinely used to examine two competing species in terms of the intrinsic rate of increase (r) of each species, and the carrying capacity (K) of a habitat in the absence of one species. When two competing species are set in an environment with two habitats, individuals, as expected, reside in the habitat where their fitness is maximized (Křivan & Sirot 2002). Thus, competition between species tends to lead to habitat specialization, and ultimately to segregation in different habitats.

Because animals can alter their foraging behavior, environmental variability and habitat mosaics might theoretically promote the coexistence of generalized and specialized species. But interactions that occur in a field biology setting at an ecological time-scale may not occur at an evolutionary time-scale. A theoretical model examining the coexistence of generalized and specialized species demonstrates that the coexistence of generalists and specialists is unlikely to occur at an evolutionary time-scale (Egas *et al.* 2004). This is especially true when trade-offs occur in the intrinsic rate of increase (r) in different habitats, rather than when trade-offs occur in carrying capacity (K). Modeling a system of four species through time tends to develop a pair of extreme specialists; if gradual evolution and speciation is allowed within the model, the system collapses to three species – two extreme specialists and one generalist (Egas *et al.* 2004: Figs 3 & 4). It is interesting that when this model is run at an evolutionary time-scale, it predicts the coexistence of generalized and specialized species only when foraging accuracy is very high, and when there is high environmental variability through time. There is some field biology confirmation for this theoretical model and its predictions. When habitat specialization is examined in coral reef fish, species with maximum growth rates are the habitat specialists (Caley & Munday 2003). Habitat specialization therefore appears to be based on trade-offs with life-history variables. Because of such trade-offs, habitat specialization can reduce the phenotypic plasticity of species, and lead to a more narrow or specialized niche. In summary, the competitive failure and unlikelihood of generalized species is the opposite of received anthropological and primatological wisdom since the nineteenth century. It was given a major impetus by the publication of Ernst Mayr's (1950) arguments about the uniquely broad and flexible nature of the hominid niche. However, it may not apply to all hominids, and it may not even apply to early members of genus *Homo* (contra Potts 1996a, 1996b, 1998).

Species can be divided into two general categories: those tolerant of stress or disruption and those that are intolerant of stress. Van Valen (1973a) hinted at the existence of a third category: species that are competitively competent, and that do well in habitats where many species with narrow niches coexist. However, the adaptive landscape is constantly changing, because of changes in the physical environment, biotic interactions, and stochastic processes. Hence, competitive competence is manifested only in an instant of evolutionary time. Van Valen (1976) later related the effects of extremely high and low selection pressure and competition to different trophic energy expended in surviving physical stresses or interacting with other species. He pointed out (1976: 203–206) that trophic energy or developmental routes devoted to adaptations to physical stresses tend to lessen adaptability to predation and competition. This is another strike against generalist species.

Phenotypic plasticity is not an exclusive hominid domain. Phenotypic plasticity pervades all nature, because it is the basis for the phenotypic variation found within populations. Empirical data derived from field biology on the costs or benefits of phenotypic plasticity have largely been produced from the study of freshwater or marine fish (e.g., Robinson & Schluter 2000, Caley & Munday

2003). Different phenotypes and degrees of phenotypic plasticity have consequences in terms of fitness. Data have been generated in experiments under natural conditions with frog tadpoles in which phenotypic traits were driven by dragonfly predation (Relyea 2002). Using survivorship as an indicator of fitness, there is no obvious pattern of costs and benefits to phenotypic plasticity across environments – the costs or benefits are dependent upon the interaction between a particular trait and local environmental circumstances. There was no correlation between increased plasticity and developmental instability, as measured by fluctuating asymmetry or phenotypic variation measured by standard deviations. Contrary to expectation, extreme phenotypes in this study were always produced by generalists, and the specialists exhibited intermediate phenotypes. The selection forces in the study were not strong enough to produce specialists with extreme phenotypes. Strangely, plasticity had a fitness benefit in predator-free environments, but it had a fitness cost in environments with predators (Relyea 2002). This surely demonstrates the degree to which predation severely constrains the phenotype of a prey species. Behavioral responses to predators were not studied.

Species that are tolerant of disruption need not require prime habitat or the best resources. They might thrive in areas where the environment is highly disturbed or disrupted. At present, human landscape alteration creates massive disruption of natural habitats, with subsequent local extinction of species. Some species, however, such as armadillos and Virginia opossums, to give two familiar North American examples, thrive under these conditions and rapidly expand their ranges. Neither one of these species would be classified as highly intelligent. Yet, they are highly adaptable.[1] This drives home the point that there is no necessary relationship between intelligence and adaptability, as often implied by reviews of human evolution. Environmental disturbance is affected by a number of general factors, irrespective of the type of disturbance (Figure 17.5).

Pleistocene climatic fluctuation also affected the tropics. It is now known that the tropics were not insulated from these climatic perturbations. Astronomical effects on climate, such as the orbital changes associated with Milankovitch cycles, must have existed, even if they were less pronounced at lower latitudes (Thompson *et al.* 2002). Yet, even a reduced climatic effect can have a profound result in tropical areas. Tropical species are more affected by seasonality than species at higher latitudes. This sensitivity, known as Rapoport's rule (Stevens 1989), means that even minor climatic oscillations can have a major effect on tropical species. Physiological studies of invertebrate species in the wild establish where the Achilles heel of climatic sensitivity lies in tropical species. It resides in an inability to acclimatize, or to adjust the body's sensitivity to temperature (Hoffmann *et al.* 2003, Roff 2003, Stillman 2003).

Factors affecting environmental disturbance

➢ **Duration of disturbance**

➢ **Frequency of disturbance**

➢ **Seasonality of disturbance (this is relatively predictable)**

➢ **Unpredictability of disturbance**

➢ **Severity of disturbance**

Figure 17.5. Any disturbance in the physical environment (e.g., volcanic eruption, windstorm, earthquake, flood) is affected by a variety of factors. These factors determine which species will thrive in disrupted environments. The timing and nature of seasonality renders it a relatively predictable disturbance. Tropical species are more affected by seasonality, and are therefore generally more susceptible to disturbance.

Hominid dispersal is affected by a number of the biological variables discussed above, such as body size, home range size, and terrestriality. In addition, a species that prospers in disrupted environments would be more tolerant of Pleistocene climatic fluctuations and habitat disturbances. *Homo erectus* accomplished the initial hominid colonization of the Old World, and it appears that this taxon was a "weed" species that thrived in disrupted environments (Cachel & Harris 1995, 1996, 1998). Widely distributed species can also be abundant, and so it is possible that this first colonization event resulted in broad areas of the Old World being seeded by respectably large hominid populations. If this were so, hominids may have survived in many areas not in minute remnant or relict populations, but as a persistent presence through long spans of time.

In addition, although the usual scenarios of hominid dispersal emphasize the rigors of life in temperate areas, movement out of the tropics would actually release hominids from the diseases and parasites that are so prevalent in tropical climates (Cachel & Harris 1995). Irrespective of climate, animals and plants encountering novel environments exhibit a reduction in disease and parasite load. This phenomenon has been best studied in introduced or invasive species. Invasive animal species have less than 20% of their usual number of parasites, and appear to acquire far fewer new parasites in the new environment (Torchin *et al.* 2003). This is also true for plants. Alien plant species have fewer fungal and viral pathogens than in their native habitat (Mitchell & Power 2003).

Tool behavior and technology

Chapter 8 detailed the contribution of tool behavior to hominid natural history intelligence. Tool behavior is widespread in the animal world, although it is rare among non-human primates (Beck 1980). Alcock's (1972) work on the origins of tool behavior in foraging animals argued that tool behavior was most likely to appear among animals whose diet was peculiar among members of their lineage. This observation (Alcock's law) agrees with archeological evidence indicating that the first stone tools were used by hominids to skin, dismember, and deflesh vertebrate carcasses, and to break open bones to remove marrow.

Imitation is responsible for the faithful reproduction and cultural transmission of human behaviors, including tool behavior. Imitation is a species-specific trait in humans, and even mentally impaired or autistic humans can demonstrate it. Some researchers argue that imitation is based on the ability to understand the intentions and goals of other individuals (Tomasello & Call 1994, 1997). This ability is called a Theory of Mind, and may support and underpin culture. Because autistic humans lack a Theory of Mind, while still being capable of imitation and culture, imitation appears to be a factor separate and independent from a Theory of Mind. A Theory of Mind may only augment imitation, and may not be necessary for culture. Other researchers argue that imitation is found in some other animals, such as small toothed whales and birds (Hauser *et al.* 2002). Culture may also exist among these species. However, imitation appears to be lacking in non-human primates, where the transmission of behavior occurs through emulation or goal-directed behavior, and faithful reproduction is absent.

Does tool behavior in non-human primates yield any clues about tool behavior in hominids? Specifically, were the earliest hominid tool users behaving like chimpanzees? So persuasive a role has chimpanzee tool behavior played in arguments about human evolution that some researchers believe that observations on chimpanzee tool behavior can be translated into reconstruction of fine details about early hominid tool behavior. For example, it is argued that the earliest hominid tool users were exploiting seasonally available resources, just as chimpanzees do (Harris & Capaldo 1993).

Povinelli *et al.* (2000) spent five years experimentally researching how captive chimpanzees use and manufacture simple tools. They ultimately conclude, contrary to anthropological expectation, that chimpanzees have no Theory of Mind. Experiments with the tool insertion problem reported in Povinelli *et al.* (2000: Chapter 8, Appendix II) are particularly instructive, because the problem was designed to test the tool insertion and termite fishing behavior of wild chimpanzees made famous by the *National Geographic* and countless anthropology textbooks. Humans, by virtue of natural history intelligence, understand that a causal relationship exists between the shape of the inserted tool and the hole through which the extracted food will emerge. Do chimpanzees understand the properties of tool shape involved in probing and extracting food items? No, they apparently do not. The concept of shape eludes them. In experiment 11, the chimpanzees initially selected tools that could not pass through the hole about 50% of the time – that is, the selection was random. Hamadryas baboons tested for tool behavior by Beck (1973) did exactly the same thing, ignoring the hooked, working end of a tool about 50% of the time during food retrieval experiments.

Yet, chimpanzee termite fishing has given rise to a whole subset of ideas on the origins of intelligence: the notion that "extractive foraging" generates intelligence (Parker & Gibson 1977, Gibson 1986). Ironically, beavers, mere lowly rodents scorned by anthropologists, appear to understand shape. When beavers construct dams, they intricately modify branches and twigs to fit snugly into any fissures and cracks that leak water (Griffin 1992). No beaver would randomly thrust twigs into a crevice until an accidental fit occurred.

When archeologists first discern stereotyped stone tools in the archeological record – e.g., single-platform cores in the Karari Industry or hand-axes in the Acheulean Industry – the ancient hominid knappers must surely have understood the concept of shape. If they did not, they would not be able to replicate the shape typical of a certain kind of tool, shapes that are recognized in formal archeological stone tool typologies. Faithful replication implies imitation. Of course, it is possible to produce stereotyped objects without possessing a Theory of Mind or a concept of shape. One can think of bees producing hexagonal brood cells, for example. They perform this task without imitation. However, when stereotyped stone tools finally appear in the archeological record at 1.7–1.5 mya, they are accompanied by undoubted indications of hominid natural history intelligence that are not found in non-human primates. Natural history intelligence is indicated in the archeological record by the practical location of sites, the selection and transport of suitable raw material for stone tools, stone tool manufacture and use, the acquisition of vertebrate meat and marrow through hunting and/or scavenging (including confrontational scavenging), the alteration of animal bones to obtain meat and marrow, the transportation of carcasses and partial carcasses, the patterning of bone and stone within a site, and distinctions between sites that are caused either by different activities, seasonality, or length of occupation. The association between the appearance of stereotyped stone tools and an increasing density and complexity of the archeological record between 1.7 and 1.5 mya marks a significant change from archeological evidence from an earlier period (see Chapter 16).

The conclusions of Beck, Povinelli, and other researchers skeptical about the hominid-like properties of chimpanzee behavior deserve to be more widely known and disseminated among primatologists and paleoanthropologists.

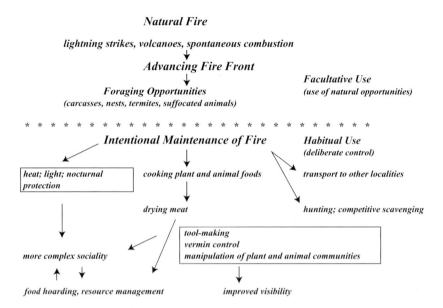

Fire and Early Hominid Behavior

Figure 17.6. The behavioral and ecological consequences of fire technology. Modified from Clark & Harris (1985: Fig. 11).

However, given the "chimpocentrism" that prevails in both groups – a mindset that actually can be traced back to Robert and Ada Yerkes and Wolfgang Köhler in the early twentieth century – it is unlikely that behavioral analogies between fossil hominids and chimpanzees will disappear anytime soon.

The ability to control fire – i.e., to maintain and transport it away from a naturally occurring source, such as a brush fire caused by a lightning strike or volcanic event – was a milestone in human evolutionary history. Careful maintenance of fire allows it to be transported across long distances, even between continents. Ethnographic evidence reveals that even some modern human groups (e.g., the indigenous people of Tasmania and the Andaman Islands) could not make fire, although they could safeguard embers and keep fire alive indefinitely. Some researchers argue that control of fire may have begun as early as 1.6 mya in East Africa (Bellomo 1994), although this date is controversial. A later, Acheulean date for control of fire comes from the Swartkrans site in South Africa (Brain & Sillen 1988), and unequivocal control of fire appears from burnt flint debitage, wood fragments, and seeds at the Israeli Acheulean site of Gesher Benot Ya'aqov, dated to about 790 000 yrs BP (Goren-Inbar *et al.* 2004). The control of fire meant that, unlike other higher primates, humans did not need to seek shelter at night in trees or cliffs, where they would be safe from nocturnal predators. Fire further permitted humans to remain active after nightfall, and provided warmth at higher altitudes or in colder habitats. Fire permitted humans to cook foods and drive hunted animals, and thus expanded human dietary range. Fire also allowed humans to modify ecosystems in a profound fashion, as they burned grasslands, cut down trees, and burned forests (Figure 17.6). The control of fire marks the beginning of human modification of the earth's surface, the signs of which are now universal. In fact, charcoal lenses appearing in pollen spectra or sediment horizons are sometimes used by archeologists as a signature of human presence, even if human skeletal material or cultural remains are absent.

Language

Modern humans are characterized by language, which is a species-specific type of vocal communication. Special neuroanatomical centers, usually located in the left cerebral hemisphere, underlie human language abilities. Human gestural or sign languages, which are non-vocal, also utilize these centers. This underscores the species-specific nature of human language. Yet, asymmetrical cerebral hemispheres also occur in other catarrhines, including Old World monkeys (Falk 1978). Furthermore, rhesus macaques (*Macaca mulatta*) process species-specific vocalizations in the left cerebral hemisphere (Poremba *et al.* 2004). Although other mammals and birds possess complex vocal communication with referential signaling, human language has the unique property of recursion (Hauser *et al.* 2002). Recursion is the ability to create an infinite number of expressions by permutations of discrete components such as words or numbers. Consequently, there is no limit to the possible communications based on language or numbers.

New World marmosets and tamarins have extremely complex natural vocal communication systems, and experimentally demonstrate a grasp of abstract linguistic rules. Untrained animals can spontaneously uncover rules relating to sequential auditory patterns. However, experiments with cotton-top tamarins (*Saguinus oedipus*) show that they cannot master hierarchically organized auditory stimuli (Fitch & Hauser 2004). They cannot learn a recursive grammar. Hence, they cannot decipher grammar in which sounds ("words") are greatly separated from each other, although dependent, as in "If-Then" constructions. This affects the whole realm of subjunctive thought and communication, dealing with possibilities and hypothesizing. This realm of the subjunctive is obviously the purview of humans, and affects the ability of humans to control future actions, and plan for unforeseen consequences. Another species-specific feature of human language is that it is constrained by a critical learning period. A critical period for the acquisition of human language occurs during infancy, and infants who are not exposed to language during this time fail to develop normal language abilities later in life, despite intensive training.

The human infant's ability to reproduce the sounds of its native language depends on imitation. Imitation is also responsible for the faithful reproduction and cultural transmission of other human behaviors, such as tool behavior. Paleoanthropologists have been consumed by the search for language origins for a long time. They have concentrated on detecting brain areas devoted to language processing on fossil endocasts (Tobias 1991); reconstructing the pharyngeal cavity and larynx placement in fossil hominids (Lieberman 1975); analyzing the hyoid bone (Arensburg 1994); investigating tongue innervation through the size of the hypoglossal canal; and estimating the likelihood of fine breath control needed for language production by studying the vertebral canals (MacLarnon 1993). There is at least one gene that is known to underlie normal human language production. This is the *FOXP2* gene. A mutant allele of this gene, inherited as an autosomal dominant, severely affects the ability to produce language, although language comprehension is unaffected. This condition resembles Broca's aphasia, and, in fact, affected individuals have a reduction in the tissue of Broca's area (Vargha-Khadem *et al.* 2005). However, the *FOXP2* gene is not unique to humans. It is among the most highly conserved of proteins. It occurs throughout the placental mammals, and may be involved in learning and performing movements of the orofacial muscles. The mutant *FOXP2* allele reveals a peculiar lack of redundancy in the genetic substrate for human language production. This may indicate intense selection pressure against *FOXP2* mutations. This selection would presumably be directed against abnormal language production itself and not orofacial movements, because affected infants have no difficulty feeding or swallowing (Vargha-Khadem *et al.* 2005).

Using all available genetic information, Cavalli-Sforza *et al.* (1994) discovered that genetic differences between human groups are frequently, but not always, associated with language differences. This implies that language can function as a reproductive barrier between humans, and can lead to a reduction in gene flow and subsequent population demarcation. However, a study of Y chromosome haplotypes in Europe (Rosser *et al.* 2000) showed that geography influences genetic diversity more than language does, at least in males. Worldwide study of factors that affect the number and distribution of human languages demonstrates that area of a country, degree to which a country is forested, and altitude affect the number of languages that are present (Sutherland 2003). The duration of human settlement in an area is unimportant in explaining the number of languages that are present, because languages change so quickly.

Because the rate of language evolution exceeds the rate of human phenotypic or genetic evolution, language does not appear to serve as a major barrier to human reproduction. In living humans, therefore, language does not function to maintain reproductive boundaries. It is not a reproductive isolating mechanism that increases genetic divergence.

Early hominid sociality

I presented a hominization model that incorporated aspects of sociality in Chapter 16. It is diagramed in Figure 16.7. Fundamental to the model is the idea that the ancestors of hominids developed a type of sociality unlike that found in other catarrhine primates. This is different from traditional models of human origins (Figure 16.6), in which changes in the physical environment trigger hominization, and changes in sociality appear long afterward, as the culmination of long ages of evolution. Yet, the "behavior first" model agrees with the accepted view in evolutionary theory that behavioral changes come first, and drive evolution.[2] Given the indeterminacy of field ecology variables, differences in life history, and the distribution and abundance of competitors and predators, behavior should be plastic. Hence, even species-specific behaviors, like geographic range, show relatively low heritability (Webb & Gaston 2003). Traditional models of human origins may place social changes last, as a finale to the hominization process, simply because human social complexity has been linked with human uniqueness and perfectibility since Greco-Roman antiquity. An initial reduction in intra-group competitive behavior – strange for catarrhine primates, especially the Old World monkeys – would lead to the origin of social foraging and anti-predator groups like those observed in other social mammals, especially the social carnivores. A major source of catarrhine intra-group competition appears derived for Old World monkeys. These behaviors involve female competition, such as female philopatry, female dominance hierarchies, female–female competition, matrilines, and competition between matrilines (Di Fiore & Rendall 1994, Rendall & Di Fiore 1995). If these behaviors are derived for Old World monkeys, it is possible that the catarrhine ancestors of the hominids never possessed such competitive behaviors. If this were so, there would be no need to reduce these behaviors – they would have only a limited manifestation among hominid ancestors. Food-sharing and sentinel behavior would make an early appearance among hominids. At the same time, natural history intelligence would arise, because of the reduction or non-appearance of intra-group competition. In this section, I provide a more detailed outline of this model, and I give specific anatomical and archeological evidence for the model.

Enamel formation tracking dental development shows accelerated growth, with a short juvenile period and early reproduction, in African and Asian *Homo erectus* and earlier hominids (Dean *et al.* 2001). Prolonged modern human-like growth periods appear only very late in the fossil record with the advent of the Neanderthals. Such accelerated growth implies the existence of self-reliant juveniles. How might this be manifested in social life? A catastrophic mortality event preserved 17 *Australopithecus afarensis* individuals at the AL 333 locality at Hadar, Ethiopia. The catastrophic event that decimated this group is unknown, despite extensive sedimentological analysis. The deaths were not caused by flooding (Behrensmeyer *et al.* 2003). This important locality establishes the existence of a social group per se, and gives the minimum number of individuals in the group, as well as their ages. Of the 17 individuals, 10 are adults, 2 are subadults or adolescents, and 5 are young – the youngest had an unworn deciduous incisor (Harmon *et al.* 2003).[3] Where are the juveniles? Their absence is not caused by taphonomic processes, because younger individuals are present. The absence of juveniles from this instantaneous time sample of a social group implies that they were spatially separated from other members of the group. In this, they appear to have resembled many non-human catarrhine species. In many of these species, juveniles establish separate play-groups, and associate most closely with each other once they are weaned. They are often physically removed from the other animals.

Accelerated growth suggests that parental investment in young and prolonged rearing of young was much more limited than in traditional reconstructions of early hominid social life, as exemplified by the theoretical work of Glynn Isaac (1978a). Modern humans formed the template for Isaac's work. A condensed period of infant and juvenile care-taking and a reduction in overlap between generations implies a very different sociality than that existing in modern humans. The Nariokotome *Homo erectus* specimen had not completed its growth, and may have been only 8 years old at the time of its death (Dean *et al.* 2001). Fast maturation, self-reliant juveniles, and abridged generational overlap also call into question the importance of matrilines and "grandmothering" emphasized in some recent work on *Homo erectus* (O'Connell *et al.* 1999). Because australopithecine species also had a pongid-like early maturation, matrilines and "grandmothering" would be similarly limited in these earlier hominids.

Isaac's (1978a) reconstruction of hominid sociality was crafted during the hey-day of Marshall Sahlins's characterization of living hunter-gatherers as "the original affluent society." Stoczkowski (2002: 23–27) argues that this characterization, and its use in Paleolithic archeology, is a classic example of conjectural history. Rowley-Conwy (2001) also inveighs against progressivist views of human societies. He believes that the mobility, fluid organization, and

egalitarianism that characterized "the original affluent society" are not components of the first typically human societies. Instead, he argues that the evidence
of archeology demonstrates that mobility, fluid organization, and egalitarianism
are specialized, late arriving aspects of human social organization.

Inferences at this level are usually fairly dubious and unconvincing. They
also generally smack of "chimpocentrism." However, I believe that it is fundamentally wrong to assume that hominids during the Plio-Pleistocene or any
earlier time were behaving like bipedal chimpanzees.

Chapter 9 ended with inferences about the basic catarrhine substrate for
sociality. This substrate included kin selection; infanticide; parental investment;
parent-offspring conflict; possible different male and female strategies of reproduction developed by sexual selection; intra-group competition; and competition between matrilines (Su 2003). These traits, which appear abstract, can be
used to infer how natural selection has molded keystone features of hominid
evolution. For example, Isbell & Young (1996) argue that bipedalism evolves
because its efficiency allows females that are competing for food to disperse
across a broader area without costs to the individual associated with fatigue.
That is, bipedalism evolves in hominids because females need to decrease food
competition. Within-group feeding competition is assumed to exist before and
after the origin of bipedalism. Isbell's model is based on work with vervet and
patas monkeys, both Old World monkey species. If phylogenetic analyses of
primate sociality are correct, it is possible that these and other Old World monkey species possess a complex of traits involving female–female competition
that is derived for primates (Di Fiore & Rendall 1994, Rendall & Di Fiore
1995). If this is so, then models of bipedal origins based on the existence of
intra-group feeding competition would not apply to the ancestors of hominids.
Furthermore, intra-group competition would not need to be reduced among the
ancestral or early hominids because it would be very limited. Yet, how is it
possible for feeding or other forms of competition to be reduced or suppressed
in primate societies?

The most complex non-human primate sociality occurs not in chimpanzees
or other pongids, but in callitrichid platyrrhines (the New World marmosets
and tamarins). Callitrichids are the smallest anthropoid primates, with a mean
adult body size of 0.418 kg (Table 11.1). Anatomical evidence (e.g., the loss
of the third molars, histological evidence that their claws are derived from
nails) indicates that callitrichids are secondarily dwarfed. That is, they evolved
from full-sized ancestors. Among callitrichids, the following intricate behaviors
occur: communal raising of young, provisioning of infants and juveniles with
high-energy foods, mental mapping of food resources, monitoring, managing,
and extracting gum and resin food resources, and complex vocal communication (Snowdon 1982, Ferrari & Martins 1992, Rylands 1993, Tardif 1994,

Savage & Baker 1996). Callitrichids are probably the only non-human primates that can be ranked as eusocial or truly social, using the formal three-fold socio-biological definition of complex sociality: (1) cooperation in care of the young; (2) reproductive division of labor; and (3) overlap of two or more generations that contribute to social life (Wilson 1975).

What factors contribute to complex sociality here? Callitrichid species have diverse social and mating systems (Savage & Baker 1996), but exist in groups with only one breeding female, and a core of highly related animals that may represent three generations. Callitrichids normally produce dizygotic twins. However, chorionic fusion of the two placentas occurs very early in pregnancy. Embryos and fetuses share the same circulatory system, because the placental arteries of the twins overlap (Wislocki 1939). This condition produces chimeras, because cells are exchanged between developing embryos and fetuses (Windle *et al.* 1999: 73). It is possible that this state – unique among primates – promotes eusociality.

Haig (1999) suggests that, if dizygotic callitrichid twins can be somatic chimeras, a twin may then have a phenotype that is more related to one of the parents than to its own germ line. This somatic make-up might increase the likelihood of an offspring suppressing its own reproduction as it cares for a younger sibling.[4] Subordinate males and females are infertile, either through physiological or behavioral suppression of reproduction (Abbott 1993). In captivity, common marmosets (*Callithrix jacchus*) enjoy improved nutrition, and can produce triplets, quadruplets, and quintuplets. A survey of the reproductive records of 12 colonies shows that triplets may sometimes be produced more often than twins (Windle *et al.* 1999). A high intrinsic rate of population growth in callitrichids therefore becomes coupled with great variation in female reproductive success. Adult body size is very small. Given the presence of twin infants, no single adult (i.e., the mother) could care for both youngsters in the wild. Alloparental care is necessary. Infants and juveniles are carried and fed on a communal basis. There is minimal striving for dominance within a callitrichid social group, and there are no competing matrilines. Physiological or behavioral suppression of reproduction also convergently evolves in other mammal species where shared care of infants is necessary. Several phylogenetically diverse examples are the naked mole-rat and dune rat, some social canids (the black-backed jackal, wolf, and Cape hunting-dog), and the dwarf mongoose. Because most of the aggression observed in primate species appears related to dominance interactions (Silk 1993), I propose that a rigid hierarchy with physiological suppression of reproduction might result in the callitrichid zenith of sociality among non-human primates. A reduction in status-striving and female–female or inter-matriline competition might also have generated complex sociality in hominids, as well.

An alternative explanation is that complex sociality may have evolved among unrelated or distantly related individuals (Chapter 9). Clutton-Brock (2002) reviews the genetic bases for cooperative behavior in vertebrates and invertebrates. Genetic relatedness may explain the origins of eusociality in haplodiploid insects, but cooperative breeding occurs without a high degree of relatedness in a number of vertebrate and invertebrate species. Because all group members experience an increase in fitness with an increase in group size, cooperative behavior may be the result of "group augmentation," rather than kin-selection (Clutton-Brock, 2002). Increasing group size can positively affect cooperative foraging, anti-predator behavior, the growth and survivorship of young, reproductive success, and competitive interactions with conspecifics in other social groups. In theory, "group augmentation" per se can maintain cooperative behavior in small groups. Hence, "group augmentation" may supplement or substitute for kin-selection as a generator of social life, or explain the presence of unrelated individuals in cooperative groups. West *et al.* (2002) review the evidence that competition between relatives within a social group may actually reduce or nullify the benefits of altruistic behavior generated by kin-selection.

In terms of the arguments developed here, a significant conclusion of this recent ethological work is that convergent evolution can also affect cooperative social life. Selection pressures act upon a myriad of different variables, and can create similar kinds and degrees of sociality in species that are phylogenetically divergent. This is homoplasy, appearing in the behavioral component of the phenotype. On the other hand, phylogenetically close species can exhibit very different social behavior.

Different social roles and a rudimentary division of labor may have appeared in hominids during the period when the archeological record is becoming more complex (1.8–1.6 mya). With the exception of callitrichids, roles and division of labor are not readily discernable in non-human primate societies. Yet, cooperative foraging and complex antipredator behavior occur in mammalian social carnivores, for example. In fact, Bednekoff (1997) first suggested that seemingly altruistic behavior might impose no costs. He argued that sentinel behavior might sometimes be an example of "safe" altruistic behavior exhibited by "selfish" individuals. Kin selection or reciprocal altruism may not be necessary for cooperation or division of labor to evolve. This suggestion has been confirmed. Clutton-Brock and his colleagues conducted a series of experiments with 16 habituated groups of wild meerkats (*Suricata suricatta*), South African desert mongooses less than 1 kg in body size. Clutton-Brock *et al.* (1999) established that the role of sentinel exists, and that complex antipredator behavior takes place among these selfish, vigilant animals. Clutton-Brock *et al.* (2001) have also experimentally shown how cooperative rearing ("babysitting" by helpers) affects weight gain in meerkat young, and results in higher survival rates during the first year, even though helpers may incur severe costs. Because they do not forage themselves, helpers may suffer a substantial loss of body weight during protracted bouts of babysitting. Helpers engage in cooperative rearing by guarding pups at the natal burrow, and feeding pups with invertebrates and small vertebrates. Helpers modify their behavior to minimize the long-term costs of helping (Russell *et al.* 2003). Helper investment is determined by their body weight, and parents increase their foraging effort after heavy parental investment, and reduce parental investment in the subsequent reproductive bout. Supplemental feeding of wild meerkats demonstrates that marked short-term costs do not translate into significant long-term energetic costs. Kin-selection alone does not explain this meerkat behavior, because both kin and non-kin are helped.

Although meerkats exhibit obligate cooperation, there are sex differences in cooperative behavior. These differences occur even though the animals are monogamous – young in the group are the offspring of a single dominant mated pair, although sometimes subordinate females may reproduce. Furthermore, male and female meerkats mature at the same rate, and exhibit very little sexual dimorphism. Female helpers contribute more to rearing and feeding of young than males do. This is thought to be caused by female philopatry, because females gain more than males when they recruit young to their natal group (Clutton-Brock *et al.* 2002). However, both males and females eventually emigrate from their natal group, so females exhibit only a qualified philopatry, in contrast to some animals (e.g., the majority of non-human social primates) identified as philopatric that remain in their natal group throughout their existence. In meerkats, the dominant breeding female forcibly ejects most females from their natal group at 2–3 years of age, and females then depart in female only groups of 2–6 animals; males of the same age leave their natal groups freely by themselves, either to join other groups or to establish their own (Clutton-Brock *et al.* 2002: 253). This distinction between true and qualified philopatry is not trivial, because Clutton-Brock *et al.* (2002) argue that philopatry in birds and mammals determines cooperative behavior per se, and the differential cooperative behavior between males and females. If both male and female meerkats disperse, however, the important factor determining differential cooperation in meerkats may be not the relative degree of philopatry between males and females, but the long term female–female associations in contrast to solitary male dispersion.

It is instructive to recognize the degree of complex cooperative behavior (sentinel behavior, provisioning of young, and babysitting or guarding of young) that occurs in these non-primate mammals. In light of the present discussion, even the appearance of one of these behaviors in hominids (sentinel behavior, provisioning, babysitting) may have been associated with an increasingly more complex archeological record. For example, increasing reliance on vertebrate meat and fat could have been affected by the appearance of sentinel behavior; long-distance foraging for lithic resources could have been affected by the appearance of helpers to guard and provision the young in a safe area.

Sentinel behavior may have been the first to appear, and could account for the early appearance of bipedal specializations in hominids. Unlike the late Miocene ape *Oreopithecus*, whose bipedal specializations appear in an island environment devoid of large mammalian carnivores, early hominids coexisted with a broad array of large sympatric carnivores. Hominid bipedalism certainly did not originate in a carnivore vacuum, and the first hominids appear to have been significantly smaller than later hominids, members of genus *Homo*. Thus, some behavioral adaptations to frequent, successful ground locomotion may pre-date the first appearance of bipedal skeletal traits. This is true even if early hominids existed in a mosaic environment that included forest habitats, as the paleoecology of hominid fossil and archeological sites documents. Sentinels who warned conspecifics of predator approach could be an important factor in establishing successful bipedalism, even in a forested or woodland environment. Because the effectiveness of sentinel warnings would be increased in more open environments – the sentinel could detect predators at a greater distance, rather than raising an alarm just as danger looms – sentinel behavior would be improved in more open country. Sentinel behavior and vigilance would need to be augmented in more open environments, given the relative paucity of trees that could offer safe havens. In any case, whether in woodland or in more open habitats, the slow rate of bipedal travel mandates more vigilance than other modes of ground locomotion. Bipedalism has an impressive cost-effectiveness in comparison to quadrupedal mammals of the same size moving at the same speed. This cost-effectiveness is especially pronounced when bipedalism is contrasted with the quadrupedalism of other primates. The energy-saving nature of bipedalism, coupled with sentinel behavior, would counteract hazards generated by slow walking speed, and the risks of being a relatively small-bodied animal on the ground. The low energy expended in bipedalism, joined with sentinel behavior, could have generated significant selection pressure benefits even if terrestrial bipedalism was merely facultative. These benefits could therefore commence at an early period in hominid evolution. They may account for the early broad dispersal abilities of hominids, documented by the discovery of *Australopithecus bahrelghazali*, 3–3.4 my old australopithecine fossils in Chad, 2500 km west of the East African Rift.

Selection for heightened vigilance and the eventual appearance of sentinel behavior can also be implicated in the origins of natural history intelligence. Vigilance and attentiveness to the world outside the social group downplays the focus on competition and intra-group social interactions that is so characteristic of non-human primate sociality. Vigilance and attention to the behavioral ecology of other animals (potential predators or prey) enhances the development of natural history intelligence. Natural history intelligence, oriented towards the landscape and observing the creation and effects of natural grassland fires, eventually allows hominids to control fire. The ability to control fire releases hominids from the dependence on sleeping trees or sleeping cliffs that is found in other catarrhines. At this point, the necessary link between hominid activities and forest or woodland areas is severed, and hominid dispersal abilities are amplified.

Like other catarrhines, hominids are diurnal, and are temporally separated from sympatric nocturnal carnivores. If hominid terrestriality were concentrated at mid-day, rather than dawn or twilight, hazardous encounters with nocturnal carnivores would be even less likely. The narrow bi-iliac breadth of australopithecines and African *Homo erectus*, a skeletal adaptation to heat, is tailored for activity in the heat of the tropical mid-day. Muscle activity is very low during bipedalism. This ensures that bipedal travel at a normal walking pace does not generate heat to the extent that quadrupedal movement does. It is also possible that the species-specific human sweating response and evaporative cooling of sweat from a hairless skin surface is very ancient, and occurs among australopithecines. This suggestion, however, is currently impossible to test. During the period between 1.64 and 1.39 mya in the Koobi Fora region of the East Turkana Basin, northern Kenya, a complex archeological record was generated by the taxon *Homo erectus* (Figure 17.7).

This archeological evidence indicates that hominid ranging was occurring over distances of about 40 km (Cachel & Harris in press, Cachel 2004). This distance far exceeds the home range sizes of other terrestrial catarrhines. It is important to note that only among carnivorous mammals does home range size increase with group size (Carbone *et al.* 2005). In other mammalian groups – including primates – there is no such relationship. Hence, if acquiring vertebrate meat and marrow necessitated group foraging by hominids, one might expect an increase in home range size associated with a dietary shift. Increased ranging is largely responsible for the increased dispersal potentials of *Homo erectus*, which was the first hominid taxon to emerge from sub-Saharan Africa to colonize other parts of the Old World, including the extreme Far East. It is reasonable to believe that increased hominid ranging and dispersal abilities in *Homo erectus* may have been associated with levels of cooperative behavior not exhibited by other catarrhine primates. Social roles and a rudimentary division of labor may already have appeared, and been accentuated by the demands of foraging and ranging over larger areas. However, philopatry, close kinship, and altruism may have been negatively impacted by increased dispersal. In this case, factors such

Figure 17.7. A summer afternoon on the bushland savannah, Karari Escarpment, East Turkana Basin, 1.6 mya. The course of an ephemeral stream is outlined in the distance by surrounding trees. Observing the flight of vultures, a party of *Homo erectus* hominids foraging for lithic resources along the streambed has located a freshly killed ungulate carcass, and has driven away the cheetah responsible for the kill. Using stone tools, a group of hominids cooperates in skinning, defleshing, and dismembering the carcass. Their relatively unhurried butchery is made possible by the vigilance of several sentinels (one positioned on a kopje) in the foreground.

as "group augmentation" might generate cooperative behavior and compensate for a reduction in kinship ties within a group.

As is true for mammalian social carnivores, it was likely that early hominids had to locate potential vertebrate food items. This necessitated the integration of cues arriving from a variety of sensory modalities when monitoring the behavior of other species, assessing the probability of successfully capturing, killing, or scavenging any individual prey animal, assessing the costs and benefits of a predatory interaction, assessing the costs and benefits of scavenging a food item (including confronting other predator species), and cooperating with other members of the social group in locating, killing, and sharing prey or scavenged items. The raucous noise of group travel and life, which is such a familiar aspect of most primate societies (Holekamp *et al.* 2000), would need to be abated if hunting took place. Furtiveness is a crucial feature of both hunting and non-confrontational scavenging in social carnivores. Hominid hunters would need to practice stealth, as well as vigilance – qualities rarely encountered among higher primates (Holekamp *et al.* 2000, Treves 2000).

Of course, predation pressure is affected by body size. The earliest hominids were small, but, even at 35 kg, they qualify as large mammals. Surveys of injuries in wild hylobatids (Schultz 1969) and pongids (Lovell 1990) document that dental pathology, arthritis, and trauma from falls are frequent. Animals face injuries from falls, not from rampaging predators. Female chimpanzees of the Mahale Mountains, Tanzania, have been studied for over 30 years (Nishida *et al.* 2003). The major cause of death is disease, accounting for 48% of the deaths. Old-age accounts for 24% of the deaths; and intra-specific aggression accounts for 16% of the deaths. Assuming that sympatric humans and human observers are the vector of disease introduction, one might infer that half of the chimpanzee deaths would be eliminated in a world without humans. On the other hand, aggression between chimpanzees accounts for a relatively high percentage of the deaths.

Food hoarding might follow killing or scavenging. If so, the location of cached food would need to be remembered and retrieved. In addition, because hominids possess no natural weapons, and cannot process a carcass without tools, the acquisition of vertebrate meat, fat, and marrow by hominids requires tool behavior. I examined the archeological record in detail in Chapter 16, and introduced a detailed outline of early hominid subsistence behavior.

In the second phase, considerably later in human evolutionary time, a social context for intelligence reappears, now separated from natural history. It is possible that the re-emergence of social intelligence took place at the end of the Paleolithic, with the advent of anatomically modern humans. At this time (best documented in European sites), population expansion shifted selection pressures on individual humans to increase social complexity and inter-group differences, and promote xenophobia (Cachel 1997).

Endnotes

1. An anecdote may illustrate this point better than abstract arguments. As a graduate student, I once witnessed the following incident. A graduate teaching assistant in the Anatomy Department was supposed to sacrifice a female Virginia opossum and her young for a demonstration in a comparative anatomy class on Monday. But it was Friday afternoon, and the weekend beckoned. Being pressed for time, he simply shoved the opossums into one of the giant horizontal freezers that held cadavers for dissection. He expected that they would freeze, and die a quick and painless death. On Monday morning, however, he opened the freezer to discover that mother and young had all survived. The mother, taking advantage of the situation, had eaten her way through a large, frozen snapping turtle carcass. Both mother and young were a little stiff, but otherwise well. I am happy to report that

the Professor of Anatomy decreed that all of the opossums deserved a reprieve. They were released out on the campus grounds, and were last seen ambling off into the shrubbery. Their many-times removed descendants undoubtedly roam the area today.

2. So well-accepted is the idea that behavioral changes drive evolution, that any contrary evidence of behavior constraining evolution becomes extremely interesting. Huey *et al.* (2003) discovered that thermoregulatory behavior restricts thermal physiology and therefore affects altitude adaptation in ectothermic *Anolis* lizards. They argue that regulatory behaviors, in particular, ought to confine or constrain evolutionary changes, and not generate them.

3. This is the age breakdown presented at the 2003 annual meeting of the Paleoanthropology Society. The online abstract, which appeared in 2002, has a different age breakdown.

4. This is theoretically possible. Such somatic chimerism may also exist in humans. As high as 8% of human fraternal twins are blood chimeras (Pearson 2002). These twins share a blood supply in the placenta. Blood stem cells pass between embryos, and come to rest in the bone marrow of the twin, becoming a continually renewed source of foreign blood cells.

18 *Final thoughts on primate and human evolution*

In summary, what special insights do non-human primates offer on the course of human evolution? I list some of these insights first in a general category, addressing evolutionary processes, and then in a particular category, addressing specific major themes in human evolution.

Speciation, extinction, and other evolutionary processes

Both human and non-human primate evolution can be examined in terms of radiations and extinctions. There are periods of time when groups are speciose and speciation is rapid, and there are periods of time when major extinction events occur. Thus, primate evolution resembles the evolution of most other mammalian orders. The timing of some primate radiations and extinctions also coincides with events occurring in other orders. A major radiation occurs in the early Paleocene, and a major extinction occurs at the end of the Miocene. The Paleocene primate radiation appears to have been triggered by the presence of an ecological vacuum created by the Cretaceous/Tertiary mass extinction. Paleocene primates diversified to occupy a series of niches for small, generalized arboreal herbivores. The contemporary condylarths diversified to occupy a series of niches for larger, generalized terrestrial herbivores. Both the plesiadapiform Paleocene primates and the condylarths, as well as other early ungulate groups, are replaced by morphologically different groups during the Eocene. A major extinction of late Miocene hominoids occurs at the same time that major extinctions occur in ungulate groups. In this case, the Miocene extinctions appear to be associated with a drop in global temperature and increasing seasonality.

Because primates are tropical mammals, they are probably more sensitive to seasonal fluctuations than temperate mammals. Late Miocene climate change, particularly increasing seasonality, therefore resulted in a major loss of primate diversity. Primates in the modern world are a mere remnant of the array of species that existed prior to the late Miocene. This probably accounts for the lack of documented competition occurring between natural populations of sympatric primate species, although competition between sympatric primate

382

species has long been thought to organize primate communities. Primate association with tropical and subtropical forests renders primates susceptible to forest loss and fragmentation. At the same time, primate persistence in isolated forest refugia can underlie the creation of new endemic species. Living cercopithecoid monkeys, for example, are speciose, and this species richness is probably generated by major Pleistocene fluctuations in tropical forest habitat that alternately expand and contract forest cover.

That primates are tropical mammals may have other consequences affecting their evolution. Tropical communities are characterized by a greater number of species within a given habitat, and therefore by a narrower niche structure caused by niche-packing. Sympatric tropical species have also been intensively examined for the existence of coevolution. Topics such as niche structure and coevolution have been relatively neglected by primatologists, although the role of primates as fruit dispersers has been examined. Hominid emergence out of the tropics was largely a result of the biological consequences of larger body size, although hominids dispersing into temperate areas of Europe and Asia made and used Oldowan (and later Acheulean) stone tools. Larger body size and obligate tool behavior allowed hominids to have a greater ecological impact on newly invaded areas than is the case for other primate species that live in temperate areas, such as the macaques. In any case, hominid invasion of new lands would both release them from competition with closely related species, and save them from the diseases and parasite loads associated with tropical areas.

Primate evolution is characterized by striking examples of convergent evolution, ranging from the superficially rodent-like or colugo-like traits of plesiadapiform primates, to the sloth-like, koala-like, or baboon-like traits of the subfossil Malagasy lemurs. A simple observation that results from this is that it is difficult to characterize primates by a laundry list of traits, and it is therefore probably unwise to exclude early species from the order simply because they do not resemble living primates. The adaptive zone of primates has shifted since the beginning of the Cenozoic, and selection pressures acting under local circumstances can create endemic primate versions of sloths or koalas. Homoplasy also runs rampant in the primate order. A recently documented example is the independent evolution of a long hand with long, curved fingers within hominoids, as implied by the hand structure of the 13–12.5 mya Spanish genus *Pierolapithecus* (Moyà-Solà *et al.* 2004). The fossil record of mammals consists chiefly of teeth. Thus, primate paleontologists rely heavily on dental evidence. However, many traits of the mammalian dentition (e.g., tooth number and size, cusp number, shape, and position, cresting) are now known to be simultaneously affected by the expression of the protein ectodysplasin in the enamel of the developing tooth crown (Kangas *et al.* 2004). Because these dental traits are not independent, coding them as independent traits in fossil teeth will obscure phylogenies. Furthermore, the fact that different dental traits do not require a separate genetic substrate, and that multiple traits are simultaneously affected by the same protein, increases the probability of convergent evolution. The widespread existence of homoplasy signals intractable problems for researchers who attempt to study evolution only in terms of cladistic phylogenies. The ultimate goal of these researchers is the recovery of the one true phylogeny produced by evolution, and thus unbridled or uncontrolled homoplasy becomes a major philosophical barrier, because it confounds the retrieval of the only real phylogeny.

Terrestrial life and bipedality

Primates originate as small arboreal mammals. However, primate evolution documents many experiments with locomotion. In fact, the variety of primate locomotion probably exceeds the range of locomotor types found in other orders of mammals. Many researchers have spent their careers in detailing the specific anatomy that underlies a particular type of primate locomotion. However, primate experimentation with terrestrial life has immediate application to ideas about human evolution. Many classic models of hominization view terrestrial life and bipedal locomotion as the fundamental hallmarks of hominid status. It is therefore important to reflect that hominids are catarrhine primates. Unlike platyrrhine primates of the New World, the Old World catarrhines have often experimented with terrestrial life. Even catarrhines that are now notably arboreal, such as the colobines, have fossil representatives that were undoubtedly terrestrial or semi-terrestrial. The failure of platyrrines to evolve some ground-dwelling species is inexplicable, but catarrhines invade the ground many times. The mere presence of open country habitats is not enough to account for this, because such habitats also occur in the New World. The generally larger body size of catarrhines can serve as a defense against predation. Larger body size is probably a major factor accounting for catarrhine terrestrial experimentation, because both living and extinct ground-dwelling species tend to be larger than closely related arboreal species. Size is not the only explanation for catarrhine terrestriality, however, as witnessed by the persistent and successful ground foraging of relatively small vervets (*Cercopithecus aethiops*). Platyrrhine inability to invade the land may also have its origins in biogeographical differences (e.g., the presence of carnivorous flightless phorusrhacoid birds that created another realm of predators, or the presence of many caviomorph rodent species that formed an array of herbivores that were more serious food competitors to primates than Old World ungulates).

Unlike other mammals, primates differentiate the functions of forelimb and hindlimb: although all four extremities are capable of grasping, the hands search, reach, seize, and manipulate, while the feet grasp strongly, hold-fast, stabilize, and propel the body. The femur is relatively long, and the hindlimb is principally responsible for both accelerating and braking the body. The primate center of gravity is located more caudad than in other mammals. In addition, catarrhine primates habitually sleep, rest, and forage with the trunk held in a vertical or orthograde position, and develop ischial callosities to support the weight of the trunk when at rest.

In Chapter 13, I argued that catarrhine orthogrady, piled atop the general primate substrate of a caudad center of gravity and hindlimb dominance in locomotion, could produce bipedal locomotion in a comparatively straightforward fashion. The anatomical transition was relatively easy and undemanding. How complicated was the behavioral transition to the ground? Bipedal locomotion was initially arboreal, as exemplified by the hylobatids, who walk and run bipedally over relatively horizontal arboreal supports. Thus, initial bipedal experimentation could occur in a familiar and protected environment. Yet, hominids evolve terrestrial bipedalism (possibly more than once, as hinted by foot anatomy), and the extinct ape *Oreopithecus* appears to have evolved facultative terrestrial bipedalism, as well. Selection pressures involving dietary change – foraging for terrestrial food items – appear to have been initially responsible for the transition to life on the ground. Bipedalism probably already existed in the behavioral repertoire of arboreal proto-hominids. However, slow bipedal walking is energy efficient, especially along a stable, horizontal terrestrial substrate, where balance and weight transfer are not problematic. Other selection pressures, involving cost-effective long-distance ranging and foraging, could be triggered once terrestrial locomotion became more than incidental. Note that *Oreopithecus* existed on a Mediterranean island that lacked potential predators, but the late Miocene of Africa abounded in dangerous terrestrial carnivores. This indicates to me that terrestrial hominids had already evolved anti-predator defenses not seen in other primates. Such defenses might include weapon use, or social behaviors such as vigilance or complex mobbing behavior.

Tool behavior

At the first appearance of the archeological record 2.5–2.6 mya, hominid tool behavior, as well as foraging and ranging behavior, is already divorced from the array of behaviors exhibited by non-human primates. What might one expect about hominid behavior before archeology? What selection pressures may have led to the origins of hominid tool behavior? Given general studies on animal tool behavior, it is reasonable to assume that hominid tool behavior arose through selection pressures involving food acquisition. Tufted capuchins (*Cebus apella*) show the most habitual use of tools among non-human primates. During the prolonged dry season of the Caatinga forests in northeastern Brazil, the normally arboreal tufted capuchins even forage on the ground in a desperate search for food items, and use tools on an almost daily basis (Moura & Lee 2004). Tools are used to acquire both plant and animal food items, and the tool behavior most often observed is cracking and battering. Stones are also used for digging. Tufted capuchins at another Brazilian dry forest site have used anvils and stone hammerstones to crack open palm nuts for about 30 years (Fragaszy *et al.* 2004). Clusters of palm nuts emerge from the ground in this area, so that the capuchins forage terrestrially. When hominid-modified animal bones first appear in the archeological record, they exhibit cut-marks and percussion marks that indicate that hominids were using stone tools to acquire vertebrate meat and marrow. However, if the behavior of living tufted capuchins has relevance to hominid behavior before the archeological record begins, then one might expect that the first stone tools (proto-Oldowan tools) would be unflaked cobbles used for digging, battering, or breaking food items. These items might be either animal or plant foods. Archeologists studying lithic technology should examine fist-sized unflaked cobbles in areas yielding hominid remains that pre-date the earliest evidence of stone tools. If characteristic signatures of digging, battering, or hammering behavior can be detected on these cobbles, hominid activity might be inferred. The hardness and texture of the item that was cracked or battered might also leave characteristic marks on unflaked cobbles. Residue analysis on suspicious cobbles might yield evidence of embedded bone spicules or plant phytoliths.

Intelligence

The new hominization model outlined in Chapter 16 is based on both anatomical and archeological evidence. It argues that changes in cognition and sociality appear early in hominid evolution, rather than being the long culmination of a series of factors, as presented in classic hominization models. How reasonable is it to assume that shifts in cognition and social behavior could occur in early hominids? Field biology indicates that behavior can be very different in closely related organisms. Hence, phylogeny need not relentlessly skew cognition and sociality.

Complex cognition has independently evolved a number of times in the animal world. The clearest example is the intelligence of corvid birds and mammals. Thus, the complex cognition of hominids is not inexplicable or the result of unknown or unknowable processes. It is the result of natural selection. However, given the 6 million year span of hominid evolution, cognitive ability underwent considerable change through time. It is possible that cognition differed substantially between sympatric species of early hominid, and resulted in different hominid minds. Relative brain size is a gross indicator of higher cognitive functions. The archeological record, as a record of human behavior, should reveal a clearer trace of cognition. Yet, archeology appears only after the first 3.5 million years of hominid history have occurred, and the existence of sympatric hominid species makes it difficult to identify the tool user or maker. It is possible that a number of hominid species experimented with tool behavior. If this were the case, contextual or spatial differences or differences in knapping technology might reveal the presence of alternative technologies, and hence differences in behavior. More comparative work on animal intelligence might yield some clues about the specific ecological factors affecting the evolution of hominid intelligence. Tool behavior did not originate de novo, and paleo-ecological conditions during the first 3.5 my of hominid evolution might reveal clues about selection pressures involved in its origin.

The first appearance of the archeological record at 2.6–2.5 mya already indicates the presence of "natural history intelligence," and hence a behavioral phenotype that differs from anything seen among non-human primates. Although hominids are good catarrhine primates when anatomy and physiology are examined, hominid behavior and sociality initially develop a different trajectory from other catarrhines.

Complex sociality

A great deal of research time over the last 40 years has been devoted to the subject of early hominid sociality. Much of this research is not grounded in evolutionary ecology or functional morphology. The result is fable and endless phylogenies. Since the early 1960s, the reigning assumption has been that non-human primate behavioral ecology has direct implications for reconstructions of early hominid behavior. Yet, the complex sociality of hominids may have had its origins in behaviors that are absent or rare in non-human primates, although they are found in several other mammal groups. In particular, field experiments demonstrate that sentinel behavior may be "penalty-free," and care taking of the young ("babysitting" of several youngsters by a single adult or subadult) may be relatively cost-free, if food is abundant and shelter is nearby. These are examples of altruistic behaviors that are usually associated with complex hominid sociality, although they occur in small carnivore species. These behaviors, as well as mutual or group defense and mobbing behaviors, have been observed in a series of mongoose species. Researchers have experimented in the field with wild meerkats to determine the variables that affect sentinel and care-taking behavior. I advocate the use of social carnivores (including small carnivore species) in assembling a "composite mammal" model – a conceptual model for the origins of early hominid social behavior.

References

Abbott, D. H. 1993. Social conflict and reproductive suppression in marmoset and tamarin monkeys. In *Primate Social Conflict*, Mason, W. A. & Mendoza, S. P., eds., pp. 331–72. Albany, NY: SUNY Press.

Abegg, C. & Thierry, B. 2002. Macaque evolution and dispersal in insular South-East Asia. *Biological Journal of the Linnean Society* 75:555–76.

Abouheif, E. & Fairbairn, D. J. 1997. A comparative analysis of allometry for sexual size dimorphism: Assessing Rensch's rule. *American Naturalist* 149:540–62.

Agnew, N. & Demas, M. 1998. Preserving the Laetoli footprints. *Scientific American* 279(3):44–55.

Agustí, J. & Antón, M. 2002. *Mammoths, Sabretooths, and Hominids: 65 Million Years of Mammalian Evolution in Europe*. New York, NY: Columbia University Press.

Agustí, J., de Siria, A. S., & Garces, M. 2003. Explaining the end of the hominoid experiment in Europe. *Journal of Human Evolution* 45:145–54.

Aiello, L. C. & Dean, C. 1990. *An Introduction to Human Evolutionary Anatomy*. San Diego, CA: Academic Press.

Aiello, L. C. & Dunbar, R. I. M. 1993. Neocortex size, group size, and the evolution of language. *Current Anthropology* 34:184–93.

Alcock, J. 1972. The evolution of the use of tools by feeding animals. *Evolution* 26:464–73.

1989. *Animal Behavior: An Evolutionary Approach*. 4th edn. Sunderland, MA: Sinauer Associates.

2001a. *Animal Behavior: An Evolutionary Approach*. 7th edn. Sunderland, MA: Sinauer Associates.

2001b. *The Triumph of Sociobiology*. Oxford: Oxford University Press.

Alexander, R. D. 1974. The evolution of social behavior. *Annual Review of Ecology and Systematics* 5:325–83.

Alexander R. McN. 1985. Body support, scaling, and allometry. In *Functional Vertebrate Morphology*, Hildebrand, M., Bramble, D. M., Liem, K. F., & Wake, D. B., eds., pp. 26–37. Cambridge, MA: Belknap Press.

1996. *Optima for Animals*. 2nd edn. Princeton, NJ: Princeton University Press.

2003a. A rodent as big as a buffalo. *Science* 301:1678–9.

2003b. *Principles of Animal Locomotion*. Princeton, NJ: Princeton University Press.

Alexander, R. McN., Jayes, A. S., Maloiy, G. M. O., & Wathuta, E. M. 1979. Allometry of limb bones of mammals from shrews (*Sorex*) to elephants (*Loxodonta*). *Journal of Zoology, London* 189:305–14.

389

1981. Allometry of the leg muscles of mammals. *Journal of Zoology, London* 194:539–52.

Allee, W. C., Emerson, A. E., Park, O., & Schmidt, K. P. 1949. *Principles of Animal Ecology*. Philadelphia, PA: W. B. Saunders Company.

Allen, P. H. 1956. *The Rain Forests of Golfo Dulce*. Gainesville, FL: University of Florida Press.

Allman, J. 1982. Reconstructing the evolution of the brain in primates through the use of comparative neurophysiological and neuroanatomical data. In *Primate Brain Evolution*. Armstrong, E., & Falk, D., eds., pp. 13–28. New York, NY: Plenum Press.

Alroy, J. 1998. Cope's rule and the dynamics of body mass evolution in North American fossil mammals. *Science* 280:731–4.

Altmann, J., Schoeller, D., Altmann, S. A., Muruthi, P., & Sapolsky, R. M. 1993. Body size and fatness of free-living baboons reflect food availability and activity levels. *American Journal of Primatology* 30:149–61.

Altmann, S. 1991. Diets of yearling female primates (*Papio cynocephalus*) predict lifetime fitness. *Proceedings of the National Academy of Sciences U.S.A.* 88:420–3.

Andrewartha, H. G. 1961. *An Introduction to the Study of Animal Populations*. Chicago, IL: University of Chicago Press.

Andrewartha, H. G. & Birch, L. C. 1954. *The Distribution and Abundance of Animals*. Chicago, IL: University of Chicago Press.

Andrews, P. J. 1978. *A Revision of the Miocene Hominoidea of East Africa. Bulletin of The British Museum of Natural History (Geology)* 30(2):85–224.

Andrews, P. 1992. Evolution and environment in the Hominoidea. *Nature* 360:641–6.

1996. Palaeoecology and hominoid palaeoenvironments. *Biological Reviews* 71:257–300.

Andrews, P., Lord, J., & Evans, E. M. N. 1979. Patterns of ecological diversity in fossil and modern mammalian faunas. *Biological Journal of the Linnean Society* 11:177–205.

Antón, S. C. 2003. Natural history of *Homo erectus. Yearbook of Physical Anthropology* 46:126–70.

Antonovics, J. 1987. The evolutionary dys-synthesis: Which bottles for which wine? *American Naturalist* 129:321–31.

Arensburg, B. 1994. Middle Paleolithic speech capabilities: A response to Dr. Lieberman. *American Journal of Physical Anthropology* 94:279–80.

Armstrong, E. 1982. Mosaic evolution in the primate brain: differences and similarities in the hominoid thalamus. In *Primate Brain Evolution*, Armstrong, E. & Falk, D., eds., pp. 131–61. New York, NY: Plenum Press.

1985. The evolved limbic system: The human revolution. *American Journal of Physical Anthropology* 66:141 (abstract).

1990. Evolution of the brain. In *The Human Nervous System*, pp. 1–16. New York, NY: Academic Press.

1991. The limbic system and culture: an allometric analysis of the neocortex and limbic nuclei. *Human Nature* 2:117–36.

Asfaw, B., White, T., Lovejoy, O., Latimer, B., Simpson, S., & Suwa, G. 1999. *Australopithecus garhi*: A new species of early hominid from Ethiopia. *Science* 284:629–34.

Ashton, E. H. 1981. The Australopithecinae: their biometrical study. *Symposia of the Zoological Society of London* 46:67–126.

Ashton, K. G., Tracy, M. C., & de Queiroz, A. 2000. Is Bergmann's rule valid for mammals? *American Naturalist* 156:390–415.

Asquith, P. J. 1995. Of monkeys and men: Cultural views in Japan and the West. In *Ape,Man, Apeman: Changing Views Since 1600*. Corbey, R. & Theunissen, B., eds., pp. 309–25. Leiden, the Netherlands: Leiden University Press.

Aureli, F. & de Waal, F. B. M. 1997. Inhibition of social behavior in chimpanzees under high-density conditions. *American Journal of Primatology* 41:213–28.

Aureli, F. & de Waal, F. B. M., eds. 2000. *Natural Conflict Resolution*. Berkeley, CA: University of California Press.

Avis, V. 1962. Brachiation: the crucial issue for man's ancestry. *Southwestern Journal of Anthropology* 18:119–48.

Ayres, J. M. & Clutton-Brock, T. H. 1992. River boundaries and species range size in Amazonian primates. *American Naturalist* 140:531–7.

Azzaroli, A., Boccaletti, M., Delson, E., Moratti, G. & Torre, D. 1986. Chronological and paleogeographical background to the study of *Oreopithecus bambolii*. *Journal of Human Evolution* 15:533–40.

Backwell, L. R. & d'Errico, F. 2001. Evidence of termite foraging by Swartkrans early hominids. *Proceedings of the National Academy of Sciences* U. S. A. 98:1358–63.

Bahn, P. G. & Vertut, J. 1988. *Images of the Ice Age*. London: Facts on File.

Balter, V., Person, A., Labourdette, N., Drucker, D., Renard, & Vandermeersch, B. 2001. Les Néanertaliens étaient-ils essentiellement carnivores? Résultats préliminaires sur les teneurs en Sr et en Ba de la paléobiocénose mammalienne de Saint-Césaire. *Comptes Rendus del'Académie des Sciences Paris, Sciences de la Terre et des planètes* 332:59–65.

Barsh, G. S., Farooqi, I. S., & O'Rahilly, S. 2000. Genetics of body-weight regulation. *Nature* 404:644–51.

Bartholomew, Jr., G. A. & Birdsell, J. B. 1953. Ecology and the protohominids. *American Anthropologist* 55:481–98.

Bartke, A., Wright, J. C., Mattison, J. A., Ingram, D. K., Miller, R. A., & Roth, G. S. 2001. Extending the lifespan of long-lived mice. *Nature* 414:412.

Barton, R. A. & Harvey, P. H. 2000. Mosaic evolution of brain structure in mammals. *Nature* 405:1055–8.

Basmajian, J. V. 1974. *Muscles Alive. Their Functions Revealed by Electromyography*, 3rd edn. Baltimore, MD: Williams & Wilkins Co.

Bayne, K., Mainzer, H., Dexter, S., Campbell, G., Yamada, F., & Suomi, S. 1991. The reduction of abnormal behaviors in individually housed rhesus monkeys (*Macaca mulatta*) with a foraging/grooming board. *American Journal of Primatology* 23:23–35.

Beard, K. C. 1990. Gliding behaviour and palaeoecology of the alleged primate family Paromomyidae (Mammalia, Dermoptera). *Nature* 345:340–1.

1998a. A new genus of Tarsiidae (Mammalia: Primates) from the Middle Eocene of Shanxi Province, China, with notes on the historical biogeography of tarsiers. In *Dawn of the Age of Mammals in Asia*, Beard, K. C. & Dawson, M. R., eds., pp. 260–77. *Bulletin of the Carnegie Museum of Natural History*, vol. 34.

1998b. East of Eden: Asia as an important center of taxonomic origination in mammalian evolution. In *Dawn of the Age of Mammals in Asia*, Beard, K. C. & Dawson, M. R., eds., pp. 5–39. *Bulletin of the Carnegie Museum of Natural History*, vol. 34.

Beard, C. 2002. East of Eden at the Paleocene/Eocene boundary. *Science* 295:2028–9.

2004. *The Hunt for the Dawn Monkey: Unearthing the Origins of Monkeys, Apes, and Humans*. Berkeley, CA: University of California Press.

Beck, B. B. 1973. Cooperative tool use by captive hamadryas baboons. *Science* 182:594–7.

1980. *Animal Tool Behavior*. New York, NY: Garland Press.

1982. Chimpocentrism: bias in cognitive ethology. *Journal of Human Evolution* 11:3–17.

Bednekoff, P. A. 1997. Mutualism among safe, selfish sentinels: a dynamic game. *American Naturalist* 150:373–92.

Beevor, A. 1999. *Stalingrad*. New York, NY: Penguin Books.

Begun, D. R. 2002. European hominoids. In *The Primate Fossil Record*, Hartwig, W. C., ed., pp. 339–68. Cambridge: Cambridge University Press.

2003. Planet of the apes. *Scientific American* 289(2):74–83.

2004. The earliest hominins – is less more? *Science* 303:1478–80.

Behrensmeyer, A. K. 1975. The taphonomy and paleoecology of Plio-Pleistocene vertebrate assemblages east of Lake Rudolf, Kenya. *Bulletin of the Museum of Comparative Zoology* 146:473–578.

1978. Taphonomic and ecologic information from bone weathering. *Paleobiology* 4:150–62.

1982. Time resolution in fluvial vertebrate assemblages. *Paleobiology* 8:211–27.

1988. Vertebrate preservation in fluvial channels. *Palaeogeography, Palaeoclimatology, Palaeoecology* 63:183–99.

Behrensmeyer, A. K. & Hill, A. P., eds. 1980. *Fossils in the Making*. Chicago, IL: University of Chicago Press.

Behrensmeyer, A. K., Damuth, J. D., DiMichele, W. A., Potts, R., Sues, H.-D., & Wing, S. L., eds. 1992. *Terrestrial Ecosystems through Time*. Chicago, IL: University of Chicago Press.

Behrensmeyer, A. K., Todd, N. E., Potts, R., & McBrinn, G. E. 1997. Late Pliocene faunal turnover in the Turkana Basin, Kenya and Ethiopia. *Science* 278:1589–94.

Behrensmeyer, A., Harmon, E. H., & Kimbel, W. H. 2003. Environmental context and taphonomy of the A. L. 333 locality, Hadar, Ethiopia. Abstracts of the annual meeting of the Paleoanthropology Society, Tempe, Arizona.

Beldade, P., Koops, K., & Brakefield, P. M. 2002. Developmental constraints versus flexibility in morphological evolution. *Nature* 416:844–6.

Belk, M. C. & Houston, D. D. 2002. Bergmann's rule in ectotherms: a test using freshwater fishes. *American Naturalist* 160:803–8.

Bell, N. H. 1990. Sarcoidosis and related disorders. In *Metabolic Bone Diseases and Clinically Related Disorders*, 2nd edn., Aviolo, L. V. & Krane, S. M., eds., pp. 804–22. Philadelphia, PA: W. B. Saunders Company.

Bellomo, R. V. 1994. Methods of determining early hominid behavioral activities associated with the controlled use of fire at FxJj 20 Main, Koobi Fora, Kenya. *Journal of Human Evolution* 27:173–95.

Bellugi, U., Wang, P. P., & Jernigan, T. L. 1994. Williams syndrome: an unusual neuropsychological profile. In *Atypical Cognitive Deficits in Developmental Disorders: Implications for Brain Function*, Broman, S. H. & Grafman, J., eds., pp. 23–56. Hillsdale, NJ: Lawrence Erlbaum.

Benefit, B. R. & McCrossin, M. L. 2002. The Victoriapithecidae, Cercopithecoidea. In *The Primate Fossil Record*, Hartwig, W. C., edn., pp. 241–53. Cambridge: Cambridge University Press.

Benton, M. J. & Ayala, F. J. 2003. Dating the tree of life. *Science* 300:1698–700.

Bergman, T. J., Beehner, J. C., Cheney, D. L., & Seyfarth, R. M. 2003. Hierarchical classification by rank and kinship in baboons. *Science* 302:1234–6.

Bergson, H. 1907. *L'Evolution Créatrice*. Genève: Éditions Albert Skira [reprinted in 1945].

Berman, J. C. 1999. Bad hair days in the Paleolithic: modern (re)constructions of the cave man. *American Anthropologist* 101:288–304.

Bicca-Marques, J. C. & Garber, P. A. 2003. Experimental field study of the relative costs and benefits to wild tamarins (*Saguinus imperator* and *S. fuscicollis*) of exploiting contestable food patches as single- and mixed-species troops. *American Journal of Primatology* 60:139–53.

Bielicki, T. & Charzewski, J. 1977. Sex differences in the magnitude of statural gains of offspring over parents. *Human Biology* 49:265–77.

Bielicki, T., Szklarska, A., Welon, Z., & Rogucka, E. 2001. Variation in body mass index among Polish adults: effects of sex, age, birth cohort, and social class. *American Journal of Physical Anthropology* 116:166–70.

Binford, L. R. 1968. Archaeological perspectives. In *New Perspectives in Archaeology*, Binford, S. R. & Binford, L. R., eds., pp. 5–32. Chicago, IL: Aldine.

    1981. *Bones: Ancient Men and Modern Myths*. New York, NY: Academic Press.

    1984. *Faunal Remains from Klasies River Mouth*. London: Academic Press.

    1989. Isolating the transition to cultural adaptations: An organizational approach. In *The Emergence of Modern Humans: Biocultural Adaptations in the Later Pleistocene*, Trinkaus, E., edn., pp. 18–41. Cambridge: Cambridge University Press.

    2001. *Constructing Frames of Reference. An Analytical Method for Archaeological Theory Building Using Ethnographic and Environmental Data Sets*. Berekeley, CA: University of California Press.

Bishop, W. W. & Clark, J. D., eds. 1967. *Background to Evolution in Africa*. Chicago, IL: University of Chicago Press.

Blankenburg, F., Taskin, B., & Ruben, J. *et al*. 2003. Imperceptible stimuli and sensory processing impediment. *Science* 299:1864.

Bloch, J. I. & Boyer, D. M. 2002. Grasping primate origins. *Science* 298:1606–10.

2003. Response to comment on "Grasping primate origins". *Science* 300:741 [and online text].

Bloch, J. I., Rose, K. D., & Gingerich, P. D. 1998. New species of *Batodonoides* (Lipotyphla, Geolabididae) from the early Eocene of Wyoming: smallest known mammal? *Journal of Mammalogy* 79:804–27.

Blomberg, S. P., Garland, T., & Ives, A. R. 2003. Testing for phylogenetic signal in comparative data: behavioral traits are more labile. *Evolution* 57:717–45.

Bloomfield, F. H., Oliver, M. H., & Hawkins, P. *et al.* 2003. A periconceptional nutritional origin for noninfectious preterm birth. *Science* 300:606.

Blumenschine, R. J. 1986. Carcass consumption sequences and the archaeological distinction of scavenging and hunting. *Journal of Human Evolution* 15:639–59.

1987. Characteristics of an early hominid scavenging niche. *Current Anthropology* 28:383–407.

1995. Percussion marks, tooth marks, and experimental determinations of the timing of hominid and carnivore access to long bones at FLK *Zinjanthropus*, Olduvai Gorge, Tanzania. *Journal of Human Evolution* 29:21–51.

Blumenschine, R. J. & Madrigal, T. G. 1993. Variability in long bone marrow yields of East African ungulates and its zooarchaeological implications. *Journal of Archaeological Science* 20:555–87.

Blumenschine, R. J., Peters, C. R., & Masao, F. T. *et al.* 2003. Late Pliocene *Homo* and hominid land use from western Olduvai Gorge, Tanzania. *Science* 299:1217–21.

Blumstein, D. T. & Armitage, K. B. 1997. Does sociality drive the evolution of communicative complexity? A comparative test with ground-dwelling sciurid alarm calls. *American Naturalist* 150:179–200.

Boas, F. 1912. Changes in bodily form of descendants of immigrants. *American Anthropologist* 14:530–62.

Bocherens, H., Billiou, D., & Patou-Mathis, M. *et al.* 1999. Palaeoenvironmental and palaeodietary implications of the isotopic biogeochemistry of last interglacial Neanderthal and mammal bones in Scladina Cave (Belgium). *Journal of Archaeological Science* 26:599–609.

Boesch, C. & Boesch, H. 1984. Possible causes of sex differences in the use of natural hammers by wild chimpanzees. *Journal of Human Evolution* 13:415–48.

Boinski, S., Quatrone, R. P., & Swartz, H. 2001. Substrate and tool use by brown capuchins in Suriname: ecological contexts and cognitive bases. *American Anthropologist* 102:741–61.

Boissinot, S. *et al.* 1998. Origins and antiquity of X-linked triallelic color vision systems in New World monkeys. *Proceedings of the National Academy of Sciences U.S.A.* 95:13749–54.

Boorman, S. A. & Levitt, P. R. 1973. A frequency-dependent natural selection model for the evolution of social cooperation networks. *Proceedings of the National Academy of Sciences U.S.A.* 70:187–9.

Borgia, G. 1994. The scandals of San Marco. *Quarterly Review of Biology* 69:373–5.

Bouchard, C., ed. 1994. *The Genetics of Obesity*. Boca Raton, FL: CRC Press.

Bourlière, F. 1963. Observations on the ecology of some large African mammals. In *African Ecology and Human Evolution*, Howell, F. C. & Bourlière, F., eds., pp. 43–54. Chicago, IL: Aldine.

Bourlière, F. 1985. Primate communities: their structure and role in tropical ecosystems. *International Journal of Primatology* 6:1–26.

Bowen, G. J., Clyde, W. C., & Koch, P. L. *et al.* 2002. Mammalian dispersal at the Paleocene/Eocene boundary. *Science* 295:2062–5.

Bower, J. M. & Parsons, L. M. 2003. Rethinking the "lesser brain". *Scientific American* 289 (2):50–7.

Bowler, P. J. 1986. *Theories of Human Evolution.* Baltimore, MD: Johns Hopkins University Press.

1995. The geography of extinction: biogeography and the expulsion of 'ape men' from human ancestry in the early twentieth century. In *Ape, Man, Apeman: Changing Views Since 1600,* Corbey, R. & Theunissen, B., eds., pp. 185–93. Leiden, the Netherlands: Leiden University Press.

Bown, T. B., Kraus, M. J., & Wing, S. L. *et al.* 1982. The Fayum forest revisited. *Journal of Human Evolution* 11:603–32.

Bown, T. M. & Rose, K. D. 1987. Patterns of dental evolution in early Eocene anaptomorphine primates (Omomyidae) from the Bighorn Basin, Wyoming. *Paleontological Society Memoir* 23.

Boyd, R., Gintis, H., Bowles, S., & Richerson, P. J. 2003. The evolution of altruistic punishment. *Proceedings of the National Academy of Sciences U.S.A.* 100:3531–5.

Boysen, S. T. 1993. Counting in chimpanzees: nonhuman principles and emergent properties of number. In *The Development of Numerical Competence. Animal and Human Models.* Boysen, S. T. & Capaldi, E. J., eds., pp. 39–59. Hillsdale, NJ: Lawrence Erlbaum Associates.

Boysen, S. T. & Berntson, G. G. 1989. Numerical competence in a chimpanzee (*Pan troglodytes*). *Journal of Comparative Psychology* 103:23–31.

Boysen, S. T., Berntson, G. G., Shreyer, T. A., & Quigley, K. S. 1993. Processing of ordinality and transitivity by chimpanzees. *Journal of Comparative Psychology* 107:208–15.

Brain, C. K. 1981. *The Hunters or the Hunted? An Introduction to African Cave Taphonomy.* Chicago, IL: University of Chicago Press.

2001. *Do We Owe Our Intelligence to a Predatory Past?* Seventieth James Arthur Lecture on the Evolution of the Human Brain. New York, American Museum of Natural History.

Brain, C. K. & Sillen, A. 1988. Evidence from the Swartkrans Cave for the earliest use of fire. *Nature* 336:464–6.

Brantingham, R. J. 1998. Mobility, competition, and Plio-Pleistocene hominid foraging groups. *Journal of Archaeological Method and Theory* 5:57–98.

Brashares, J. S., Arcese, P., Sam, M. K., Coppolillo, P. B., Sinclair, A. R. E., & Balmford, A. 2004. Bushmeat hunting, wildlife declines, and fish supply in West Africa. *Science* 306:1180–3.

Brauer, G. W. 1982. Size sexual dimorphism and secular trend: Indicators of subclinical malnutrition? In *Sexual Dimorphism in* Homo sapiens. *A Question of Size,* Hall, R. L., ed., pp. 245–59. New York, NY: Praeger Publishers.

Britten, R. J. 2002. Divergence between samples of chimpanzee and human DNA sequences is 5%, counting indels. *Proceedings of the National Academy of Sciences U.S.A.* 99: 13633–5.

Broadfield, D. C., Holloway, R. L., Mowbray, K., Silvers, A., Yuan, M. S., & Marquez, S. 2001. Endocast of Sambungmacan 3 (Sm 3): A new *Homo erectus* from Indonesia. *Anatomical Record* 262:369–79.

Bronowski, B. & Long, W. M. 1952. Statistics of discrimination in anthropology. *American Journal of Physical Anthropology* 10:385–94.

Brooks, R. A. 1991. New approaches to robotics. *Science* 253:1227–32.

Brosnan, S. F. & de Waal, F. B. M. 2003. Monkeys reject unequal pay. *Nature* 425:297–9.

Brown, J. S. 1989. Coexistence on a seasonal resource. *American Naturalist* 133:168–82.

Brown, P., Sutikna, T., & Morwood, M. J. *et al.* 2004. A new small-bodied hominin from the Late Pleistocene of Flores, Indonesia. *Nature* 431:1055–61.

Browne, J. 2002. *Charles Darwin. The Power of Place.* New York, NY: Alfred A. Knopf, Inc.

Brűne, M., Brűne-Cohrs, U., & McGrew, W. 2004. Psychiatric treatment for great apes? *Science* 306:2039.

Brunet, M., Guy, F., & Pilbeam, D. *et al.* 1995. The first australopithecine 2500 kilometres west of the Rift Valley (Chad). *Nature* 378:273–5.

Brunet, M., Beauvilain, A., Coppens, Y., Heintz, E., Moutaye, A., & Pilbeam, D. 2002. A new hominid from the Upper Miocene of Chad, Central Africa. *Nature* 418:145–51.

Buchardt, B. 1979. Oxygen isotope paleotemperatures from the Tertiary Period in the North Sea area. *Nature* 275:121–3.

Bunn, H. T. 1994. Early Pleistocene hominid foraging strategies along the ancestral Omo River at Koobi Fora, Kenya. *Journal of Human Evolution* 27:247–66.

   1997. The bone assemblages from the excavated sites. In Koobi Fora Research Project, Vol. 5. *Plio-Pleistocene Archaeology*, Isaac, G. L. & Isaac, B., eds., pp. 402–44. Oxford: Clarendon Press.

Burckle, L. H. 1995. A critical review of the micropaleontological evidence used to infer a major drawdown of the East Antarctic ice sheet during the early Pliocene. In *Paleoclimate and Evolution with Emphasis on Human Origins*, Vrba, E. S., Denton, G. H., Partridge, T. C., & Burckle, L. H., eds., pp. 230–41. New Haven, CT: Yale University Press.

Burdick, A. 2004. Gross anatomy. *Discover* 25 (3):46–51.

Burness, G. P., Diamond, J., & Flannery, T. 2001. Dinosaurs, dragons, and dwarfs: The evolution of maximal body size. *Proceedings of the National Academy of Sciences U.S.A.* 98:14518–23.

Burney, D. A., Burney, L. P., & Godfrey, L. R. *et al.* 2004. A chronology for late prehistoric Madagascar. *Journal of Human Evolution* 47:25–63.

Burney, D. A. & Ramilisonina. 1999. The *kilopilopitsofy*, *kidoky*, and *bokyboky*: Accounts of strange animals from Belo-sur-mer, Madagascar, and the megafaunal "extinction window". *American Anthropologist* 100:957–66.

Bush, G. L. 1993. A reaffirmation of Santa Rosalia, or why are there so many kinds of small animals? In *Evolutionary Patterns and Processes*, Lees, D. R. & Edwards, D., eds., pp. 229–49. London: Academic Press.

Butzer, W. K. 1971. *Environment and Archeology, An Ecological Approach to Prehistory*, 2nd edn. Chicago, IL: Aldine.

Buzas, M. A. & Culver, S. J. 1994. Species pool and dynamics of marine paleocommunities. *Science* 264:1439–41.

Byers, J. A. 1997. *The American Pronghorn: Social Adaptations and the Ghosts of Predators Past*. Chicago, IL: University of Chicago Press.

Byrne, R. 1995. *The Thinking Ape. The Evolutionary Origins of Intelligence*. Oxford: Oxford University Press.

1996. Machiavellian intelligence. *Evolutionary Anthropology* 5:172–80.

Byrne, R. W. & Whiten, A., eds. 1988. *Machiavellian Intelligence. Social Expertise and the Evolution of Intellect in Monkeys, Apes, and Humans*. Oxford: Clarendon Press.

Cachel, S. 1975a. A new view of speciation in *Australopithecus*. In *Paleoanthropology, Morphology and Paleoecology*, Tuttle, R. H., ed., pp. 183–201. The Hague, the Netherlands: Mouton Press.

1975b. The beginnings of the Catarrhini. In *Primate Functional Morphology and Evolution*, Tuttle, R. H., ed., pp. 23–36. The Hague, the Netherlands: Mouton Press.

1979a. A functional analysis of the primate masticatory system and the origin of the anthropoid post-orbital septum. *American Journal of Physical Anthropology* 50:1–18.

1979b. A paleoecological model for the origin of higher primates. *Journal of Human Evolution* 8:351–9.

1981. Plate tectonics and the problem of anthropoid origins. *Yearbook of Physical Anthropology* 24:139–72.

1984. Growth and allometry in primate masticatory muscles. *Archives of Oral Biology* 29:287–93.

1986. *The Growth of Biological Thought* revisited. *American Anthropologist* 88:452–4.

1990. Partisan primatology. *American Journal of Primatology* 22:139–42.

1992. The theory of punctuated equilibria and evolutionary anthropology. In *The Dynamics of Evolution. The Punctuated Equilibrium Debate in the Natural and Social Sciences*, Somit, A. & Peterson, S. A., eds., pp. 187–220. Ithaca, NY: Cornell University Press.

1994. The natural history origin of human intelligence: A new perspective on the origin of human intelligence. *Social Neuroscience Bulletin* 7(2):25–30.

1996a. Phylogeny triumphant. *American Journal of Primatology* 36:365–8.

1996b. Megadontia in the teeth of early hominids. *Kaupia. Darmstadter Beiträge zür Naturgeschichte* 6:119–28. M. D. Leakey Festschrift volume.

1997. Dietary shifts and the European Upper Palaeolithic transition. *Current Anthropology* 38:579–603.

1998. Review of *Paleoclimate and Evolution, with Emphasis on Human Origins*. *American Journal of Physical Anthropology* 105:97–9.

1999. Dietary shifts and the origins of the European Upper Palaeolithic. In *Lifestyles and Survival Strategies in Pliocene and Pleistocene Hominids*, Ullrich, H., edn., pp. 494–504. Schwelm, Germany: Edition Archaea.

2000a. Review of *Structure and Contingency. Evolutionary Processes in Life and Human Society*, Bintloff, J., ed., Leicester University Press, 1999. *American Anthropologist* 102:648–9.

2000b. Subsistence factors among Arctic peoples and the reconstruction of social organization from prehistoric human diet. In *Animal Bones, Human Societies*, Rowley-Conwy, P., ed., pp. 39–48. Oxford: Oxbow Books.

2001. The impact of dietary constraints on social organization in high latitudes. In *On Being First: Cultural Innovation and Environmental Consequences of First Peopling*, deMille, C. *et al.*, eds., pp. 61–80. Calgary: Department of Archaeology, University of Calgary.

2004. The paleobiology of *Homo erectus* and early hominid dispersal. *Athena Review* 4(1):23–31.

Cachel, S. & Harris, J. W. K. 1995. Ranging patterns, land-use and subsistence in *Homo erectus* from the perspective of evolutionary ecology. In *Proceedings of the Pithecanthropus Centennial, 1893–1993; Vol. I, Palaeoanthropology: Evolution & Ecology of* Homo erectus, Bower, J. R. F. & Sartono, S., eds., pp. 51–66. Leiden, the Netherlands: Leiden University Press.

1996. The paleobiology of *Homo erectus*: implications for understanding the adaptive zone of this species. In *Aspects of African Archaeology. Papers from the 10th Congress of the PanAfrican Association for Prehistory and Related Studies*, Pwiti, G. & Soper, R., eds., pp. 3–9. Harare: University of Zimbabwe Press.

1998. The lifeways of *Homo erectus* inferred from archaeology and evolutionary ecology: A perspective from East Africa. In *Early Human Behaviour in Global Context. The Rise and Diversity of the Lower Palaeolithic Record*, Petraglia, M. D. & Korisettar, R., eds., pp. 108–32. London: Routledge Press.

in press. The behavioural ecology of early Pleistocene hominids in the Koobi Fora region, east Lake Turkana Basin, northern Kenya. In *An Odyssey of Space*, Robertson, E. *et al.*, eds. Calgary: University of Calgary Press.

Cachel, S., Harris, J. W. K., Monahan, C. M., & Rogers, M. J. 2000. Inferring hominid behavioral adaptations during Okote Member times in the Koobi Fora region. *American Journal of Physical Anthropology* Supplement 30:116–17.

Calder, W. A., III. 1984. *Size, Function, and Life History*. Cambridge, MA: Harvard University Press.

Caley, M. J. & Munday, P. L. 2003. Growth trades off with habitat specialization. *Proceedings of the Royal Society of London B* 270 (supplement):S175–7.

Cantalupo, C. & Hopkins, W. D. 2001. Asymmetric Broca's area in great apes. *Nature* 414:505.

Carbone, C., Mace, G. M., Roberts, S. C., & Macdonald, D. W. 1999. Energetic constraints on the diet of terrestrial carnivores. *Nature* 402:286–8.

Carbone, C., Cowlishaw, G., Isaac, N. J. B., & Rowcliffe, J. M. 2005. How far do animals go? Determinants of day range in mammals. *American Naturalist* 165:290–7.

Carothers, J. H. 1986. Homage to Huxley: On the conceptual origin of minimum size ratios among competing species. *American Naturalist* 128:440–2.

Carpenter, C. R. 1934. *A Field Study of the Behavior and Social Relations of Howling Monkeys. Comparative Psychology Monographs*, Vol. 10, no. 2.

  1940. *A Field Study in Siam of the Behavior and Social Relations of the Gibbon* (Hylobates lar). *Comparative Psychology Monographs*, Vol. 16, No. 5.

  1965. *Naturalistic Behavior of Nonhuman Primates*. University Park, PA: Pennsylvania State University Press.

Cartmill, M. 1972. Arboreal adaptations and the origin of the Order Primates. In *The Functional and Evolutionary Biology of Primates*, Tuttle, R. H., ed., pp. 97–122. Chicago, IL: Aldine.

  1974. Rethinking primate origins. *Science* 184:436–43.

  1993. *A View to a Death in the Morning: Hunting and Nature Through History.* Cambridge, MA: Harvard University Press.

Cavalieri, P. & Singer, P. 1993. *The Great Ape Project: Equality Beyond Humanity.* New York: St. Martin's Press.

Cavalli-Sforza, L. L., Menozzi, P., & Piazza, A. 1994. *The History and Geography of Human Genes*. Princeton, NJ: Princeton University Press.

Cavallo, J. A. & Blumenschine, R. J. 1989. Tree-stored leopard kills: expanding the hominid scavenging niche. *Journal of Human Evolution* 18:393–9.

Cela-Conde, C. J. & Altaba, C. R. 2002. Multiplying genera versus moving species: a new taxonomic proposal for the Family Hominidae. *South African Journal of Science* 98:229–32.

Cerling, T. E. 1992. Development of grasslands and savannas in East Africa during the Neogene. *Palaeogeography, Palaeoclimatology, Palaeoecology* 97:241–7.

Chaimanee, Y., Jolly, D., & Beammi, M. *et al.* 2003. A Middle Miocene hominoid from Thailand and orangutan origins. *Nature* 422:61–5.

Chaimanee, Y., Suteethorn, V., Jintasakul, P., Vidthayanon, C., Marandat, B., & Jaeger, J. J. 2004. A new orang-utan relative from the Late Miocene of Thailand. *Nature* 427:439–41.

Chapman, C. A., Chapman, L. J., Bjorndal, K. A., & Onderdonk, D. A. 2002. Application of protein-to-fiber ratios to predict colobine abundance on different spatial scales. *International Journal of Primatology* 23:283–310.

Charles-Dominique, P. & Martin, R. D. 1970. Evolution of lorises and lemurs. *Nature* 227:257–60.

Chehab, F. F., Mounzih, K., Lu, R. H., & Lim, M. E. 1997. Early onset of reproductive function in normal female mice treated with leptin. *Science* 275:88–90.

Chen, Y. S., Olckers, A., Schurr, T. G., Kogelnik, A. M., Huoponen, K., & Wallace, D. C. 2000. Mitochondrial DNA variation in the South African Kung and Khwe and their genetic relationships to other African populations. *American Journal of Human Genetics* 66:1362–83.

Cheney, D. L. & Seyfarth, R. M. 1990. *How Monkeys see the World*. Chicago, IL: University of Chicago Press.

Cheney, D. L., Seyfarth, R. M., & Smuts, B. B. 1986. Social relationships and social cognition in nonhuman primates. *Science* 234:1361–6.

Chenn, A. 2002. Making a bigger brain by regulating cell cycle exit. *Science* 298:766–7.

Chenn, A. & Walsh, C. A. 2002. Regulation of cerebral cortical size by control of cell cycle exit in neural precursors. *Science* 297:365–9.

Chiu, C.-H. & Hamrick, M. W. 2002. Evolution and development of the primate limb skeleton. *Evolutionary Anthropology* 11:94–107.

Chivers, D. J. 1984. Feeding and ranging in gibbons: A summary. In *The Lesser Apes. Evolutionary and Behavioural Ecology*, Preuschoft, H. *et al.*, eds., pp. 267–81. Edinburgh: Edinburgh University Press.

Churchfield, S. 1990. *The Natural History of Shrews*. Ithaca, NY: Comstock Publishing Associates, Cornell University Press.

Clark, D. A., Mitra, P. P., & Wang, S. S. H. 2001. Scalable architecture in mammalian brains. *Nature* 411:189–93.

Clark, J. D. & Harris, J. W. K. 1985. Fire and its roles in early hominid lifeways. *The African Archaeological Review* 3:3–27.

Clark, J. D., Beyene, Y., & WoldeGabriel, G. *et al.* 2003. Stratigraphic, chronological and behavioural contexts of Pleistocene *Homo sapiens* from Middle Awash, Ethiopia. *Nature* 423:747–52.

Clarke, R. J. 1998. First ever discovery of a well-preserved skull and associated skeleton of *Australopithecus*. *South African Journal of Science* 94:460–3.

  1999. Discovery of complete arm and hand of the 3.3 million-year-old *Australopithecus* skeleton from Sterkfontein. *South African Journal of Science* 95:477–80.

Clarke, R. J. & Tobias, P. V. 1995. Sterkfontein Member 2 foot bones of the oldest South African hominid. *Science* 269:521–4.

Clayton, N. S. 1998. Memory and hippocampus in food-storing birds: a comparative approach. *Neuropharmacology* 37:441–52.

Clubb, R. & Mason, G. 2003. Captivity effects on wide-ranging carnivores. *Nature* 425:473–4.

Clutton-Brock, T., ed. 1977. *Primate Ecology*. New York, NY: Academic Press.

  2002. Breeding together: kin selection and mutualism in cooperative vertebrates. *Science* 296:69–72.

Clutton-Brock, T., Harvey, P., & Rudder, B. 1977. Sexual dimorphism, socioeconomic sex ratio and body weight in primates. *Nature* 269:797–800.

Clutton-Brock, T. H., O'Riain, M. J., & Brotherton, P. N. M. *et al.* 1999. Selfish sentinels in cooperative mammals. *Science* 284:1640–4.

Clutton-Brock, T. H., Russell, A. F., & Sharpe, L. L. *et al.* 2001. Effects of helpers on juvenile development and survival in meerkats. *Science* 293:2446–9.

Clutton-Brock, T. H., Russell, A. F., Sharpe, L. L., Young, A. J., Balmforth, Z., & McIlrath, G. M. Evolution and development of sex differences in cooperative behavior in meerkats. *Science* 297:253–6.

Collard, M. & Wood, B. 2000. How reliable are human phylogenetic hypotheses? *Proceedings of the National Academy of Sciences, U.S.A.* 97:5003–6.

Collins, K. J. 1996. Heat stress and related disorders. In *Manson's Tropical Diseases*, 20th edn., Cook, G., ed., pp. 421–32. London: W.B. Saunders.

Collins, S., Raina, A., Tedrake, R., & Wisse, M. 2005. Efficient bipedal robots based on passive-dynamic walkers. *Science* 307:1082–5.

Coltman, D. W., O'Donoghue, P., Jorgenson, J. T., Hogg, J. T., Strobeck, C., & Festa Bianchet, M. 2003. Undesirable evolutionary consequences of trophy hunting. *Nature* 426:655–8.

Conroy, G. C. 1990. *Primate Evolution*. New York, NY: W. W. Norton & Company.

2002. Speciosity in the early *Homo* lineage: too many, too few, or just about right? *Journal of Human Evolution* 43:759–66.

2003. The inverse relationship between species diversity and body mass: do primates play by the "rules"? *Journal of Human Evolution* 45:783–95.

Conway Morris, S. 1998. *The Crucible of Creation. The Burgess Shale and the Rise of Animals*. New York, NY: Oxford University Press.

Coolidge, H. J. 1933. *Pan paniscus*: Pygmy chimpanzee from south of the Congo River. *American Journal of Physical Anthropology* 18:1–57.

Coon, C. S. 1962. *The Origin of Races*. New York, NY: Knopf.

1965. *The Living Races of Man*. New York, NY: Knopf.

1982. *Racial Adaptations*. Chicago, IL: Nelson-Hall.

Cope, E. D. 1871. The method of creation of organic forms. *Proceedings of the American Philosophical Society* 12:229–63.

Correia, H. R., Balseiro, S. C., Correia, E. R., Mota, P. G., & de Areia, M. L. 2004. Why are human newborns so fat? Relationship between fatness and brain size at birth. *American Journal of Human Biology* 16:24–30.

Corvelo, T. C. O., Schneider, H., & Harada, M. L. 2002. ABO blood groups in the primate species of *Cebidae* from the Amazon region. *Journal of Medical Primatology* 31:136–41.

Couzin, I. D., Krause, J., Franks, N. R., & Levin, S. A. 2005. Effective leadership and decision-making in animal groups on the move. *Nature* 433:513–16.

Cowlishaw, G. & Hacker, J. E. 1997. Distribution, density, and latitude in African primates. *American Naturalist* 150:505–11.

Crampton, J. S., Beu, A. G., Cooper, R. A., Jones, C. M., Marshall, B., & Maxwell, P. A. 2003. Estimating the rock volume bias in paleobiodiversity studies. *Science* 301:358–60.

Crandall, K. A., Olaf, R., Bininde-Emonds, P., Mace, G. M., & Wayne, R. K. 2000. Considering evolutionary processes in conservation biology. *Trends in Ecology and Evolution* 15:290–5.

Critser, G. 2003. *Fat Land: How Americans Became the Fattest People in the World*. Boston, MA: Houghton Mifflin Co.

Crook, J. H. & Gartlan, J. S. 1966. Evolution of primate societies. *Nature* 210:1200–3.

Cruciani, F., Santolamazza, P., & Shen, P. D. *et al.* 2002. A back migration from Asia to sub-Saharan Africa is supported by high resolution analysis of human Y chromosome haplotypes. *American Journal of Human Genetics* 70:1197–214.

Cunningham, D. L., Cole III, T. M., Ward, C. V., & Wescott, D. J. 2004. Postcranial sexual dimorphism at the A. L. 333 site. *American Journal of Physical Anthropology* 123 (Supplement 38):80–1.

Damuth, J. D. 1993. *Evolution of Terrestrial Ecosystems Database Manual*. Washington, DC: Smithsonian Institution.

Damuth, J. & MacFadden, B. J., eds. 1990. *Body Size in Mammalian Paleobiology: Estimation and Biological Implications.* Cambridge: Cambridge University Press.

Dart, R. A. 1925. *Australopithecus africanus*: The man-ape of South Africa. *Nature* 115:195–9.

1949. The predatory implemental technique of *Australopithecus. American Journal of Physical Anthropology* 7:1–38.

1957. *The Osteodontokeratic Culture of* Australopithecus prometheus. Transvaal Museum Memoir No. 10, Pretoria.

Darwin, C. 1859. *On the Origin of Species by Means of Natural Selection or the Preservation of Favoured Races in the Struggle for Life.* Amherst, NY: Prometheus Books [1991 reprint].

1871. *The Descent of Man and Selection in Relation to Sex.* Amherst, NY: Prometheus Books [1998 reprint].

Dausmann, K. H., Glos, J., Ganzhorn, J. V., & Heldmaier, G. 2004. Hibernation in a tropical primate. *Nature* 429:825–6.

Davis, M., Hut, P., & Muller, R. A. 1984. Extinction of species by periodic comet showers. *Nature* 308:715–17.

de Chardin, P. T. 1953. The idea of fossil man. In *Anthropology Today. An Encyclopedic Inventory*, Kroeber, A. L., edn., pp. 93–100. Chicago, IL: University of Chicago Press.

1955. *Le Phénomene Humain.* Paris: Editions du Seuil.

de Heinzelin, J., Clark, J. D., White, T. *et al.* 1999. Environment and behavior of 2.5-million-year-old Bouri hominids. *Science* 284:625–9.

de Menocal, P. 1995. Plio-Pleistocene African climate. *Science* 270:53–9.

de Menocal, P. & Bloemendal, J. 1995. Plio-Pleistocene climatic variability in subtropical Africa and the paleoenvironment of hominid evolution: a combined data-model approach. In *Paleoclimate and Evolution with Emphasis on Human Origins*, Vrba, E. S., Denton, G. H., Partridge, T. C., & Burckle, L. H. eds., pp. 262–88. New Haven, CT: Yale University Press.

d'Errico, F. & Backwell, L. R. 2003. Possible evidence of bone tool shaping by Swartkrans early hominids. *Journal of Archaeological Science* 30:1559–76.

de Waal, F. B. M. 1982. *Chimpanzee Politics. Power and Sex Among Apes.* New York, NY: Harper & Row.

1989. *Peacemaking among Primates.* Cambridge, MA: Harvard University Press.

1996. *Good Natured.* Cambridge, MA: Harvard University Press.

de Winter, W. & Oxnard, C. E. 2001. Evolutionary radiations and convergences in the structural organization of mammalian brains. *Nature* 409:710–14.

Dean, C., Leakey, M. G., & Reid, D. 2001. Growth processes in teeth distinguish modern humans from *Homo erectus* and earlier hominins. *Nature* 414:628–31.

Deloison, Y. 1991. Les australopithèques marchaient-ils comme nous? In *Origine(s) de la Bipédie chez les Hominidés*, Coppens, Y. & Senut, B., eds., pp. 177–86. Cahiers de Paléoanthropologie. Paris: Editions du Centre National de la Recherche Scientifique.

Delson, E. 1975. Evolutionary history of the Cercopithecidae. *Contributions to Primatology* 5:167–217.

1986. An anthropoid enigma: historical introduction to the study of *Oreopithecus bambolii*. *Journal of Human Evolution* 15:523–31.

Delson, E. & Rosenberger, A. 1984. Are there any anthropoid primate living fossils? In *Living Fossils*, Eldredge, N. & Stanley, S. M., eds., pp. 50–61. New York, NY: Springer Verlag.

Dermitzakis, E. T., Reymond, A., & Scamuffa, N. *et al.* 2003. Evolutionary discrimination of mammalian conserved non-genic sequences (CNGs). *Science* 302:1033–5.

Desmond, A. 1984. *Archetypes and Ancestors. Palaeontology in Victorian London, 1850–1875.* Chicago, IL: University of Chicago Press.

DeVore, I. 1963. Comparative ecology and behavior of monkeys and apes. In *Classification and Human Evolution*, Washburn, S. L., ed., 301–19. New York, NY: Wenner-Gren Foundation.

ed. 1965. *Primate Behavior: Field Studies of Monkeys and Apes.* New York, NY: Holt, Rinehart and Winston.

DeVore, I. & Washburn, S. L. 1963. Baboon ecology and human evolution. In *African Ecology and Human Evolution*, Howell, F. C. & Bourlière, F., eds., pp. 335–67. New York, NY: Wenner-Gren Foundation.

Diamond, J. M. 1984. "Normal" extinctions of isolated populations. In *Extinctions*, Nitecki, M., ed., pp. 191–246. Chicago, IL: University of Chicago Press.

1996. Competition for brain space. *Nature* 382:756–7.

Dieckmann, U. & Doebeli, M. 1999. On the origin of species by sympatric speciation. *Nature* 400:354–7.

Dietl, G. P., Herbert, G. S., & Vermeij, G. J. 2004. Reduced competition and altered feeding behavior among marine snails after a mass extinction. *Science* 306:2229–31.

Di Fiore, A. 2003. Molecular genetic approaches to the study of primate behavior, social organization, and reproduction. *Yearbook of Physical Anthropology* 46:62–99.

Di Fiore, A. & Rendall, D. 1994. Evolution of social organization: a reappraisal for primates by using phylogenetic methods. *Proceedings of the National Academy of Sciences U.S.A.* 91:9941–5.

Dixson, A. 2002. Sexual selection by cryptic female choice and the evolution of primate sexuality. *Evolutionary Anthropology* 11(S1):195–9.

Doebeli, M. & Dieckmann, U. 2003. Speciation along environmental gradients. *Nature* 421:259–64.

Domb, L. G. & Pagel, M. 2001. Sexual swellings advertise female quality in wild baboons. *Nature* 410:204–6.

2002. Evolutionary biology (communication arising): Significance of primate sexual swellings. *Nature* 420:143.

Duarte-Quiroga, A. & Estrada, A. 2003. Primates as pets in Mexico City: an assessment of the species involved, source of origin, and general aspects of treatment. *American Journal of Primatology* 61:53–60.

Dubois, E. 1895. Dr. Dubois' 'Missing Link'. *Nature* 53:115–16.

Duggen, S., Hoernle, K., van den Bogaard, P., Rupke, L., & Morgan, J. P. 2003. Deep roots of the Messinian salinity crisis. *Nature* 422:602–6.

Dunbar, R. I. M. 1992. Neocortex size as a constraint on group size in primates. *Journal of Human Evolution* 22:469–93.

Duncan, J., Seitz, R. J., Kolodny, J. *et al.* 2000. A neural basis for general intelligence. *Science* 289:457–60.

Eadie, J. M., Broekhoven, L., & Colgan, P. 1987. Size ratios and artifacts: Hutchinson's rule revisited. *American Naturalist* 129:1–17.

Eckhardt, R. B. 2000. *Human Paleobiology*. Cambridge: Cambridge University Press.

Egas, M., Dieckmann, U., & Sabelis, M. W. 2004. Evolution restricts the coexistence of specialists and generalists: the role of trade-off structure. *American Naturalist* 163:518–31.

Eisenberg, J. F. 1981. *The Mammalian Radiations: An Analysis of Trends in Evolution, Adaptation, and Behavior*. Chicago, IL: University of Chicago Press.

Eisenberg, J. F., Muckenhirn, N., & Rudran, R. 1972. The relation between ecology and social structure in primates. *Science* 176:863–74.

Eldredge, N. & Cracraft, J. 1980. *Phylogenetic Patterns and the Evolutionary Process*. New York, NY: Columbia University Press.

Eldredge, N. & Gould, S. J. 1972. Punctuated equilibria: an alternative to phyletic gradualism. In *Models in Paleobiology*, Schopf, T. J. M., ed., pp. 82–115. San Francisco, CA: Freeman, Cooper.

Emerson, S. B. & Hastings, P. A. 1998. Morphological correlations in evolution: consequences for phylogenetic analysis. *Quarterly Review of Biology* 73:141–62.

Emery, N. J. & Clayton, N. S. 2004. The mentality of crows: convergent evolution of intelligence in corvids and apes. *Science* 306:1903–7.

Emmons, L. H. & Gentry, A. H. 1983. Tropical forest structure and the distribution of gliding and prehensile-tailed vertebrates. *American Naturalist* 121:513–24.

Enard, W., Khaitovich, P., Klose, J. *et al.* 2002. Intra- and interspecific variation in primate gene expression patterns. *Science* 296:340–3.

Endler, J. A. 1986. *Natural Selection in the Wild*. Princeton, NJ: Princeton University Press.

Erikson, G. E. 1963. Brachiation in New World monkeys and in anthropoid apes. *Symposia of the Zoological Society of London* 10:135–63.

Eronen, J. T. & Rook, L. 2004. The Mio-Pliocene European primate fossil record: dynamics and habitat tracking. *Journal of Human Evolution* 47:323–41.

Erwin, D. H. 2001. Lessons from the past: biotic recoveries from mass extinctions. *Proceedings of the National Academy of Sciences U.S.A.* 98:5399–403.

Eysenck, H. J. 1993. The biological basis of intelligence. In *Biological Approaches to the Study of Human Intelligence*, Vernon, P. A., ed., pp. 1–32. Norwood, NJ: Ablex Publishing Co.

Falk, D. 1978. Cerebral asymmetry in Old World monkeys. *Acta Anatomica* 101:334–9.

1987. Hominid paleoneurology. *Annual Review of Anthropology* 16:13–30.

2001. The evolution of sex differences in primate brains. In *Evolutionary Anatomy of the Primate Cerebral Cortex*, Falk, D. & Gibson, K. R., eds., pp. 98–112. Cambridge: Cambridge University Press.

Fehr, E. & Fischbacher, U. 2003. The nature of human altruism. *Nature* 425:785–91.

Feibel, C. S. 1997. Debating the environmental factors in human evolution. *GSA Today* 7:1–7.

Feibel, C. S., Brown, F. H., & McDougall, I. 1989. Stratigraphic context of fossil hominids from the Omo Group deposits: Northern Turkana Basin, Kenya and Ethiopia. *American Journal of Physical Anthropology* 78:595–622.

Feng, D.-F., Cho. G., & Doolittle, R. F. 1997. Determining divergence times with a protein clock: update and reevaluation. *Proceedings of the National Academy of Sciences U.S.A.* 94:13028–33.

Fennell, K. J. & Trinkaus, E. 1997. Bilateral femoral and tibial periostitis in the La Ferrassie 1 Neanderthal. *Journal of Archaeological Science* 24:985–95.

Ferrari, S. F. & Martins, E. S. 1992. Gummivory and gut morphology in two sympatric callitrichids (*Callithrix emiliae* and *Saguinus fuscicollis weddelli*) from Western Brazilian Amazonia. *American Journal of Physical Anthropology* 88:97–103.

Fimbel, C. 1992. *Use of Regenerating Farm Clearings and Older Secondary Forest by Primates and a Forest Antelope at Tiwai, Sierra Leone.* Ph.D. dissertation, Department of Environment and Natural Resources, Rutgers University, New Brunswick, NJ.

1994. Ecological correlates of species success in modified habitats may be disturbance- and site-specific: The primates of Tiwai Island. *Conservation Biology* 8:106–13.

Finlay, B. L. & Darlington, R. B. 1995. Linked regularities in the development and evolution of mammalian brains. *Science* 268:1578–84.

Fischman, J. 1993. New clues surface about the making of mind. *Science* 262:1517.

Fitch, W. T. & Hauser, M. D. 2004. Computational constraints on syntactic processing in a nonhuman primate. *Science* 303:377–80.

Fleagle, J. G. 1978. Locomotion, posture, and habitat utilization in two sympatric, Malaysian leaf-monkeys (*Presbytis obscura* and *Presbytis melalophos*). In *The Ecology of Arboreal Folivores*, Montgomery, G. G., ed., pp. 243–51. Washington, DC: Smithsonian Institution Press.

Fleagle, J. G. & Reed, K. E. 1996. Comparing primate communities: a multivariate approach. *Journal of Human Evolution* 30:489–510.

Fleagle, J. G., Stern, Jr., J. T., Jungers, W. L., Susman, R. L., Vangor, A. K., & Wells, J. P. 1981. Climbing: a biomechanical link with brachiation and with bipedalism. *Symposia of the Zoological Society of London* 48:359–75.

Foley, R. A. 1989. The evolution of hominid social behavior. In *Comparative Socioecology*, Standen, V. & Foley, R. A., eds., pp. 473–94. Oxford: Blackwell Scientific.

1999. Hunting down the hunter-gatherers. *Evolutionary Anthropology* 8(4):115–17.

Foley, R. A. & Lee, P. C. 1989. Finite social space, evolutionary pathways, and reconstructing hominid behavior. *Science* 243:901–6.

Foote, M. 1996. Ecological controls on the evolutionary recovery of post-Paleozoic crinoids. *Science* 274:1492–5.

Foote, M., Hunter, J. P., Janis, C. M., & Sepkoski, J. J. 1999. Evolutionary and preservational constraints on origins of biologic groups: divergence times of eutherian mammals. *Science* 283:1310–14.

Ford, S. M. 1994. Evolution of sexual dimorphism in body weight in platyrrhines. *American Journal of Primatology* 34:221–44.

Forey, P. L., Humphries, C. J., Kitching, I. L., Scotland, R. W., Siebert, D. J., & Williams, D. M. 1992. *Cladistics. A Practical Course in Systematics.* Oxford: Clarendon Press.

Fortelius, M. 1990. Problems with using fossil teeth to estimate body sizes of extinct mammals. In *Body Size in Mammalian Paleobiology: Estimation and Biological Implications*, Damuth, J. & MacFadden, B. J., eds., pp. 207–28. Cambridge: Cambridge University Press.

Fragaszy, D. M. & Visalberghi, E. 1989. Social influences on the acquisition of tool-using behaviors in tufted capuchin monkeys (*Cebus apella*). *Journal of Comparative Psychology* 103:159–70.

Fragaszy, D., Izar, P., Visalberghi, E., Ottoni, E. B., & De Oliveira, M. G. 2004. Wild capuchin monkeys (*Cebus libidinosus*) use anvils and stone pounding tools. *American Journal of Primatology* 64:359–66.

Frakes, L. A., Francis, J. E., & Syktus, J. I. 1992. *Climate Modes of the Phanerozoic. The History of the Earth's Climate over the Past 600 Million Years.* Cambridge: Cambridge University Press.

Frank, S. A. 2003. Perspective: Repression of competition and the evolution of cooperation. *Evolution* 57:693–705.

Freckleton, R. P., Harvey, P. H., & Pagel, M. 2003. Bergmann's rule and body size in mammals. *American Naturalist* 161:821–5.

Fredriks, A. M., Van Buuren, S., Burgmeijer, R. J. F. *et al.* 2000. Continuing positive secular growth change in the Netherlands 1955–1997. *Pediatric Research* 47:316–23.

Frisancho, A. R. 1993. *Human Adaptation and Accommodation.* Ann Arbor, MI: University of Michigan Press.

Frisch, R. E. & McArthur, J. W. 1974. Menstrual cycles: fatness as a determinant of minimum weight necessary for their maintenance or onset. *Science* 185:949–51.

Gabunia, L., Vekua, A., Lordkipanidze, D. *et al.* 2000. Earliest Pleistocene hominid cranial remains from Dmanisi, Republic of Georgia: taxonomy, geological setting, and age. *Science* 288:1019–25.

Gagnon, M. 1997. Ecological diversity and community ecology in the Fayum sequence (Egypt). *Journal of Human Evolution* 32:133–60.

Gaillard, J.-M., Festa-Bianchet, M., Delorme, D., & Jorgenson, J. 2000. Body mass and individual fitness in female ungulates: bigger is not always better. *Proceedings of the Royal Society of London B* 267:471–7.

Galik, K. *et al.* 2004. External and internal morphology of the BAR 1002′ 00 *Orrorin tugenensis* femur. *Science* 305:1450–3.

Ganzhorn, J. U. 1999. Body mass, competition and the structure of primate communities. In *Primate Communities*, Fleagle, J. G., Janson, C., & Reed, K., eds., pp. 141–57. Cambridge: Cambridge University Press.

Garber, P. A. 1990. Tamarins: body size/weight. *AnthroQuest* 41:12.

Gardner, M. B. 2003. Simian AIDS: an historical perspective. *Journal of Medical Primatology* 32:180–6.

Garland, Jr., T. 1983. Scaling the ecological cost of transport to body mass in terrestrial mammals. *American Naturalist* 121:571–587.

Gaudzinski, S. 2003. Subsistence patterns of Early Pleistocene hominids in the Levant – taphonomic evidence from the 'Ubeidiya Formation (Israel). *Journal of Archaeological Science* 31:65–75.

Gautier-Hion, A. 1988. Polyspecific associations among forest guenons: ecological, behavioural and evolutionary aspects. In *A Primate Radiation. Evolutionary Biology of the African Guenons*, Gautier-Hion, A., Bourlière, F., Gautier, J. P., & Kingdon, J., eds., pp. 452–76. Cambridge: Cambridge University Press.

Gautier-Hion, A., Bourlière, F., Gautier, J. P., & Kingdon, J., eds. 1988. *A Primate Radiation: Evolutionary Biology of the African Guenons*. Cambridge: Cambridge University Press.

Gebo, D. L. 2004. A shrew-sized origin for primates. *Yearbook of Physical Anthropology* 47:40–62.

Gebo, D. L., Dagosto, M., Beard, K. C., & Qi, T. 2000. The smallest primates. *Journal of Human Evolution* 38:585–94.

Gendron, R. P. & Reichman, O. J. 1995. Food perishability and inventory management: a comparison of three caching strategies. *American Naturalist* 145:948–68.

Gervais, P. 1872. Sur un singe fossile, d'espèce non encore décrite, qui a été découvert au Monte Bamboli (Italie). *Comptes rendus de l'Académie des sciences de Paris* 74:1217–23.

Gibbs, S., Collard, M., & Wood, B. 2000. Soft-tissue characters in higher primate phylogenetics. *Proceedings of the National Academy of Sciences U.S.A.* 97:11130–2.

Gibson, K. R. 1986. Cognition, brain size and the extraction of embedded food resources. In *Primate Ontogeny, Cognition and Social Behavior*, Else, J. G. & Lee, P. C., eds., pp. 93–103. Cambridge: Cambridge University Press.

Gilad, Y., Man, O., Paabo, S., & Lancet, D. 2003. Human specific loss of olfactory receptor genes. *Proceedings of the National Academy of Sciences U.S.A.* 100:3324–7.

Gilbert, K. 1994. *Endoparasitic Infections in Red Howler Monkeys (Alouatta seniculus) in the Central Amazon Basin: A Cost of Sociality*. Ph. D. dissertation, Department of Anthropology, Rutgers University, New Brunswick, NJ.

  1997. Red howling monkey use of specific defecation sites as a parasite avoidance strategy. *Animal Behaviour* 54:451–5.

Gilchrist, G. W. 1995. Specialists and generalists in changing environments. I. Fitness landscapes of thermal sensitivity. *American Naturalist* 146:252–70.

Gingerich, P. D. 1976. Paleontology and phylogeny: patterns of evolution at the species level in early Tertiary mammals. *American Journal of Science* 276:1–28.

  1983. Rates of evolution: effects of time and temporal scaling. *Science* 222:159–61.

  1984. Pleistocene extinctions in the context of origination-extinction equilibria in Cenozoic mammals. In *Quaternary Extinctions: A Prehistoric Revolution*. Martin, P. S. & Klein, R. G., eds., pp. 211–22. Tucson, AZ: University of Arizona Press.

  1990. Prediction of body mass in mammalian species from long-bone lengths and diameters. *Contributions from the University of Michigan Museum of Paleontology* 28:79–92.

  ed. 1993a. Functional Morphology and Evolution. *American Journal of Science* 293-A (special supplementary issue in honor of John Ostrom).

1993b. Quantification and comparison of evolutionary rates. *American Journal of Science* 293-A:453–478. (special supplementary issue in honor of John Ostrom: *Functional Morphology and Evolution*, edited by P. D. Gingerich).

2001. Rates of evolution on the time scale of the evolutionary process. *Genetica* 112–113:127–44.

Gingerich, P. D. & Smith, B. H. 1985. Allometric scaling in the dentition of primates and insectivores. In *Size and Scaling in Primate Biology*, Jungers, W. L., ed., pp. 257–72. New York, NY: Plenum.

Gingerich, P. D., Smith, B. H., & Rosenberg, K. 1982. Allometric scaling in the dentition of primates and prediction of body weight from tooth size in fossils. *American Journal of Physical Anthropology* 58:81–100.

Ginsberg, B. E. & Laughlin, W. S. 1966. The multiple bases of human adaptability and achievement: a species point of view. *Eugenics Quarterly* 13:240–57.

Ginsberg, J. R., Mace, G. M., & Albon, S. 1995. Local extinction in a small and declining population: wild dogs in the Serengeti. *Proceedings of the Royal Society of London* B 262:221–8.

Girondot, M. & Laurin, M. 2003. Bone profiler: a tool to quantify, model, and statistically compare bone-section compactness profiles. *Journal of Vertebrate Paleontology* 23:458–61.

Glander, K. E. 1982. The impact of plant secondary compounds on primate feeding behavior. *Yearbook of Physical Anthropology* 25:1–18.

Godfray, H. C. J. & Blythe, S. P. 1990. Complex dynamics in multispecies communities. *Philosophical Transactions of the Royal Society of London B* 330:221–33.

Goldschmidt, R. 1940. *The Material Basis of Evolution*. New Haven, CT: Yale University Press.

Goodall, J. 1964. Tool-using and aimed throwing in a community of free-living chimpanzees. *Nature* 201:1264–6.

1971. *In the Shadow of Man*. Boston, MA: Houghton Mifflin.

1977. Infant killing and cannibalism in free-living chimpanzees. *Folia Primatologica* 28:259–82.

1979. Life and death at Gombe. *National Geographic* 155:592–621.

1986. *The Chimpanzees of Gombe: Patterns of Behavior*. Cambridge, MA: Belknap Press.

Goodman, M. 1963. Man's place in the phylogeny of the primates as reflected in serum proteins. In *Classification and Human Evolution*, Washburn, S. L., ed., pp. 204–34. Chicago, IL: Aldine.

1973. The chronicle of primate phylogeny contained in proteins. *Symposia of the Zoological Society of London* 33:339–75.

Goodman, M. & Moore, G. W. 1971. Immunodiffusion systematics of the primates. I. The Catarrhini. *Systematic Zoology* 20:19–62.

Gopnik, A., Meltzoff, A. N., & Kuhl, P. K. 2001. *The Scientist in the Crib. What Early Learning Tells Us About The Mind*. New York, NY: Perennial.

Goren-Inbar, N., Feibel, C. S., Verosub, K. L. *et al.* 2000. Pleistocene milestones on the Out-of-Africa corridor at Gesher Benot Ya'aqov, Israel. *Science* 289: 944–7.

Goren-Inbar, N., Alperson, N., Kisley, M. E. *et al.* 2004. Evidence of hominin control of fire at Gesher Benot Ya'aqov, Israel. *Science* 304:725–7.

Gould, S. J. 1978. Sociobiology: The art of story-telling. *New Scientist* 80:530–3.

1980. Is a new and general theory of evolution emerging? *Paleobiology* 6:119–30.

1983. Irrelevance, submission, and partnership: the changing role of palaeontology in Darwin's three centennials, and a modest proposal for macroevolution. In *Evolution from Molecules to Men*, Bendall, D. S., ed., pp. 347–66. Cambridge: Cambridge University Press.

1984. Toward the vindication of punctuational change. In *Catastrophes and Earth History*, Berggren, W. A. & Van Couvering, J. A., eds., pp. 9–34. Princeton, NJ: Princeton University Press.

1989. *Wonderful Life. The Burgess Shale and the Nature of History*. New York, NY: W. W. Norton & Company, Inc.

Gould, S. J. & Eldredge, N. 1977. Punctuated equilibria: the tempo and mode of evolution reconsidered. *Paleobiology* 3:115–51.

Gould, S. J. & Eldredge, N. 1993. Punctuated equilibrium comes of age. *Nature* 366:223–7.

Gould, S. J. & Lewontin, R. C. 1979. The spandrels of San Marco and the Panglossian paradigm. A critique of the adaptationist program. *Proceedings of the Royal Society of London B* 205:581–98.

Gradstein, F. M., Ogg, J. G., Smith, A. G. *et al.* 2004. *A Geologic Time Scale 2004.* Cambridge: Cambridge University Press.

Grand, T. I. 1978. Adaptations of tissue and limb segments to facilitate moving and feeding in arboreal folivores. In *The Ecology of Arboreal Folivores*, Montgomery, G. G., ed., pp. 231–41. Washington, DC: Smithsonian Institution Press.

Grant, P. R. & Grant, B. R. 1992. Hybridization of bird species. *Science* 256:193–7.

Gray, J. R., Braver, T. S., & Raichle, M. E. 2002. Integration of emotion and cognition in the lateral prefrontal cortex. *Proceedings of the National Academy of Sciences U.S.A.* 99(6):4115–20.

Grayson, D. E. & Delpech, F. 2003. Ungulates and the Middle-to-Upper Paleolithic transition at Grotte XVI (Dordogne, France). *Journal of Archaeological Science* 30:1633–48.

Grayson, D. K. 1983. *The Establishment of Human Antiquity*. New York, NY: Academic Press.

Gregory, W. K. 1910. *The Orders of Mammals. Bulletin of the American Museum of Natural History* 27:1–524.

1916. Studies on the evolution of the primates. *Bulletin of the American Museum of Natural History* 35:239–355.

1920. *On the Structure and Relations of* Notharctus, *an American Eocene Primate. Memoirs of the American Museum of Natural History*, n. s. Vol. 3, Part 2.

1927. The origin of man from the anthropoid stem – when and where? *Proceedings of the American Philosophical Society* 66:439–63.

1928. Were the ancestors of man primitive brachiators? *Proceedings of the American Philosophical Society* 67:129–50.

1930. The origin of man from a brachiating anthropoid stock. *Science* 71:645–50.

1934. *Man's Place Among the Anthropoids*. Oxford: Oxford University Press.

1949. The bearing of the Australopithecinae upon the problem of man's place in nature. *American Journal of Physical Anthropology* 7:485–512.

1951. *Evolution Emerging. A Survey of Changing Patterns from Primeval Life to Man*. 2 vols. New York, NY: Macmillan.

Gregory, W. K. & Hellman, M. 1926. The dentition of *Dryopithecus* and the origin of man. *Anthropological Papers of the American Museum of Natural History*, Vol. 28, Part 1.

Griffin, A. S. & West, S. A. 2003. Kin discrimination and the benefit of helping in cooperatively breeding vertebrates. *Science* 302:634–6.

Griffin, D. R. 1992. *Animal Minds*. Chicago, IL: University of Chicago Press.

Groce, N. E. & Marks, J. 2001. The great ape project and disability rights: ominous undercurrents of eugenics in action. *American Anthropologist* 102:818–22.

Groves, C. P. 2001a. *Primate Taxonomy*. Washington, DC: Smithsonian Institution Press.

2001b. Why taxonomic stability is a bad idea, or why are there so few species of primates (or are there?). *Evolutionary Anthropology* 10(6):192–8.

2004. The what, why and how of primate taxonomy. *International Journal of Primatology* 25:1105–26.

Gunther, M. M. & Boesch, C. 1993. Energetic cost of nut-cracking behaviour in wild chimpanzees. In *Hands of Primates*, Preuschoft, H. & Chivers, D. J., eds., pp. 109–29. Wien: Springer-Verlag.

Guo, Z. T., Ruddiman, W. F., Hao, Q. Z. *et al*. 2002. Onset of Asian desertification by 22 Myr ago inferred from loess deposits in China. *Nature* 416:159–63.

Gur, R. C., Mozley, L. H., Mozley, P. D. *et al*. 1995. Sex differences in regional cerebral glucose metabolism during a resting state. *Science* 267:528–31.

Hagelin, J., Carlsson, H. E., Suleman, M. A., & Hau, J. 2000. Swedish and Kenyan medical and veterinary students accept nonhuman primate use in biomedical research. *Journal of Medical Primatology* 29:431–2.

Hahn, E. 1988. *Eve and the Apes*. New York, NY: Weidenfeld and Nicholson.

Haig, D. 1999. What is a marmoset? *American Journal of Primatology* 49:285–96.

Haile-Selassie, Y. 2001. Late Miocene hominids from the Middle Awash, Ethiopia.*Nature* 412:178–81.

Haile-Selassie, Y., Suwa, G., & White, T. D. 2004. Late Miocene teeth from Middle Awash, Ethiopia, and early hominid dental evolution. *Science* 303:1503–5.

Haldane, J. B. S. 1949. Suggestions as to quantitative measurement of rates of evolution. *Evolution* 3:51–6.

Hall, J. G., Flora, C., Scott, Jr., C. I., Pauli, R. H., & Tanaka, K. I. 2004. Majewski osteodysplastic primordial dwarfism type II (MOPD II): natural history and clinical findings. *American Journal of Medical Genetics* 130A:55–72.

Hall, K. R. L. 1963. Variations in the ecology of the chacma baboon (*P. ursinus*). *Symposia of the Zoological Society of London* 10:1–28.

1965. Social organization of the Old-World monkeys and apes. *Symposia of the Zoological Society of London* 14:265–89.

Hall, R. L. 1982. Sexual dimorphism for size in seven nineteenth-century Northwest Coast populations. In *Sexual Dimorphism in* Homo sapiens. *A Question of Size*, Hall, R. L., ed., pp. 231–243. New York, NY: Praeger Publishers.

Hallam, A. 1989. *Great Geological Controversies*. 2nd edn. Oxford: Oxford University Press.

Hamilton, I. M. 2000. Recruiters and joiners: using optimal skew theory to predict group size and the division of resources within groups of social foragers. *American Naturalist* 155:684–95.

Hamilton, W. D. 1964. The genetical theory of social behavior, I & II. *Journal of Theoretical Biology* 7:1–52.

1971. Geometry for the selfish herd. *Journal of Theoretical Biology* 31:295–311.

Haraway, D. J. 1989. *Primate Visions. Gender, Race, and Nature in the World of Modern Science*. New York, NY: Routledge.

Harcourt, A. H. 1991. Sperm competition and the evolution of nonfertilizing sperm in mammals. *Evolution* 48:314–28.

1997. Sperm competition in primates. *American Naturalist* 149:189–94.

Harcourt, A. H. & Schwartz, M. W. 2001. Primate evolution: a biology of Holocene extinction and survival on the Southeast Asian Sunda Shelf islands. *American Journal of Physical Anthropology* 114:4–17.

Harcourt, A. H., Harvey, P. H., Larson, S. G., & Short, R. V. 1981. Testis weight, body weight and breeding system in primates. *Nature* 293:55–7.

Harcourt-Smith, W. E. H. & Aiello, L. C. 2004. Fossils, feet and the evolution of human bipedal locomotion. *Journal of Anatomy* 204:403–16.

Hardin, G. 1968. The tragedy of the commons. *Science* 162:1243–8.

Harland, W. B., Armstrong, R. L., Cox, A. V., Craig, L. E., Smith, A. G., & Smith, D. G. 1989. *A Geologic Time Scale 1989*. Cambridge: Cambridge University Press.

Harlow, H. F. & Harlow, M. K. 1962. Social deprivation in monkeys. *Scientific American* 207:136–46.

Harlow, H. F., Harlow, M. K., Dodsworth, R. O., & Arling, G. L. 1966. Maternal behavior of rhesus monkeys deprived of mothering and peer associations in infancy. *Proceedings of the American Philosophical Society* 110: 58–66.

Harlow, H. F., Harlow, M. K., & Suomi, S. J. 1971. From thought to therapy: lessons from a primate laboratory. *American Scientist* 59:538–49.

Harmon, E. H., Behrensmeyer, A. K., Kimbel, W. K., & Johanson, D. C. 2003. Preliminary taphonomic analysis of hominin remains from A. L. 333, Hadar Formation, Ethiopia. Paper presented at the annual meeting of the Paleoanthropology Society, Tempe, Arizona.

Harmon, L. J., Schulte, J. A., Larson, J. A., & Losos, J. B. 2003. Tempo and mode of evolutionary radiation in iguanian lizards. *Science* 301:961–4.

Harris, J. A. & Benedict, F. G. 1919. *A Biometric Study of Basal Metabolism in Man*. Washington, D.C: Carnegie Institution of Washington.

Harris, J. W. K. 1978. *The Karari Industry: Its Place in East African Prehistory*. Ph. D. Dissertation. Department of Anthropology, University of California, Berkeley.

Harris, J. W. K. & Capaldo, S. D. 1993. The earliest stone tools: their implications for an understanding of the activities and behaviour of late Pliocene hominids. In *The Use of Tools by Human and Non-Human Primates*, Berthelet, A. & Chavaillon, J., eds., pp. 196–223. Oxford: Clarendon Press.

Harrison, T. 1987. The phylogenetic relationships of the early catarrhine primates: a review of the current evidence. *Journal of Human Evolution* 16:41–80.

1993. Cladistic concepts and the species problem in hominoid evolution. In *Species, Species Concepts and Primate Evolution*, Kimbel, W. H. & Martin, L. B., eds., pp. 345–71. New York, NY: Plenum.

Hart, J., Jr. & Gordon, B. 1992. Neural subsystems for object knowledge. *Nature* 359:60–4.

Harvey, P. H. & Krebs, J. R. 1990. Comparing brains. *Science* 249:140–6.

Harvey, P. H. & Pagel, M. D. 1991. *The Comparative Method in Evolutionary Biology*. Oxford: Oxford University Press.

Hastings, I. M. 1996. The genetics and physiology of size reduction in mice. *Symposia of the Zoological Society of London* 69:129–42.

Hauser, M. D., Chomsky, N., & Fitch, W. T. 2002. The faculty of language: what is it, who has it, and how did it evolve? *Science* 298:1569–79.

Hausfater, G. & Hrdy, S. B., eds. 1984. *Infanticide: Comparative and Evolutionary Perspectives*. New York, NY: Aldine.

Hawks, J. 2004. How much can cladistics tell us about early hominid relationships? *American Journal of Physical Anthropology* 125:207–19.

Hediger, H. 1964. *Wild Animals in Captivity*. New York, NY: Dover Publications.

1968. *The Psychology and Behaviour of Animals in Zoos and Circuses*. New York, NY: Dover Publications.

1969. *Man and Animal in the Zoo. Zoo Biology*. New York, NY: Delacorte Press.

Heinrich, B. 2000. *Mind of the Raven. Investigations and Adventures with Wolf-Birds*. New York, NY: Cliff Street Books, Harper Collins.

Hendrickx, A. G. 1971. *Embryology of the Baboon*. Chicago, IL: University of Chicago Press.

Hendry, A. P. & Kinnison, M. T. 1999. Perspective: The pace of modern life: measuring rates of contemporary microevolution. *Evolution* 53:1637–53.

Henig, R. M. 2000. *The Monk in the Garden. The Lost and Found Genius of Gregor Mendel, the Father of Genetics*. Boston, MA: Houghton Mifflin.

Henneberg, M. 1988. Decrease of human skull size in the Holocene. *Human Biology* 60:395–405.

Hennig, W. 1966. *Phylogenetic Systematics*. Urbana, IL: University of Illinois Press.

Hershkovitz, P. 1977. *Living New World Monkeys (Platyrrhini) with an Introduction to Primates*. Chicago, IL: University of Chicago Press.

Heyes, C. M. 1993. Anecdotes, training, trapping and triangulating: do animals attribute mental states? *Animal Behavior* 46:177–88.

Hildebrand, M. 1985. Walking and running. In *Functional Vertebrate Morphology*. Hildebrand, M., Bramble, D. M., Liem, K. F., & Wake, D. B., eds., pp. 38–57. Cambridge, MA: Belknap Press.

Hildebrand, M., Bramble, D. M., Liem, K. F., & Wake, D. B., eds. 1985. *Functional Vertebrate Morphology*. Cambridge, MA: Belknap Press.

Hirasaki, E., Kumakura, H., & Matano, S. 2000. Biomechanical analysis of vertical climbing in the spider monkey and the Japanese macaque. *American Journal of Physical Anthropology* 113:455–72.

Hirasaki, E., Ogihara, N., Hamada, Y., & Nakatsukasa, M. 2002. Kinematics of bipedal locomotion in bipedally-trained Japanese macaques (monkey performance monkeys). *American Journal of Physical Anthropology* 117(S34):85.

Hoffman, A. 1985. Patterns of family extinction depend on definition and geological timescale. *Nature* 315:659–62.

Hoffmann, A. A., Hallas, R. J., Dean, J. A., & Schiffer, M. 2003. Low potential for climatic stress adaptation in a rainforest *Drosophila* species. *Science* 301:100–2.

Holekamp, K. E. *et al.* 2000. Group travel in social carnivores. In *On the Move*, Boinski, S. & Garber, P., eds., pp. 587–627. Chicago, IL: University of Chicago Press.

Holland, E. L., Cran, G. W., Elwood, J. H., Pinkerton, J. H. M., & Thompson, W. 1982. Associations between pelvic anatomy, height, and year of birth of men and women in Belfast. *Annals of Human Biology* 9:113–20.

Hoogenraad, C. C., Akhmanova, A., Galjart, N., & De Zeeuw. 2004. LIMK1 and CLIP-115: linking cytoskeletal defects to Williams Syndrome. *BioEssays* 26:141–50.

Hooton, E. A. 1942. *Man's Poor Relations*. Garden City, NY: Doubleday, Doran & Company, Inc.

Hopwood, A. T. 1933. Miocene primates from Kenya. *Zoological Journal of the Linnean Society, London* 38:437–64.

Howell, F. C. 1969. Remains of Hominidae from the Pliocene/Pleistocene formations in the lower Omo basin, Ethiopia. *Nature* 223:1234–9.

1978. Hominidae. In *Evolution of African Mammals*, Maglio, V. J. & Cooke, H.B.S., eds., pp. 154–248. Cambridge, MA: Harvard University Press.

Howell, S. 1999. An assessment of primatology in the 1990s. http://www.primate.wisc.edu/pin/careers/howell.html.

Howells, W. W. 1973. *Cranial Variation in Man. A Study by Multivariate Analysis of Patterns of Difference Among Recent Human Populations*. Papers of the Peabody Museum of Archaeology and Ethnology, Harvard University, Vol. 67.

1995. *Who's Who in Skulls. Ethnic Identification of Crania from Measurements*. Papers of the Peabody Museum of Archaeology and Ethnology, Harvard University, Vol. 82.

1997. *Getting Here. The Story of Human Evolution*, 2nd edn. Washington, DC: Compass Press.

Hrdy, S. B. 1977. *The Langurs of Abu: Female and Male Strategies of Reproduction*. Cambridge, MA: Harvard University Press.

1981. *The Woman That Never Evolved*. Cambridge, MA: Harvard University Press.

Hsü, K. J. 1983. *The Mediterranean was a Desert. A Voyage of the Glomar Challenger*. Princeton, NJ: Princeton University Press.

1986. *The Great Dying*. San Diego, CA: Harcourt Brace Jovanovich.

Hu, Y., Meng, J., Wang, Y. Q., & Li, C. K. 2005. Large Mesozoic mammals fed on young dinosaurs. *Nature* 433:149–52.

Hubbell, S. P. 2001. *The Unified Neutral Theory of Biodiversity and Biogeography*. Princeton, NJ: Princeton University Press.

Huey, R. B., Hertz, P. E., & Sinervo, B. 2003. Behavioral drive versus behavioral inertia in evolution: a null model approach. *American Naturalist* 161:357–66.

Hughes, A. R. 1954. Hyaenas versus australopithecines as agents of bone accumulations. *American Journal of Physical Anthropology* 12:467–86.

Hulbert, A. H. & Haskell, J. P. 2003. The effect of energy and seasonality on avian species richness and community composition. *American Naturalist* 161:83–97.

Humphrey, N. K. 1976. The social function of intellect. In *Growing Points in Ethology*, Bateson, P. G. & Hinde, R. A., eds., pp. 303–17. Cambridge: Cambridge University Press.

Hunt, G. 1996. Manufacture and use of hook-tools by New Caledonian crows. *Nature* 379:249–51.

Hunt, K. D. 1992. Positional behavior of *Pan troglodytes* in the Mahale Mountains and Gombe Stream National Parks, Tanzania. *American Journal of Physical Anthropology* 87:83–105.

  1994. The evolution of human bipedality: ecology and functional morphology. *Journal of Human Evolution* 2:183–202.

Hürzeler, J. 1958. *Oreopithecus bambolii* Gervais: A preliminary report. *Verhandlungen der naturforschenden Gesellschaft Basel* 69:1–47.

Hutchinson, G. E. 1959. Homage to Santa Rosalia, or why are there so many kinds of animals? *American Naturalist* 93:145–59.

Hutchinson, G. E. & MacArthur, R. H. 1959. A theoretical ecological model of size distributions among species of animals. *American Naturalist* 93:117–25.

Huxley, J. S. 1958. Evolutionary processes and taxonomy with special reference to grades. *Uppsala University Årss.* 1:21–39.

Huxley, T. H. 1863. *Evidence as to Man's Place in Nature*. Ann Arbor, MI: University of Michigan Press [reprinted in 1959].

  1874. On the hypothesis that animals are automata, and its history. *Nature* 10:362–6.

  1894. *Evolution and Ethics–Prolegomena*. From The Huxley File, online URL: http://aleph0.clarku.edu/huxley/.

Hyodo, M., Nakaya, H., Urabe, A. *et al.* 2002. Paleomagnetic dates of hominid remains from Yuanmou, China, and other Asian sites. *Journal of Human Evolution* 43:27–41.

Isaac, G. L. 1971. The diet of early man: aspects of archaeological evidence from Lower and Middle Pleistocene sites in Africa. *World Archaeology* 2:278–98.

  1975. Early hominids in action: a commentary on the contribution of archeology to understanding the fossil record in East Africa. *Yearbook of Physical Anthropology* 19:19–35.

  1978a. The food-sharing behavior of proto-human hominids. *Scientific American* 238:90–108.

  1978b. Food-sharing and human evolution: archaeological evidence from the Plio-Pleistocene of East Africa. *Journal of Anthropological Research* 34:311–25.

  1983. Bones in contention: competing explanations for the juxtaposition of Early Pleistocene artifacts and faunal remains. In *Animals and Archaeology: Hunters and Their Prey*, Clutton-Brock, J. & Grigson, C., eds., pp. 3–20. Oxford: BAR International Series, no. 163.

Isaac, G. L. & Crader, D. 1981. To what extent were early hominids carnivorous? An archaeological perspective. In *Omnivorous Primates: Gathering and Hunting in*

Human Evolution, Harding, R. S. O. & Teleki, G., eds., pp. 37–103. New York, NY: Columbia University Press.

Isaac, G. L. & Harris, J. W. K. 1975. The scatter between the patches. Paper presented at the Kroeber Anthropological Society Meeting, Berkeley, California.

Isaac, N. J. B., Mallet, J. & Mace, G. M. 2004. Taxonomic inflation: its influence on macroecology and conservation. *Trends in Ecology and Evolution* 19:464–9.

Isbell, L. A. 1990. Sudden short-term increase in mortality of vervet monkeys (*Cercopithecus aethiops*) due to leopard predation in Amboseli National Park, Kenya. *American Journal of Primatology* 21:41–52.

Isbell, L. A. & Young, T. P. 1996. The evolution of bipedalism in hominids and reduced group size in chimpanzees: alternative responses to decreasing resource availability. *Journal of Human Evolution* 30:389–97.

Iturralde-Vinent, M. A. & MacPhee, R. D. E. 1999. Paleogeography of the Caribbean region: implications for Cenozoic biogeography. *Bulletin of the American Museum of Natural History* 238:1–95.

Jablonski, D. 1997. Body-size evolution in Cretaceous molluscs and the status of Cope's rule. *Nature* 385:250–2.

1998. Geographic variation in the molluscan recovery from the end Cretaceous extinction. *Science* 279:1327–30.

1999. The future of the fossil record. *Science* 284:2114–16.

2001. Lessons from the past: evolutionary impacts of mass extinctions. *Proceedings of the National Academy of Sciences U.S.A.* 98:5393–8.

2004. Extinction: past and present. *Nature* 427:589.

Jablonski, D. & Bottjer, D. J. 1991. Environmental patterns in the origins of higher taxa: the post-Paleozoic record. *Science* 252:1831–3.

Jablonski, D., Roy, K., Valentine, J. W., Price, R. M., & Anderson, P. S. 2003. The impact of the pull of the recent on the history of marine diversity. *Science* 300:1133–5.

Jablonski, N. G., ed. 1993. Theropithecus: *Rise and Fall of a Primate Genus*. Cambridge: Cambridge University Press.

Janzen, D. H., ed. 1983. *Costa Rican Natural History*. Chicago, IL: University of Chicago Press.

Jaquish, C. E., Dyer, T., Williams Blangero, S., Dyke, B., Leland, M., & Blangero, J. 1997. Genetics of adult body mass and maintenance of adult body mass in captive baboons (*Papio hamadryas* subspecies). *American Journal of Primatology* 42:281–8.

Jasieńska, G., Thune, I., & Ellison, P. T. 2004a. Energy status, energy balance and energy expenditure in relation to ovarian function in rural and urban women from Poland. *American Journal of Physical Anthropology* 123(S38):120.

Jasieńska, G., Ziomkiewicz, A., Ellison, P. T., Lipson, S. F., & Thune, I. 2004b. Large breasts and narrow waists indicate high reproductive potential in women. *Proceedings of the Royal Society of London B* 271:1213–17.

Jenike, M. R. 2001. Nutritional ecology: Diet, physical activity and body size. In *Hunter Gatherers: An Interdisciplinary Perspective*, Panter-Brick, C., Layton, R. H., & Rowley Conwy, P., eds., pp. 205–38. Cambridge: Cambridge University Press.

Jerison, H. J. 1973. *Evolution of the Brain and Intelligence.* New York, NY: Academic Press.

1975. Fossil evidence for the evolution of the human brain. *Annual Review of Anthropology* 4:27–58.

1991. *Brain Size and the Evolution of Mind.* Fifty-Ninth James Arthur Lecture on the Evolution of the Human Brain. New York, American Museum of Natural History.

2001. The study of primate brain evolution: Where do we go from here? In *Evolutionary Anatomy of the Primate Cerebral Cortex*, Falk, D. & Gibson, K. R., eds., pp. 305–37. Cambridge: Cambridge University Press.

Jernvall, J., Hunter, J. P., & Fortelius, M. 1996. Molar tooth diversity, disparity, and ecology in Cenozoic ungulate radiations. *Science* 274:1489–92.

Jernvall, J. & Jung, H.-S. 2000. Genotype, phenotype, and developmental biology of molar tooth characters. *Yearbook of Physical Anthropology* 43:171–90.

Jernvall, J. & Wright, P. C. 1998. Diversity components of impending primate extinctions. *Proceedings of the National Academy of Sciences U.S.A.* 95:11279–83.

Johanson, D. C., Lovejoy, C. O., Kimbel, W. H. *et al.* 1982. Morphology of the Pliocene partial hominid skeleton (A.L. 288–1) from the Hadar Formation, Ethiopia. *American Journal of Physical Anthropology* 57:403–52.

Johnson, K. R. & Ellis, B. 2002. A tropical rainforest in Colorado 1.4 million years after the Cretaceous-Tertiary boundary. *Science* 296:2379–83.

Jolly, A. 1966. Lemur social behavior and primate intelligence. *Science* 153:501–6.

Jolly, C. J. 1970a. The large African monkeys as an adaptive array. In *Old World Monkeys – Evolution, Systematics, and Behavior*, Napier, J. & Napier, P., eds., pp. 227–62. New York, NY: Academic Press.

1970b. The seed-eaters: a new model of hominid differentiation based on a baboon analogy. *Man* 5:1–26.

1993. Species, subspecies, and baboon systematics. In *Species, Species Concepts, and Primate Evolution*, Kimbel, W. H. & Martin, L. B., eds., pp. 67–107. New York, NY: Plenum Press.

2001. A proper study for mankind: analogies from the papionin monkeys and their implications for human evolution. *Yearbook of Physical Anthropology* 44:17–204.

Jungers, W. L. 1985. Body size and scaling of limb proportions in primates. In *Size and Scaling in Primate Biology*, Jungers, W. L., ed., pp. 345–81. New York, NY: Plenum.

Kaiser, J. 2003. Ebola, hunting push ape populations to the brink. *Science* 300:232.

Kamil, A. C. & Jones, J. E. 1997. The seed-storing corvid Clark's nutcracker learns geometric relationships among landmarks. *Nature* 390:276–9.

Kangas, A. T., Evans, A. R., Thesleff, I., & Jernvall, J. 2004. Nonindependence of mammalian dental characters. *Nature* 432:211–14.

Kappeler, P. M. & Pereira, M. E., eds. 2003. *Primate Life Histories and Socioecology.* Chicago, IL: University of Chicago Press.

Karlsson, C. *et al.* 1997. Expression of functional leptin receptors in the human ovary. *Journal of Clinical Endocrinology and Metabolism* 82:4144–8.

Katz, M. E., Pak, D. K., Dickens, G. R., & Miller, K. G. 1999. The source and fate of massive carbon input during the latest Paleocene thermal maximum. *Science* 286:1531–3.

Katzmarzyk, P. T. & Leonard, W. R. 1998. Climatic influences on human body size and proportions: ecological adaptations and secular trends. *American Journal of Physical Anthropology* 106:483–503.

Kawai, M. 1965. Newly acquired pre-cultural behavior of the natural troop of Japanese monkeys on Koshima Island. *Primates* 6:1–30.

Kawamura, S. 1959. The process of sub-culture propagation among Japanese macaques. *Primates* 2:43–60.

Kay, R. F. & Couvert, H. H. 1984. Anatomy and behaviour of extinct primates. In *Food Acquisition and Processing in Primates*, Chivers, D. J., Wood, B. A., & Bilsborough, A., eds., pp. 467–508. New York, NY: Plenum.

Kay, R. F., Rossie, J. B., Colbert, M. W., & Rowe, T. B. 2004. Observations on the olfactory system of *Tremacebus harringtoni* (Platyrrhini, early Miocene, Sacanana, Argentina) based on high resolution X-ray CT scans. *American Journal of Physical Anthropology* 123 (Supplement 38):123–4.

Kearney, M. & Clark, J. M. 2003. Problems due to missing data in phylogenetic analyses including fossils: a critical review. *Journal of Vertebrate Paleontology* 23:263–74.

Keith, A. 1926. *The Engines of the Human Body*. Philadelphia, PA: Lippincott Company.

   1940. Fifty years ago. *American Journal of Physical Anthropology* 26:251–67.

   1950. *An Autobiography*. London: Watts.

Kenneally, C. 2003. AIBO as research tool. *Discover* 24 (3):46–53.

Kennett, J. P. 1995. A review of polar climatic evolution during the Neogene, based on the marine sediment record. In *Paleoclimate and Evolution with Emphasis on Human Origins*, Vrba, E. S., Denton, G. H., Partridge, T. C., & Burckle, L. H., eds., pp. 49–64. New Haven, CT: Yale University Press.

Kenward, B., Weir, A. A. S., Rutz, C., & Kacelnik, A. 2005. Tool manufacture by naïve juvenile crows. *Nature* 433:121.

Keverne, E., Martel, F. L., & Nevison, C. M. 1996. Primate brain evolution: genetic and functional considerations. *Proceedings of the Royal Society of London B* 262:689–96.

Keys, A. & the Laboratory of Physiological Hygiene, University of Minnesota 1950. *The Biology of Human Starvation*. Vol. I. Minneapolis, MN: University of Minnesota Press.

Keyser, A. W. 2000. The Drimolen skull: the most complete australopithecine cranium and mandible to date. *South African Journal of Science* 96:189–93.

Keyser, A. W., Menter, C. G., Moggi-Cecchi, J., Pickering, T. R., & Berger, L. R. 2000. Drimolen: A new hominid-bearing site in Gauteng, South Africa. *South African Journal of Science* 96:193–7.

Kibunjia, M. 1994. Pliocene archaeological occurrences in the Lake Turkana Basin. *Journal of Human Evolution* 27:159–71.

   2002. *Plio-Pleistocene Archaeological Sites in the West Turkana Basin*. Ph. D. dissertation, Department of Anthropology, Rutgers University, New Brunswick, NJ.

Kibunjia, M., Roche, H., Brown, F. H., & Leakey, R. E. 1992. Pliocene and Pleistocene archaeological sites west of Lake Turkana, Kenya. *Journal of Human Evolution* 23:431–8.

Kiers, E. T., Rousseau, R. A., West, S. A., & Denison, R. F. 2003. Host sanctions and the legume-rhizobium mutualism. *Nature* 425:78–81.

Kihlstrom, J. F. 1987. The cognitive unconscious. *Science* 237:1445–52.

Kim, S.-G., Ugurbil, K., & Strick, P. L. 1994. Activation of a cerebellar output nucleus during cognitive processing. *Science* 265:949–51.

Kimbel, W. H., Walter, R. C., Johanson, D. C. *et al.* 1996. Late Pliocene *Homo* and Oldowan tools from the Hadar Formation (Kada Hadar Member), Ethiopia. *Journal of Human Evolution* 31:549–61.

Kimura, M. 1983. *The Neutral Theory of Molecular Evolution*. Cambridge: Cambridge University Press.

Kindlmann, P., Dixon, A. F. G., & Dostalkova, I. 1999. Does body size optimization result in skewed body size distribution on a logarithmic scale? *American Naturalist* 153:445–7.

Kingdon, J. 1988. What are face patterns and do they contribute to reproductive isolation in guenons? In *A Primate Radiation: Evolutionary Biology of the African Guenons*, Gautier-Hion, A., Bourlière, F., Gautier, J. P., & Kingdon, J., eds., pp. 227–45. Cambridge: Cambridge University Press.

  1989. *Island Africa: The Evolution of Africa's Rare Animals and Plants*. Princeton, NJ: Princeton University Press.

  1997. *Kingdon Field Guide to African Mammals*. New York, NY: Academic Press.

  2003. *Lowly Origin: Where, When, and Why Our Ancestors First Stood Up*. Princeton, NJ: Princeton University Press.

Kingston, J. D., Marino, B. D., & Hill, A. 1994. Isotopic evidence for Neogene hominid paleoenvironments in the Kenya Rift Valley. *Science* 264:955–9.

Kinnison, M. T. & Hendry, A. P. 2001. The pace of modern life II: from rates of contemporary microevolution to pattern and process. *Genetica* 112–113: 145–64.

Kinzey, W. G., ed. 1987. *The Evolution of Human Behavior: Primate Models*. Albany, NY: State University of New York Press.

Kirk, E. C. 2004. Effects of activity pattern on eye and orbit morphology in primates. *American Journal of Physical Anthropology* 123(S38):126.

Kirk, E. C. *et al.* 2003. Comment on "Grasping primate origins". *Science* 300:741 [and online text].

Kleiber, M. 1961. *The Fire of Life. An Introduction to Animal Energetics*. New York, NY: Wiley.

Klein, R. G. 1989. Biological and behavioural perspectives on modern human origins in Southern Africa. In *The Human Revolution*, Mellars, P. & Stringer, C., eds., pp. 530–46. Edinburgh: Edinburgh University Press.

  1999. *The Human Career. Human Biological and Cultural Origins*, 2nd edn. Chicago, IL: University of Chicago Press.

Kmita, M., Fraudeau, N., Herault, Y., & Duboule, D. 2002. Serial deletions and duplications suggest a mechanism for the collinearity of *Hoxd* genes in limbs. *Nature* 420:145–50.

Knight, A., Underhill, P. A., Mortenson, H. M. *et al.* 2003. African Y chromosome and mtDNA divergence provides insight into the history of click languages. *Current Biology* 13:464–73.

Knowler, W. C., Pettitt, D. J., Bennett, P. H., & Williams, R. C. 1983. Diabetes mellitus in the Pima Indians: genetic and evolutionary considerations. *American Journal of Physical Anthropology* 62:107–14.

Köhler, M. & Moyà-Solà, S. 1997. Ape-like or hominid-like? The positional behavior of *Oreopithecus bambolii* reconsidered. *Proceedings of the National Academy of Sciences U.S.A.* 94:11747–50.

Kopelman, P. G. 2000. Obesity as a medical problem. *Nature* 404:635–43.

Kornack, D. R. & Rakic, P. 2001. Cell proliferation without neurogenesis in adult primate neocortex. *Science* 294:2127–30.

Kortlandt, A. 1972. *New Perspectives on Ape and Human Evolution*. Amsterdam: Stichting Voor Psychobiologie, University of Amsterdam.

1980. The Fayum primate forest: did it exist? *Journal of Human Evolution* 9:277–97.

Kozlowski, J. & Weiner, J. 1997. Interspecific allometries are by-products of body size optimization. *American Naturalist* 149:352–80.

Krebs, J. R., Clayton, N. S., Healy, S. D., Cristol, D. A., Patel, S. N., & Jolliffe, A. R. 1996. The ecology of the brain: food-storing and the hippocampus. *Ibis* 138:34–46.

Krijgsman, W., Hilgen, F. J., Raffi, I., Sierro, F. J., & Wilson, D. S. 1999. Chronology, causes and progression of the Messinian salinity crisis. *Nature* 400:652–5.

Křivan, V. & Sirot, E. 2002. Habitat selection by two competing species in a two-habitat environment. *American Naturalist* 160:214–34.

Kühmer, W. 1965. Communal food distribution and division of labour in African hunting dogs (*Lycaon pictus lupinus*). *Nature* 205:443–4.

1966. Freilanstudien zur soziologie des Hyänenhundes (*Lycaon pictus lupinus* Thomas 1902). *Zeitung fur Tierpsychologie* 22:495–541.

Kuhn, S. L. & Stiner, M. C. 2001. The antiquity of hunter-gatherers. In *Hunter-Gatherers: An Interdisciplinary Perspective*, Panter-Brick, C., Layton, R. H., & Rowley-Conwy, P., eds., pp. 99–142. Cambridge: Cambridge University Press.

Kummer, H. 1968. *Social Organization of Hamadryas Baboons: A Field Study*. Chicago, IL: University of Chicago Press.

1971. *Primate Societies: Group Techniques of Ecological Adaptation*. Chicago, IL: Aldine-Atherton.

1995. *In Quest of the Sacred Baboon. A Scientist's Journey*. Princeton, NJ: Princeton University Press.

2002. Topics gained and lost in primate social behavior. *Evolutionary Anthropology* 11 (supplement 1):73–4.

Kummer, H., Dasser, V., & Hoyningen-Huene, P. 1990. Exploring primate social cognition: some critical remarks. *Behaviour* 112:84–98.

Kuper, A. 1988. *The Invention of Primitive Society: Transformations of an Illusion*. New York, NY: Routledge.

Kurita, H., Shiniomura, T., & Fujita, T. 2002. Temporal variation in Japanese macaque body mass. *International Journal of Primatology* 23:411–28.

Kuzawa, C. W. 1998. Adipose tissue in human infancy and childhood: an evolutionary perspective. *Yearbook of Physical Anthropology* 41:177–209.

Lacy, R. C. & Sherman, P. W. 1983. Kin recognition by phenotype matching. *American Naturalist* 121:489–512.

Lande, R. 1993. Risks of population extinction from demographic and environmental stochasticity and random catastrophes. *American Naturalist* 142:911–27.

Lane, M. A., Baer, D. J., Rampler, W. V. *et al.* 1996. Calorie restriction lowers body temperature in rhesus monkeys, consistent with a postulated anti-aging mechanism in rodents. *Proceedings of the National Academy of Sciences U.S.A.* 93:4159–64.

Lauder, G. V. 1995. On the inference of function from structure. In *Functional Morphology in Vertebrate Paleontology*, Thomason, J. J., ed., pp. 1–18. Cambridge: Cambridge University Press.

Laughlin, W. S. 1968. Hunting: an integrating biobehavior system and its evolutionary importance. In *Man the Hunter*, Lee, R. B. & DeVore, I., eds., pp. 304–20. Chicago, IL: Aldine Publishing Company.

Leakey, L. S. B. 1959. A new fossil skull from Olduvai. *Nature* 184:491–3.

 1960. *Adam's Ancestors. The Evolution of Man and His Culture*, 5th edn. New York, NY: Harper Torchbooks.

Leakey, L. S. B., Evernden, J. F., & Curtis, G. H. 1961. Age of Bed I, Olduvai Gorge, Tanganyika. *Nature* 191:478–9.

Leakey, L. S. B., Tobias, P. V., & Napier, J. R. 1964. A new species of the genus *Homo* from Olduvai Gorge. *Nature* 202:7–9.

Leakey, M. D. 1971. *Olduvai Gorge. Excavations in Beds I and II, 1960–1963*. Cambridge: Cambridge University Press.

Leakey, M. D. & Hay, R. L. 1979. Pliocene footprints in the Laetolil Beds at Laetoli, Northern Tanzania. *Nature* 278:317–23.

Leakey, M. G., Feibel, C. S., McDougall, J., & Walker, A. 1995. New four-million-year-old hominid species from Kanapoi and Alia Bay, Kenya. *Nature* 376:565–71.

Leakey, R. E. F. 1973a. Australopithecines and hominines: a summary on the evidence from the Early Pleistocene of Eastern Africa. *Symposia of the Zoological Society of London* 33:53–69.

 1973b. Evidence for an advanced Plio-Pleistocene hominid from East Rudolf, Kenya. *Nature* 242:447–50.

Lebovitz, R. M. 2002. Only vital need justifies primate experiments. *Nature* 418:273.

Lee, P. C. & Hauser, M. D. 1998. Long-term consequences of changes in territory quality on feeding and reproductive strategies of vervet monkeys. *Journal of Animal Ecology* 67:347–58.

Lee, R. B. & DeVore, I., eds. 1968. *Man the Hunter*. Chicago, IL: Aldine Publishing Company.

Lee-Thorp, J. A., Thackeray, J. F., & van der Merwe, N. 2000. The hunters and the hunted revisited. *Journal of Human Evolution* 39:565–76.

Lee-Thorp, J. A., Sponheimer, M., & Van der Merwe, N. H. 2003. What do stable isotopes tell us about hominid dietary and ecological niches in the Pliocene? *International Journal of Osteoarchaeology* 13:104–13.

Legendre, S. 1986. Analysis of mammalian communities from the late Eocene and Oligocene of southern France. *Palaeovertebrata* 16(4):191–212.

Le Gros Clark, W. E. 1947. Observations on the anatomy of the fossil Australopithecinae. *Journal of Anatomy* 81:300–33.

1950. Hominid characters of the australopithecine dentition. *Journal of the Royal Anthropological Institute* 80:37–54.

1967. *Man-Apes or Ape-Men? The Story of Discoveries in Africa.* New York, NY: Holt, Rinehart and Winston.

1971. *The Antecedents of Man. An Introduction to the Evolution of The Primates,* 3rd edn. Chicago, IL: Quadrangle Books.

Leigh, Jr., E. G., Rand, A. S., & Windsor, D. M., eds. 1982. *The Ecology of a Tropical Forest: Seasonal Rhythms and Long-Term Changes.* Washington, DC: Smithsonian Institution Press.

Leigh, Jr., E. G. 1999. *Tropical Forest Ecology: A View from Barro Colorado Island.* New York, NY: Oxford University Press.

Leinders-Zufall, T., Lane, A. P., Puche, A. C. *et al.* 2000. Ultrasensitive pheromone detection by mammalian vomeronasal neurons. *Nature* 405:792–6.

Lenski, R. E., Ofria, C., Pennock, R. T., & Adami, C. 2003. The evolutionary origin of complex features. *Nature* 423:139–45.

Leonard, W. R. & Robertson, M. L. 1997. Comparative primate energetics and hominid evolution. *American Journal of Physical Anthropology* 102:265–81.

2001. Locomotor economy and the origin of bipedality: Reply to Steudel-Numbers. *American Journal of Physical Anthropology* 116:174–6.

Leroy, E. M., Rouquet, P., Formenty, P. *et al.* 2004. Multiple Ebola virus transmission events and rapid decline of Central African wildlife. *Science* 303:387–90.

Lewontin, R. C. 1974. *The Genetic Basis of Evolutionary Change.* New York, NY: Columbia University Press.

1982. *Human Diversity.* New York, NY: W. H. Freeman.

2000. *It Ain't Necessarily So: The Dream of the Human Genome and Other Illusions.* New York, NY: New York Review of Books.

Lewontin, R. C., Rose, S., & Kamin, L. J. 1984. *Not In Our Genes: Biology, Ideology, and Human Nature.* New York, NY: Pantheon Books.

Lieberman, P. 1975. *On the Origins of Language. An Introduction to the Evolution of Human Speech.* New York, NY: Macmillan.

Liem, K. F. 1989. Milton Hildebrand: architect of the re-birth of vertebrate morphology. *American Zoologist* 29:191–4.

Liman, E. R. & Innan, H. 2003. Relaxed selection pressure on an essential component of pheromone transduction in primate evolution. *Proceedings of the National Academy of Sciences U.S.A.* 100:3328–32.

Lister, A. M. 1996. Dwarfing in island elephants and deer: processes in relation to time of isolation. *Symposia of the Zoological Society of London* 69:277–92.

Little, M. A. 1989. Human biology of African pastoralists. *Yearbook of Physical Anthropology* 32:215–47.

Lockwood, C. A. *et al.* 2004. Morphometrics and hominoid phylogeny: support for a chimpanzee-human clade and differentiation among great ape subspecies. *Proceedings of the National Academy of Sciences U.S.A.* 101:4356–60.

Lonsdorf, E. V., Eberly, L. E., & Pusey, A. E. 2004. Sex differences in learning in chimpanzees. *Nature* 428:715.

Lorenz, K. 1966. *On Aggression*. New York, NY: Harcourt, Brace and World.

Lotan, E. 2000. Feeding the scavengers. Actualistic taphonomy in the Jordan Valley, Israel. *International Journal of Osteoarchaeology* 10:407–25.

Lovejoy, A. O. 1936. *The Great Chain of Being. A Study of the History of an Idea*. New York, NY: Harper & Row. [reprinted in 1960].

Lovejoy, C. O. 1981. The origin of man. *Science* 211:341–50.

1988. Evolution of human walking. *Scientific American* 259:118–25.

Lovejoy, C. O. *et al.* 1999. Morphological analysis of the mammalian postcranium: a developmental perspective. *Proceedings of the National Academy of Sciences U.S.A.* 96:13247–52.

Lovell, N. C. 1990. *Patterns of Injury and Illness in Great Apes. A Skeletal Analysis*. Washington, DC: Smithsonian Institution Press.

Lucas, P. W. 2004. *Dental Functional Morphology: How Teeth Work*. Cambridge: Cambridge University Press.

Lummaa, V., Haukioja, E., Lemmetyinen, R., & Pikkola, M. 1998. Natural selection on human twinning. *Nature* 394:533–4.

Luo, M., Fee, M. S., & Katz, L. C. 2003. Encoding pheromonal signals in the accessory olfactory bulb of behaving mice. *Science* 299:1196–201.

Luo, Z.-X., Crompton, A. W., & Sun, A. L. 2001. A new mammaliaform from the Early Jurassic and evolution of mammalian characteristics. *Science* 292:1535–40.

Lutz, C., Well, A., & Novak, M. 2003. Stereotypic and self-injurious behavior in rhesus macaques: a survey and retrospective analysis of environment and early experience. *American Journal of Primatology* 60:1–15.

Lyman, R. L. 1994. *Vertebrate Taphonomy*. Cambridge: Cambridge University Press.

Mac Arthur, R. H. & Levins, R. 1967. The limiting similarity, convergence and divergence of coexisting species. *American Naturalist* 101:377–85.

Mac Arthur, R. H. & Wilson, E. O. 1967. *The Theory of Island Biogeography*. Princeton, NJ: Princeton University Press.

Macdonald, D. 1984. *The Encyclopedia of Mammals*. New York, NY: Facts on File Publications.

Mace, G. M., Gittleman, J. L., & Purvis, A. 2003. Preserving the tree of life. *Science* 300:1707–9.

MacKenzie, D. 1999. New clues as to why size equals destiny. *Science* 284:1607–8.

MacKinnon, J. 1974a. The behaviour and ecology of wild orang-utans (*Pongo pygmaeus*). *Animal Behavior* 22:3–74.

1974b. *In Search of the Red Ape*. New York, NY: Holt, Rinehart and Winston.

MacKinnon, J. R. & MacKinnon, K. S. 1980. Niche differentiation in a primate community. In *Malayan Forest Primates*, Chivers, D. J., ed., pp. 167–90. New York, NY: Plenum Press.

MacLarnon, A. 1993. The vertebral canal. In *The Nariokotome* Homo erectus *Skeleton*, Walker, A. & Leakey, R., eds., pp. 359–90. Cambridge, MA: Harvard University Press.

MacLatchy, L., Gebo, D., Kityo, R., & Pilbeam, D. 2000. Postcranial functional morphology of *Morotopithecus bishopi*, with implications for the evolution of modern ape locomotion. *Journal of Human Evolution* 39:159–83.

MacPhee, R. D. E., ed. 1993. *Primates and Their Relatives in Phylogenetic Perspective*. New York, NY: Plenum.

MacPhee, R. D. E. & Iturralde-Vinent, M. A. 1995. Earliest monkey from Greater Antilles. *Journal of Human Evolution* 28:197–200.

MacPhee, R. D. E., Cartmill, M., & Gingerich, P. D. 1983. New Paleogene primate basicrania and the definition of the Order Primates. *Nature* 301:509–11.

Maggioncalda, A. N., Sapolsky, R. M., & Czekala, N. M. 1999. Reproductive hormone profiles in captive male orangutans: implications for understanding developmental arrest. *American Journal of Physical Anthropology* 109:19–32.

Maggioncalda, A. N., Czekala, N. M., & Sapolsky, R. M. 2002. Male orangutan subadulthood: a new twist on the relationship between chronic stress and developmental arrest. *American Journal of Physical Anthropology* 118:25–32.

Maiorana, V. C. 1978. An explanation of ecological and developmental constants. *Nature* 273:375–7.

  1990. Evolutionary strategies and body size in a guild of mammals. In *Body Size in Mammalian Paleobiology*, Damuth, J., MacFadden, B. J., eds., pp. 69–102. Cambridge: Cambridge University Press.

Marivaux, L., Chaimanee, Y., Ducrocq, S. *et al.* 2003. The anthropoid status of a primate from the late middle Eocene Pondaung Formation (Central Myanmar): tarsal evidence. *Proceedings of the National Academy of Sciences U.S.A.* 100:13173–8.

Markowitz, H. & Stevens, V. J., eds. 1978. *Behavior of Captive Wild Animals*. Chicago, IL: Nelson-Hall.

Marlowe, F. W. 2004. Body size and fat predict fertility and reproductive success among Hadza hunter-gatherers. *American Journal of Physical Anthropology* 123(S38):142.

Marroig, G. & Cheverud, J. M. 2004. Did natural selection or genetic drift produce the cranial diversification of Neotropical monkeys? *American Naturalist* 163:417–28.

Marsh, C. W. & Mittermeier, R. A., eds. 1987. *Primate Conservation in the Tropical Rain Forest*. New York, NY: A. R. Liss.

Martin, G. T. 1991. *The Hidden Tombs of Memphis. New Discoveries from the Time of Tutankhamun and Ramesses the Great*. London: Thames and Hudson.

Martin, L. B. & Andrews, P. 1993. Species recognition in Middle Miocene hominoids. In *Species, Species Concepts and Primate Evolution*, Kimbel, W. H. & Martin, L. B., eds., pp. 393–427. New York, NY: Plenum.

Martin, R. D. 1983. *Human Brain Evolution in an Ecological Context*. Fifty-Second James Arthur Lecture on the Evolution of the Human Brain. American Museum of Natural History, New York.

  1986a. Are fruit bats primates? *Nature* 320:482–3.

  1986b. Primates: A definition. In *Major Topics in Primate and Human Evolution*, Wood, B., Martin, L. & Andrews, P., eds., pp. 1–31. Cambridge: Cambridge University Press.

  1990. *Primate Origins and Evolution. A Phylogenetic Reconstruction*. Princeton, NJ: Princeton University Press.

Martin, R. E. 1999. *Taphonomy: A Process Approach*. Cambridge: Cambridge University Press.

Maslin, T. P. 1952. Morphological criteria of phylogenetic relationships. *Systematic Zoology* 1:49–70.

Masuzaki, H., Paterson, J., Shinyama, H. *et al.* 2001. A transgenic model of visceral obesity and the metabolic syndrome. *Science* 294:2166–70.

Matano, S. & Hirasaki, E. 1997. Volumetric comparisons in the cerebellar complex of anthropoids, with special reference to locomotor types. *American Journal of Physical Anthropology* 103:173–83.

Matano, S., Baron, G., Stephan, H., & Frahm, H. 1985. Volume comparisons in the cerebellar complex of primates. II. Cerebellar nuclei. *Folia primatologica* 44:182–203.

Matthew, W. D. 1914. Climate and evolution. *Annals of the New York Academy of Science* 24:171–318.

Maynard Smith, J. 1982. *Evolution and the Theory of Games*. Cambridge: Cambridge University Press.

1992. *Did Darwin Get It Right?* New York, NY: Chapman and Hall.

Maynard Smith, J. & Brown, R. L. W. 1986. Competition and body size. *Theoretical Population Biology* 30:166–79.

Mayr, E. 1950. Taxonomic categories in fossil hominids. *Cold Spring Harbor Symposia on Quantitative Biology* 15:109–18.

1963. *Animal Species and Evolution*. Cambridge, MA: Harvard University Press.

1982. *The Growth of Biological Thought. Diversity, Evolution, and Inheritance*. Cambridge, MA: Belknap Press.

1983. How to carry out the adaptationist program? *American Naturalist* 121:324–34.

McCollum, M. A. 1999. The robust australopithecine face: a morphogenetic perspective. *Science* 284:301–5.

McCrossin, M. L. & Benefit, B. R. 1993. Recently recovered *Kenyapithecus* mandible and its implications for great ape and human origins. *Proceedings of the National Academy of Sciences U.S.A.* 90:1962–6.

McFarland, D. & Bösser, T. 1993. *Intelligent Behavior in Animals and Robots*. Cambridge, MA: M. I. T. Press.

McGhee, G. R., Jr. 1999. *Theoretical Morphology*. New York, NY: Columbia University Press.

McGrew, W. C. 1992. *Chimpanzee Material Culture*. Cambridge: Cambridge University Press.

McHenry, H. M. 1974. How large were the australopithecines? *American Journal of Physical Anthropology* 40:329–40.

1992. Body size and proportions in early hominids. *American Journal of Physical Anthropology* 87:407–31.

1994. Behavioral ecological implications of early hominid body size. *Journal of Human Evolution* 27:77–87.

McKenna, M. C. 1980. Eocene paleolatitude, climate, and mammals of Ellesmere Island. *Palaeogeography, Palaeoclimatology, Palaeoecology* 30:349–62.

McKenna, M. C. & Bell, S. K. 1997. *Classification of Mammals Above the Species Level*. New York, NY: Columbia University Press.

McNab, B. K. 1980. Food habits, energetics and the population biology of mammals. *American Naturalist* 116:106–24.

Mendoza, S. P., Reeder, D. M., & Mason, W. A. 2002. Nature of proximate mechanisms underlying primate social systems: Simplicity and redundancy. *Evolutionary Anthropology* 11 (Supplement 1):112–16.

Menzel, C. R. & Beck, B. A. 2000. Homing and detour behavior in golden lion tamarin social groups. In *On the Move. How and Why Animals Travel in Groups*, Boinski, S. & Garber, P. A., eds., pp. 299–326. Chicago, IL: University of Chicago Press.

Mercader, J., Panger, M., & Boesch, C. 2002. Excavation of a chimpanzee stone tool site in the African rainforest. *Science* 296:1452–5.

Merrick, H. V. & Merrick, J. P. S. 1976. Archaeological occurrences of earlier Pleistocene age from the Shungura Formation. In *Earliest Man and Environments in the Lake Rudolf Basin: Stratigraphy, Paleoecology, and Evolution*, Coppens, Y., Howell, F. C., Isaac, G. L., & Leakey, R. E. F., eds., pp. 574–84. Chicago, IL: University of Chicago Press.

Merritt, S. 2000. *A Quantitative Analysis of Cut Marks Experimentally Produced by Large Bifacial Tools and Small Flakes*. Henry Rutgers undergraduate research thesis, Department of Anthropology, Rutgers University, New Brunswick, NJ.

Mervis, C. B. & Bertrand, J. 1997. Developmental relations between cognition and language: evidence from Williams syndrome. In *Communication and Language Acquisition: Discoveries from Atypical Development*, Adamson, L. B. & Romski, M. A., eds. Baltimore, MD: Brooks.

Meyer, J. M. & Stunkard, A. J. 1994. Twin studies of human obesity. In *The Genetics of Obesity*, Bouchard, C., ed., pp. 63–78. Boca Raton, FL: CRC Press.

Middleton, F. A. & Strick, P. L. 1994. Anatomical evidence for cerebellar and basal ganglia involvement in higher cognitive function. *Science* 266:458–61.

Mifflin, M. D., St. Jeor, S. T., Hill, L. A., Scott, B. J., Daugherty, S. A., & Koh, Y. O. 1990. A new predictive equation for resting energy expenditure in healthy individuals. *American Journal of Clinical Nutrition* 51:241–7.

Miller, A. I. & Foote, M. 2003. Increased longevities of post-Paleozoic marine genera after mass extinctions. *Science* 302:1030–2.

Miller, G. 2003. Hungry ewes deliver offspring early. *Science* 300:561–2.

Miller, G. H., Magee, J. W., Johnson, B. J. *et al.* 1999. Pleistocene extinction of *Genyornis newtoni*: human impact on Australian megafauna. *Science* 283:205–8.

Miller, K. G., Sugerman, P. J., Browning, J. V., Kominz, M. A., Olsson, R. K., Feigenson, M. D., & Hernández. 2004. Upper Cretaceous sequences and sea-level history, New Jersey Coastal Plain. *GSA Bulletin* 116:368–93.

Miller, L. E., ed. 2002. *Eat or be Eaten. Predator Sensitive Foraging Among Primates*. Cambridge: Cambridge University Press.

Milton, K. 1980. *The Foraging Strategy of Howler Monkeys. A Study in Primate Economics*. New York, NY: Columbia University Press.

1981. Distribution patterns of tropical plant foods as an evolutionary stimulus to primate mental development. *American Anthropologist* 83:554–8.

Minelli, A. 2003. *The Development of Animal Form. Ontogeny, Morphology, and Evolution*. Cambridge: Cambridge University Press.

Mitchell, C. E. & Power, A. G. 2003. Release of invasive plants from fungal and viral pathogens. *Nature* 421:625–7.

Mithen, S. 1996. *The Prehistory of the Mind. The Cognitive Origins of Art, Religion and Science.* New York, NY: Thames and Hudson.

Moffett, M. W. 1993. *The High Frontier: Exploring the Tropical Rainforest Canopy.* Cambridge, MA: Harvard University Press.

Moore, J. A. 1993. *Science as a Way of Knowing.* Cambridge, MA: Harvard University Press.

Moore, S. L. & Wilson, K. 2002. Parasites as a viability cost of sexual selection in natural populations of mammals. *Science* 297:2015–18.

Morden. J. 1991. *Hominid Taphonomy: Density, Fluvial Transport, and Carnivore Consumption of Human Remains with Application to Three Plio/Pleistocene Hominid Sites.* Ph. D. dissertation, Department of Anthropology, Rutgers University, New Brunswick, NJ.

Morgan, M. E., Kingston, J. D., & Marino, B. D. 1994. Carbon isotopic evidence for the emergence of C4 plants in the Neogene from Pakistan and Kenya. *Nature* 367:162–5.

Morwood, M. J., Soejono, R. P., Roberts, R. G. *et al.* 2004. Archaeology and age of a new hominin from Flores in eastern Indonesia. *Nature* 431:1087–91.

Morris, R. & Morris, D. 1968. *Men and Apes.* New York, NY: Bantam Books.

Morton, D. J. 1924. Evolution of the human foot. II. *American Journal of Physical Anthropology* 7:1–52.

   1927. Human origin. Correlation of previous studies of primate feet and posture with other morphologic evidence. *American Journal of Physical Anthropology* 10:173–203.

Moss, M. L. 1971. Ontogenetic aspects of cranio-facial growth. In *Cranio-facial Growth in Man*, Moyers, R. E. & Krogman, W. M., eds., pp. 109–24. Oxford: Pergamon Press.

Moss, M. L. & Young, R. W. 1960. A functional approach to craniology. *American Journal of Physical Anthropology* 18:281–92.

Moura, A. C. de A. & Lee, P. C. 2004. Capuchin stone tool use in Caatinga dry forest. *Science* 306:1909.

Mowbray, K. M. 2002. *Growth of the Occipital in Human Evolution: Understanding the Developmental Reasons for Morphological Variability through Experimental Models and Surface Bone Histology.* Ph. D. dissertation, Department of Anthropology, Rutgers University, New Brunswick, N. J.

Moyà-Solà, S. & Köhler, M. 1996. The first *Dryopithecus* skeleton: origins of great ape locomotion. *Nature* 379:156–9.

Moyà-Solà, S., Köhler, M., Alba, D. M., Casanovas-Vilar, I., & Galindo, J. 2004. *Pierolapithecus catalaunicus*, a new middle Miocene great ape from Spain. *Science* 306:1339–44.

Muragaki,Y., Mundlos, S., Upton, J., & Olsen, B. R. 1996. Altered growth and branching patterns in synpolydactyly caused by mutations in Hoxd13. *Science* 272:548–51.

Musiba, C. M., Tuttle, R. H., Hallgrimsson, B., & Webb, D. M. 1997. Swift and sure-footed on the savanna: a study of Hadzabe gaits and feet in northern Tanzania. *American Journal of Human Biology* 9:303–21.

Myers, N. & Knoll, A. H. 2001. The biotic crisis and the future of evolution. *Proceedings of the National Academy of Sciences U. S. A.* 98:5389–92.

Nagano, A., Umberger, B. R., Marzke, M. W., & Gerritsen, K. G. M. 2005. Neuromusculoskeletal computer modeling and simulation of upright, straight-legged, bipedal locomotion of *Australopithecus afarensis* (A.L. 288–1). *American Journal of Physical Anthropology* 126:2–13.

Nakatsukasa, M. 2004. Acquisition of bipedalism: the Miocene hominoid record and modern analogues for bipedal protohominids. *Journal of Anatomy* 204:385–402.

Napier, J. R. & Napier, P. H. 1967. *A Handbook of Living Primates*. London: Academic Press.

1985. *The Natural History of the Primates*. Cambridge, MA: MIT Press.

Napier, J. R. & Walker, A. C. 1967. Vertical clinging and leaping: a newly recognized category of locomotor behavior in primates. *Folia primatologica* 6:180–203.

Naples, V. L. 1995. The artificial generation of wear patterns on tooth models as a means to infer mandibular movement during feeding in mammals. In *Functional Morphology in Vertebrate Paleontology*, Thomason, J. J., ed., pp. 136–50. Cambridge: Cambridge University Press.

Naughton-Treves, L., Treves, A., Chapman, C., & Wrangham, R. 1998. Temporal patterns of crop raiding by primates: linking food availability in croplands and adjacent forest. *Journal of Applied Ecology* 35:596–606.

Navarro, A. & Barton, N. H. 2003. Chromosomal speciation and molecular divergence – accelerated evolution in rearranged chromosomes. *Science* 300:321–4.

Neel, J. V. 1962. Diabetes mellitus: a "thrifty" genotype rendered detrimental by "progress?" *American Journal of Human Genetics* 14:353–62.

1982. The "thrifty" genotype revisited. In *The Genetics of Diabetes Mellitus*, Kobberling, J. & Tattersall, J., eds., pp. 49–60. New York, NY: Academic Press.

Nelson, R. 1997. *Heart and Blood: Living with Deer in America*. New York, NY: Knopf.

Nelson, S. V. 2003. *The Extinction of* Sivapithecus: *Faunal and Environmental Changes Surrounding the Disappearance of a Miocene Hominoid in the Siwaliks of Pakistan*. Boston, MA: Brill Academic Publishers, Inc.

Newmark, W. D. 1996. Insularization of Tanzanian parks and the local extinction of large mammals. *Conservation Biology* 10:1549–56.

Ni, X., Wang, Y., Hu, Y. & Li, C. 2004. A euprimate skull from the early Eocene of China. *Nature* 427:65–8.

Nishida, T., Corp, N., Hamai, M. *et al.* 2003. Demography, female life history, and reproductive profiles among the chimpanzees of Mahale. *American Journal of Primatology* 59:99–121.

Norell, M. A. & Wheeler, W. C. 2003. Missing entry replacement data analysis: a replacement approach to dealing with missing data in paleontological and total evidence data sets. *Journal of Vertebrate Paleontology* 23:275–83.

Novak, M. A. & Petto, A. J. 1991. *Through the Looking Glass: Issues of Psychological Well-Being in Captive Nonhuman Primates*. Washington, DC: American Psychological Association.

Nowak, M. A. & Sigmund, K. 2004. Evolutionary dynamics of biological games. *Science* 303:793–9.

Nowak, M. A., Sasaki, A., Taylor, C., & Fudenberg, D. 2004. Emergence of cooperation and evolutionary stability in finite populations. *Nature* 428:646–50.

Nowak, R. M. 1999. *Walker's Mammals of the World*, 2 vols., 6th edn. Baltimore, MD: Johns Hopkins University Press.

Nunn, C. L. & Barton, R. A. 2000. Allometric slopes and independent contrasts: a comparative test of Kleiber's law in primate ranging patterns. *American Naturalist* 156:519–33.

Nunn, C. L., Altizer, S., Jones, K.E., & Sechrest, W. 2003. Comparative tests of parasite species richness in primates. *American Naturalist* 162:597–614.

Nunn, C. L., Altizer, S., Sechrest, W., Jones, K. E., Barton, R. A., & Gittleman, J. L. 2004. Parasites and the evolutionary diversification of primate clades. *American Naturalist* 164 (Supplement):S90–103.

Oakley, K. 1957. Tools makyth man. *Antiquity* 31:199–209.

Ober, C., Hyslop, T., Elias, S., Weitkamp, L. R., & Hauck, W. W. 1998. Human leucocyte antigen matching and fetal loss: a result of a 10-year prospective study. *Human Reproduction* 13:33–8.

O'Connell, J. F., Hawkes, K., & Jones, N. G. B. 1999. Grandmothering and the evolution of *Homo erectus*. *Journal of Human Evolution* 36:461–85.

Ohnuki-Tierney, E. 1995. Representations of the monkey (*saru*) in Japanese culture. In *Ape, Man, Apeman: Changing Views Since 1600*, Corbey, R. & Theunissen, B., eds., pp. 297–308. Leiden, the Netherlands: Leiden University Press.

Olsen, E. M., Heino, M., Lilly, G. R., Morgan, M. J., Brattey, J., Ernande, B., & Dieckmann, U. 2004. Maturation trends indicative of rapid evolution preceded the collapse of northern cod. *Nature* 428:932–5.

Olson, E. C. 1980. Taphonomy: its history and role in community evolution. In *Fossils in The Making*, Behrensmeyer, A. K. & Hill, A. P., eds., pp. 5–19. Chicago, IL: University of Chicago Press.

Olson, E. & Miller, R. 1958. *Morphological Integration*. Chicago, IL: University of Chicago Press.

Osborn, H. F. 1924. *Men of the Old Stone Age. Their Environment, Life and Art*, 3rd edn. New York, NY: Charles Scribner's Sons.

1927. *Man Rises to Parnassus. Critical Epochs in the Prehistory of Man*. Princeton, NJ: Princeton University Press.

1934. Aristogenesis, the creative principle in the origin of species. *American Naturalist* 68:193–235.

Ostrom, J. J. 1969. Osteology of *Deinonychus antirrhopus*, an unusual theropod from the Lower Cretaceous of Montana. *Peabody Museum of Natural History Bulletin* 30:1–165.

O'Sullivan, P. B., Morwood, M., Hobbs, D. *et al.* 2001. Archaeological implications of the geology and chronology of the Soa basin, Flores, Indonesia. *Geology* 29:607–10.

Otte, D. & Endler, J. A., eds. 1989. *Speciation and Its Consequences*. Sunderland, MA: Sinauer Associates.

Owen-Smith, N. 1998. *Megaherbivores: The Influence of Very Large Body Size on Ecology*. Cambridge: Cambridge University Press.

Owens, I. P. F. 2002. Sex differences in mortality rate. *Science* 297:2008–9.

Oxnard, C. E. 1969. Evolution of the human shoulder: some possible pathways. *American Journal of Physical Anthropology* 30:319–32.

1975. *Uniqueness and Diversity in Human Evolution. Morphometric Studies of Australopithecines*. Chicago, IL: University of Chicago Press.

1984. *The Order of Man. A Biomathematical Anatomy of the Primates*. New Haven, CI: Yale University Press.

2004. Brain evolution: mammals, primates, chimpanzees, and humans. *International Journal of Primatology* 25:1127–58.

Ozanne, C. M. P., Anhuf, D., Boulter, S. L. *et al.* 2003. Biodiversity meets the atmosphere: a global view of forest canopies. *Science* 301:183–6.

Packer, C., Pusey, A. E., & Eberly, L. E. 2001. Egalitarianism in female African lions. *Science* 293:690–3.

Padian, K. 1995. Form versus function: the evolution of a dialectic. In *Functional Morphology in Vertebrate Paleontology*, Thomason, J. J., ed., pp. 264–77. Cambridge: Cambridge University Press.

Palombit, R. A. 1999. Infanticide and the evolution of pair bonds in nonhuman primates. *Evolutionary Anthropology* 7:117–28.

Palombit, R. A., Seyfarth, R. M., & Cheney, D. L. 1997. The adaptive value of 'friendships' to female baboons: experimental and observational evidence. *Animal Behaviour* 54:599–614.

Palombit, R. A., Cheney, D. L., & Seyfarth, R. M. 2001. Female-female competition for male 'friends' in wild chacma baboons, *Papio cynocephalus ursinus*. *Animal Behaviour* 61:1159–71.

Panger, M. A., Brooks, A. S., Richmond, B. G., & Wood, B. 2002. Older than the Oldowan? Rethinking the emergence of hominin tool use. *Evolutionary Anthropology* 11:235–45.

Papageorgiou, S. 2004. A cluster translocation model may explain the colinearity of *Hox* gene expressions. *BioEssays* 26:189–95.

Parker, S. T. & Gibson, K. R. 1977. Object manipulation, tool use and sensorimotor intelligence as feeding adaptations in *Cebus* monkeys and great apes. *Journal of Human Evolution* 6:623–41.

Partridge, T. C., Bond, G. C., Hartnady, C. J. H., de Menocal, P. B., & Ruddiman, W. F. 1995. Climatic effects of late Neogene tectonism and volcanism. In *Paleoclimate and Evolution with Emphasis on Human Origins*, Vrba, E. S., Denton, G. H., Partridge, T. C., & Burckle, L. H., eds., pp. 8–23. New Haven, CT: Yale University Press.

Partridge, T. C., Granger, D. E., Caffee, M. W., & Clarke, R. J. 2003. Lower Pliocene hominid remains from Sterkfontein. *Science* 300:607–12.

Passarino, G., Semino, O., Quintana-Murci, L., Excoffier, L., Hummer, M., & Santachiara-Benerecetti, A. S. 1998. Different genetic components in the Ethiopian populations, identified by mtDNA and Y-chromosome polymorphisms. *American Journal of Human Genetics* 62:420–34.

Paterson, H. E. H. 1993. *Evolution and the Recognition Concept of Species: Collected Writings*. Baltimore, MD: Johns Hopkins University Press.

Patricelli, G. L., Jac, U., Walsh, G., & Borgia, G. 2002. Sexual selection: male displays adjusted to female's response. *Nature* 415:279–80.

Patterson, B. 1949. Rates of evolution in taeniodonts. In *Genetics, Paleontology, and Evolution*, Jepsen, G. L. *et al.*, eds., pp. 243–78. Princeton, NJ: Princeton University Press.

Pearson, H. 2002. Dual identities. *Nature* 417:10–11.

Pearson, O. M. 2000. Postcranial remains and the origin of modern humans. *Evolutionary Anthropology* 9(6):229–47.

Penniman, T. K. 1965. *A Hundred Years of Anthropology*. 3rd edn. New York, NY: International Universities Press.

Penny, D. & Phillips, M. J. 2004. The rise of birds and mammals: are microevolutionary processes sufficient for macroevolution? *Trends in Ecology and Evolution* 19:516–22.

Perkins, S. 2003. Learning from the present. Fresh bones could provide insight into earth's patchy fossil record. *Science News* 164:42–4.

Perry, D. 1986. *Life Above the Forest Floor*. New York, NY: Simon & Schuster.

Perry, S. & Manson, J. H. 2003. Traditions in monkeys. *Evolutionary Anthropology* 12:71–81.

Perry, S., Manson, J. H., Dower, G., & Wikberg, E. 2003. White-faced capuchins cooperate to rescue a groupmate from a *Boa constrictor*. *Folia Primatologica* 74:109–11.

Peters, R. H. 1983. *The Ecological Implications of Body Size*. Cambridge: Cambridge University Press.

Petersen, D. 2003. *Eating Apes*. Berkeley, CA: University of California Press.

Pettigrew, J. D. 1986. Flying primates? Megabats have the advanced pathway from eye to midbrain. *Science* 231:1304–6.

Pfeiffer, S. 2001. Patterns of skeletal trauma during the Holocene Later Stone Age, Southern Africa. *Abstracts of the 11th PanAfrican Congress for Prehistory and Related Fields*, pp. 46–47. Bamako, Mali.

Piantadosi, C. A. 2003. *The Biology of Human Survival. Life and Death in Extreme Environments*. New York, NY: Oxford University Press.

Pickford, M. 1983. Sequence and environments of the Lower and Middle Miocene hominoids of Western Kenya. In *New Interpretations of Ape and Human Ancestry*, Ciochon, R. L. & Corruccini, R. S., eds., pp. 421–39. New York, NY: Plenum.

Pickford, M., Senut, B., Gommery, D., & Treil, J. 2002. Bipedalism in *Orrorin tugenensis* revealed by its femora. *Comptes Rendus Palevol* 1:191–203.

Pickford, M. & Senut, B. 2001. The geological and faunal context of Late Miocene hominid remains from Lukeino, Kenya. *Comptes Rendus de l'Académie des Sciences Paris, Sciences de la Terre et des planètes* 332:145–52.

Pieczarka, J. C., de Souza-Barros, R. M., de Faria, Jr., F. M., & Nagamachi, C. Y. 1993. *Aotus* from the southwestern Amazon region is geographically and chromosomally intermediate between *A. azarae boliviensis* and *A. infulatus*. *Primates* 34:197–204.

Pilbeam, D. R. 1969. *Tertiary Pongidae of East Africa: Evolutionary Relationships and Taxonomy. Bulletin Peabody Museum of Natural History Yale University*, no. 31.

Pimm, S. 2003. Expiry dates. *Nature* 426:235–6.

Pimm, S. L., Jones, H. L., & Diamond, J. 1998. On the risk of extinction. *American Naturalist* 132:757–85.

Plavcan, J. M. 2000. Inferring social behavior from sexual dimorphism in the fossil record. *Journal of Human Evolution* 39:327–44.

2001. Sexual dimorphism in primate evolution. *Yearbook of Physical Anthropology* 44:25–53.

2002. Taxonomic variation in the patterns of craniofacial dimorphism in primates. *Journal of Human Evolution* 42:579–608.

2003. Scaling relationships between craniofacial sexual dimorphism and body mass dimorphism in primates: implications for the fossil record. *American Journal of Physical Anthropology* 120:38–60.

Plomin, R. & Kosslyn, S. M. 2001. Genes, brain and cognition. *Nature Neuroscience* 4:1153–5.

Plummer, T. 2004. Flaked stones and old bones: biological and cultural evolution at the dawn of technology. *Yearbook of Physical Anthropology* 47:118–64.

Plummer, T., Bishop, L. C., Ditchfield, P., & Hicks, J. 1999. Research on late Pliocene Oldowan sites at Kanjera South, Kenya. *Journal of Human Evolution* 36:151–70.

Poldrack, R. A., Clark, J., Pare-Blagoev, E. J. *et al.* 2001. Interactive memory systems in the human brain. *Nature* 414:546–50.

Pond, C. M. 1977. The significance of lactation in the evolution of mammals. *Evolution* 31:177–99.

1978. Morphological aspects and the ecological and mechanical consequences of fat deposition in wild vertebrates. *Annual Review of Ecology and Systematics* 9:519–70.

1992a. An evolutionary and functional view of mammalian adipose tissue. *Proceedings of the Nutrition Society* 51:367–77.

1992b. The structure and function of adipose tissue in humans with comments on the evolutionary origin and physiological consequences of sex differences. *Collegium Antropologicum* 16:135–43.

Poremba, A., Malloy, M., Saunders, R. C., Carson, R. E., Herscovitch, P., & Mishkin, M. 2004. Species-specific calls evoke asymmetric activity in the monkey's temporal poles. *Nature* 427:448–51.

Potts, R. 1988. *Early Hominid Activities at Olduvai*. New York, NY: Aldine de Gruyter.

1996a. Evolution and climatic variability. *Science* 273:922–3.

1996b. *Humanity's Descent. The Consequences of Ecological Instability*. New York, NY: William Morrow and Company, Inc.

1998. Variability selection in hominid evolution. *Evolutionary Anthropology* 7:81–96.

Potts, R. & Shipman, P. 1981. Cutmarks made by stone tools on bones from Olduvai Gorge, Tanzania. *Nature* 291:577–80.

Potts, R., Behrensmeyer, A. K., & Ditchfield, P. 1999. Paleolandscape variation and Early Pleistocene hominid activities: Members 1 and 7, Olorgesailie Formation, Kenya. *Journal of Human Evolution* 37:747–88.

Povinelli, D. J. & Vonk, J. 2003. Chimpanzee minds: suspiciously human? *Trends in Cognitive Sciences* 7:157–60.

Povinelli, D. J., Reaux, J. E., Theall, L. A., & Giambrone, S. 2000. *Folk Physics for Apes. The Chimpanzee's Theory of How the World Works.* Oxford: Oxford University Press.

Preuschoft, H. 2004. Mechanisms for the acquisition of habitual bipedality: are there biomechanical reasons for the acquisition of upright bipedal posture? *Journal of Anatomy* 204:363–4.

Preuss, T. 2001. The discovery of cerebral diversity: an unwelcome scientific revolution. In *Evolutionary Anatomy of the Primate Cerebral Cortex*, Falk, D. & Gibson, K. R., eds., pp. 138–64. Cambridge: Cambridge University Press.

Price, P. W. 2003. *Macroevolutionary Theory on Macroecological Patterns.* Cambridge: Cambridge University Press.

Price, R. A., Charles, M. A., Pettitt, D. J., & Knowler, W. C. 1993. Obesity in Pima Indians: large increases among post-World War II birth cohorts. *American Journal of Physical Anthropology* 92:473–9.

Prothero, D. R. & Berggren, W. A., eds. 1992. *Eocene-Oligocene Climatic and Biotic Evolution.* Princeton, NJ: Princeton University Press.

Purdue, J. R. & Reitz, E. J. 1993. Decrease in body size of white-tailed deer (*Odocoileus virginianus*) during the late Holocene in South Carolina and Georgia. In *Morphological Change in Quaternary Mammals of North America*, Martin, R. A. & Barnosky, A. D., eds., pp. 281–98. Cambridge: Cambridge University Press.

Purvis, A. & Harvey, P. H. 1997. The right size for a mammal. *Nature* 386:332–3.

Purvis, A., Agapow, P. M., Gittleman, J. L., & Mace, G. M. 2000. Nonrandom extinction and the loss of evolutionary history. *Science* 288:328–30.

Quade, J., Levin, N., Semaw, S. *et al.* 2004. Paleoenvironments of the earliest stone toolmakers, Gona, Ethiopia. *Geological Society of America Bulletin* 116:1529–44.

Queller, D. C. 1995. The spaniels of St. Marx and the Panglossian paradox: a critique of a rhetorical programme. *Quarterly Review of Biology* 70:485–9.

Quick, D. L. F. 1986. Activity budgets and the consumption of human food in two troops of baboons, *P. anubis*, at Gilgil, Kenya. In *Primate Ecology and Conservation*, Lee, P. C. & Else, J. G., eds., pp. 223–8. Cambridge: Cambridge University Press.

Radinsky, L. 1972. Endocasts and studies of primate brain evolution. In *The Functional and Evolutionary Biology of Primates*, Tuttle, R., ed., pp. 175–84. Chicago, IL: Aldine-Atherton, Inc.

  1979. *The Fossil Record of Primate Brain Evolution.* Forty-Ninth James Arthur Lecture on the Evolution of the Human Brain. American Museum of Natural History, New York.

Raemaekers, J. 1984. Large versus small gibbons: relative roles of bioenergetics and competition in their ecological segregation in sympatry. In *The Lesser Apes. Evolutionary and Behavioural Ecology*, Preuschoft, H. *et al.*, eds., pp. 209–18. Edinburgh: Edinburgh University Press.

Raichle, M. E., MacLeod, A. M., Snyder, A. Z., Powers, W. J., Gusnard, D. A., & Shulman, G. L. 2001. A default mode of brain function. *Proceedings of the National Academy of Sciences U.S.A.* 98(2):676–82.

Rainger, R. 1989. What's the use: William King Gregory and the functional morphology of fossil vertebrates. *Journal of the History of Biology* 22:103–39.

1991. *An Agenda for Antiquity: Henry Fairfield Osborn and Vertebrate Paleontology at the American Museum of Natural History 1890–1935.* Tuscaloosa, AL: University of Alabama Press.

Ralls, K. 1976. Mammals in which females are larger than males. *Quarterly Review of Biology* 51:245–76.

Ramnani, N. & Miall, R. C. 2004. A system in the human brain for predicting the actions of others. *Nature Neuroscience* 7:85–90.

Ratnieks, F. L. W. & Wenseleers, T. 2005. Policing insect societies. *Science* 307:54–6.

Raup, D. M. 1986. *The Nemesis Affair: A Story of the Death of the Dinosaurs and the Ways of Science.* New York, NY: Norton.

1991. *Extinction: Bad Genes or Bad Luck?* New York, NY: Norton.

Raup, D. M. & Sepkoski, Jr., J. J. 1984. Periodicity of extinctions in the geologic past. *Proceedings of the National Academy of Sciences U.S.A.* 81:801–5.

1986. Periodic extinctions of families and genera. *Science* 231:833–6.

1988. Testing for periodicity of extinctions. *Science* 241:94–6.

Rawlins, R. G. & Kessler, M. J., eds. 1986. *The Cayo Santiago Macaques. History, Behavior and Biology.* Albany, NY: State University of New York Press.

Reader, S. M. & Laland, K. N. 2002. Social intelligence, innovation, and enhanced brain size in primates. *Proceedings of the National Academy of Sciences U.S.A.* 99:4436–41.

Real, L. A. 1991. Animal choice behavior and the evolution of cognitive architecture. *Science* 253:980–6.

1992. Information processing and the evolutionary ecology of cognitive architecture. *American Naturalist* 140 (supplement):S108–45.

Redfield, T. F., Wheeler, W. H., & Often, M. 2003. A kinematic model for the development of the Afar Depression and its paleogeographic implications. *Earth and Planetary Science Letters* 216:383–98.

Rees, G., Frith, C. D., & Lavie, N. 1997. Modulating irrelevant motion perception by varying attentional load in an unrelated task. *Science* 278:1616–18.

Reichard, U. & Boesch, C., eds. 2003. *Monogamy. Mating Strategies and Partnerships in Birds, Humans and Other Mammals.* Cambridge: Cambridge University Press.

Relyea, R. A. 2002. Costs of phenotypic plasticity. *American Naturalist* 159:272–82.

Rendall, D. & Di Fiore, A. 1995. The road less traveled: phylogenetic perspectives in primatology. *Evolutionary Anthropology* 4:43–52.

Reno, P., Lovejoy, C. O., McCollum, M. A., Hamrick, M. W., Meindl, R. S., & Cohn, M. J. 2001. Ontogenetic data suggest the presence of HOXD targets that act as growth scalars in the hominoid forearm and hand. *American Journal of Physical Anthropology* 32 (Supplement):125.

Reno, P. L., Meindl, R. S., McCollum, M. A., & Lovejoy, C. O. 2003. Sexual dimorphism in *Australopithecus afarensis* was similar to that of modern humans. *Proceedings of the National Academy of Sciences U.S.A.* 100:9404–9.

Rensberger, J. M. 1995. Determination of stresses in mammalian dental enamel and their relevance to the interpretation of feeding behaviors in extinct taxa. In *Functional Morphology in Vertebrate Paleontology*, Thomason, J. J., ed., pp. 151–72. Cambridge: Cambridge University Press.

Retallack, G. J. 1991. *Miocene Paleosols and Ape Habitats of Pakistan and Kenya.* Oxford: Oxford University Press.

Reynolds, T. R. 1981. *Mechanics of Interlimb Weight Redistribution in Primates.* Unpublished Ph. D. dissertation, Department of Anthropology, Rutgers University, New Brunswick, NJ.

 1983. Stride lengths of mammals, primates, humans and early hominids. *American Journal of Physical Anthropology* 60(2):244.

 1985a. Mechanics of increased support of weight by the hindlimbs in primates. *American Journal of Physical Anthropology* 67:335–49.

 1985b. Stresses on the limbs of quadrupedal primates. *American Journal of Physical Anthropology* 67:351–62.

Reznick, D. N., Shaw, F. H., Rodd, F. H., & Shaw, R. G. 1997. Evaluation of the rate of evolution in natural populations of guppies (*Poecilia reticulata*). *Science* 275:1934–7.

Rich, T. H., Hopson, J. A., Musser, A. M., Flannery, T. F., & Vickers-Rich, P. 2005. Independent origins of middle ear bones in monotremes and therians. *Science* 307:910–14.

Richard, A. F. 1981. Changing assumptions in primate ecology. *American Anthropologist* 83:517–33.

 1985. *Primates in Nature.* New York, NY: Freeman.

Richard, A. F., Goldstein, S. J., & Dewar, R. E. 1989. Weed macaques: the evolutionary implications of macaque feeding ecology. *International Journal of Primatology* 10:569–94.

Richards, P. W. 1996. *The Tropical Rain Forest: An Ecological Study*, 2nd edn. Cambridge: Cambridge University Press.

Richmond, B. G. & Strait, D. S. 2000. Evidence that humans evolved from a knuckle-walking ancestor. *Nature* 404:382–5.

Ridley, M. 1986. The number of males in a primate group. *Animal Behaviour* 34:1848–58.

Rieseberg, L. H. & Livingstone, K. 2003. Chromosomal speciation in primates. *Science* 300:267–8.

Roberts, D. F. 1953. Body weight, race, and climate. *American Journal of Physical Anthropology* 11:533–58.

Roberts, D. L. & Solow, A. R. 2003. When did the dodo become extinct? *Nature* 426:245.

Roberts, R. G., Flannery, T. F., Ayliffe, L. K. *et al.* 2001. New ages for the last Australian megafauna: continent-wide extinction about 46,000 years ago. *Science* 292:1888–92.

Robinson, B. W. & Schluter, D. 2000. Natural selection and the evolution of adaptive genetic variation in northern freshwater fishes. In *Adaptive Genetic Variation in the Wild*, Mousseau, T. A., Sinervo, B., & Endler, J., eds., pp. 65–94. New York, NY: Oxford University Press.

Robinson, J. T. 1954. Prehominid dentition and hominid evolution. *Evolution* 8:324–34.

 1956. *The Dentition of the Australopithecinae.* Transvaal Museum Memoir No. 9, Pretoria.

1972. *Early Hominid Posture and Locomotion*. Chicago, IL: University of Chicago Press.

Roche, H., Delagnes, A., Brugel, J. P. *et al.* 1999. Early hominid stone tool production and technical skill 2.34 myr ago in West Turkana, Kenya. *Nature* 399:57–60.

Rodier, P. M. 2000. The early origins of autism. *Scientific American* 282(2):56–63.

Roebroeks, W. 1995. 'Policing the boundary'? Continuity of discussions in 19th and 20th century palaeoanthropology. In *Ape, Man, Apeman: Changing Views Since 1600*, Corbey, R. & Theunissen, B., eds., pp. 173–9. Leiden, the Netherlands: Leiden University Press.

Roff, D. 2003. Evolutionary danger for rainforest species. *Science* 301:58–9.

Rogers, M. J. 1997. *A Landscape Archaeological Study at East Turkana, Kenya*. Ph. D. Dissertation. Department of Anthropology, Rutgers University, New Brunswick, NJ.

Rogers, M. J., Harris, J. W. K., & Feibel, C. S. 1994. Changing patterns of landuse by Plio/Pleistocene hominids in the Lake Turkana Basin. *Journal of Human Evolution* 27:139–58.

Rooijakkers, G. 1995. European apelore in popular prints, 17th–19th centuries. In *Ape, Man, Apeman: Changing Views Since 1600*, Corbey, R. & Theunissen, B., eds., pp. 327–35. Leiden, the Netherlands: Leiden University Press.

Rook, L., Bondioli, L., Köhler, M., Moyà-Solà, S., & Macchiavelli, R. 1999. *Oreopithecus* was a bipedal ape after all: evidence from the iliac cancellous architecture. *Proceedings of the National Academy of Sciences U.S.A.* 96:8795–9.

Rose, K. D. & Bown, T. M. 1993. Species concepts and species recognition in Eocene primates. In *Species, Species Concepts, and Primate Evolution*, Kimbel, W. H. & Martin, L. B., eds., pp. 299–330. New York, NY: Plenum Press.

Rose, M. D. 1991. The process of bipedalization in hominids. In *Origine(s) de la Bipédie chez les Hominidés*, Coppens, Y. & Senut, B. eds., pp. 37–48. Cahiers de Paléoanthropologie. Paris: Editions du Centre National de la Recherche Scientifique.

Rose, M. D., Leakey, M. G., Leakey, R. E. F., & Walker, A. C. 1992. Postcranial specimens of *Simiolus enjiessi* and other primitive catarrhines from the early Miocene of Lake Turkana, Kenya. *Journal of Human Evolution* 22:171–237.

Rose, M. R. & Lauder, G. V., eds. 1996a. *Adaptation*. San Diego, CA: Academic Press.
1996b. Post-Spandrel adaptationism. In *Adaptation*, Rose, M. R. & Lauder, G. V., eds., pp. 1–8. San Diego, CA: Academic Press.

Rosenzweig, M. L. 1995. *Species Diversity in Space and Time*. Cambridge: Cambridge University Press.
2001. Loss of speciation rate will impoverish future diversity. *Proceedings of the National Academy of Sciences U.S.A.* 98:5404–10.

Rosser, Z. H., Zerjal, T., Hurles, M. E. *et al.* 2000. Y chromosomal diversity in Europe is clinal and influenced primarily by geography rather than language. *American Journal of Human Genetics* 67:1526–43.

Roth, V. L. 1981. Constancy in the size ratios of sympatric species. *American Naturalist* 118:394–404.

Rowley-Conwy, P. 2001. Time, change and the archaeology of hunter-gatherers: how original is the 'Original Affluent Society'? In *Hunter-Gatherers: An*

Interdisciplinary Perspective, Panter-Brick, C., Layton, R. H., & Rowley-Conwy, P., eds., pp. 39–72. Cambridge: Cambridge University Press.

Rudwick, M. J. S. 1972. *The Meaning of Fossils*. New York, NY: American Elsevier.
1992. *Scenes from Deep Time: Early Pictorial Representations of the Prehistoric World*. Chicago, IL: University of Chicago Press.

Ruff, C. B. 1994. Morphological adaptation to climate in modern and fossil hominids. *Yearbook of Physical Anthropology* 37:65–107.

Ruff, C. B. & Hayes, W. C. 1983. Cross-sectional geometry of Pecos Pueblo femora and tibiae – a biomechanical investigation: 1. Method and general patterns of variation. *American Journal of Physical Anthropology* 60:359–81.

Ruff, C. B. & Walker, A. 1993. Body size and body shape. In *The Nariokotome* Homo erectus *Skeleton*, Walker, A. & Leakey, R., eds., pp. 234–65. Cambridge, MA: Harvard University Press.

Ruse, M. 1999. *Mystery of Mysteries. Is Evolution a Social Construction?* Cambridge, MA: Harvard University Press.

Russell, A. F., Sharpe, L. L., Brotherton, P. N. M., & Clutton-Brock, T. H. 2003. Cost minimization by helpers in cooperative vertebrates. *Proceedings of the National Academy of Sciences U.S.A.* 100:3333–8.

Rutherford, S. & D'Hondt, S. 2000. Early onset and tropical forcing of 100,000-year Pleistocene glacial cycles. *Nature* 408:72–4.

Rylands, A. B., ed. 1993. *Marmosets and Tamarins: Systematics, Behaviour, and Ecology*. Oxford: Oxford University Press.

Sahlins, M. D. 1959. The social life of monkeys, apes and primitive man. *Human Biology* 31:54–73.

Sakai, S. & Harada, Y. 2001. Why do large mothers produce large offspring? Theory and a test. *American Naturalist* 157:348–59.

Sanderson, I. T. 1957. *The Monkey Kingdom. An Introduction to the Primates*. Garden City, NY: Doubleday & Company, Inc.

Sanderson, M. J. & Hufford, L., eds. 1996. *Homoplasy*. New York, NY: Academic Press.

Sarich, V. M. & Wilson, A. C. 1967. Immunological time scale for hominoid evolution. *Science* 158:1200–3.

Sarmiento, E. E. 1998. Generalized quadrupeds, committed bipeds, and the shift to open habitats: an evolutionary model of hominid divergence. *American Museum Novitates* 3250:1–78.

Saunders, W. B., Work, D. M., & Nikolaeva, S. V. 1999. Evolution of complexity in Paleozoic ammonoid sutures. *Science* 286:760–3.

Savage, A. & Baker, A. J., eds. 1996. Callitrichid social structure and mating systems: Evidence from field studies. Vol. 38, No. 1, *American Journal of Primatology* (special issue).

Savage-Rumbaugh, E. S. 1986. *Ape Language*. New York, NY: Columbia University Press.

Savage-Rumbaugh, S. & Lewin, R. 1994. *Kanzi. The Ape at the Brink of the Human Mind*. New York, NY: John Wiley & Sons, Inc.

Sawada, Y., Pickford, M., & Senut, B. *et al.* 2002. The age of *Orrorin tugenensis*, an early hominid from the Tugen Hills, Kenya. *Comptes Rendus Palevol* 1:293–303.

Schaal, S. & Ziegler, W., eds. 1992. *Messel: An Insight into the History of Life and of the Earth*. Oxford: Clarendon Press.

Schaller, G. B. 1963. *The Mountain Gorilla: Ecology and Behavior*. Chicago, IL: University of Chicago Press.

Schiermeier, Q. 2003. Setting the record straight. *Nature* 424:482–3.

Schluter, D. 2000. Ecological character displacement in adaptive radiation. *American Naturalist* 156 (supplement):S4–16.

Schmidt, G. A. & Shindell, D. T. 2003. Atmospheric composition, radiative forcing, and climate change as a consequence of a massive methane release from gas hydrates. *Paleooceanography* 18 doi: 10.1029/2002PA000757.

Schmidt-Nielsen, K. 1984. *Scaling: Why is Animal Size So Important?* Cambridge: Cambridge University Press.

Schmitt, D. & Lemelin, P. 2002. Origins of primate locomotion: gait mechanics of the woolly opossum. *American Journal of Physical Anthropology* 118:231–8.

Schoener, T. W. 1984. Size differences between sympatric, bird-eating hawks: a worldwide survey. In *Ecological Communities. Conceptual Issues and the Evidence*, Strong, D. R., Jr. *et al.*, eds., pp. 254–81. Princeton, NJ: Princeton University Press.

Schultz, A. H. 1933. Die Körperproportionen der erwachsenen catarrhinen Primaten, mit spezieller Berücksichtigung der Menschenaffen. *Anthrop. Anz.* 10:154–85.

    1936a. Characters common to higher primates and characters specific for man. *Quarterly Review of Biology* 11:259–83.

    1936b. Characters common to higher primates and characters specific for man (continued). *Quarterly Review of Biology* 11:425–55.

    1950. The physical distinctions of man. *Proceedings of the American Philosophical Society* 94:428–49.

    1961. Some factors influencing the social life of primates in general and of early man in particular. In *The Social Life of Early Man*, Washburn, S. L., ed., pp. 58–90. Chicago, IL: Aldine.

    1968. The recent hominoid primates. In *Perspectives on Human Evolution*, Washburn, S. L. & Jay, P. C., eds., pp. 122–95. New York, NY: Holt, Rinehart and Winston.

    1969. *The Life of Primates*. New York, NY: Universe Books.

    1976. The rise of primatology in the twentieth century. *Folia primatologica* 26:5–23.

Schwartz, J. H., Tattersall, I., & Eldredge, N. 1978. Phylogeny and classification of the primates revisited. *Yearbook of Physical Anthropology* 21:95–133.

Schwartz, J. H., Collard, M., & Cela-Conde, C. J. 2001. Systematics of "humankind". *Evolutionary Anthropology* 10:1–3.

Scott-Ram, N. R. 1990. *Transformed Cladistics, Taxonomy, and Evolution*. Cambridge: Cambridge University Press.

Sealy, J. & Pfeiffer, S. 2000. Diet, body size, and landscape use among Holocene people in the Southern Cape, South Africa. *Current Anthropology* 41:642–55.

Searle, J. B. 1996. Speciation in small mammals. *Symposia of the Zoological Society of London* 69:143–56.

Segerstråle, U. 2000. *Defenders of the Truth. The Battle for Science in the Sociobiology Debate and Beyond*. Oxford: Oxford University Press.

Semaw, S. 1997. *Earliest Stone Age Sites in the Gona Region, Ethiopia*. Ph.D. dissertation, Department of Anthropology, Rutgers University, New Brunswick, NJ.

Semaw, S., Renne, P., Harris, J. W. K. *et al.* 1997. 2.5-million-year-old stone tools from Gona, Ethiopia. *Nature* 385:333–6.

Semaw, S., Rogers, M. J., Quade, J. *et al.* 2003. 2.6-million-year-old stone tools and associated bones from OGS-6 and OGS-7, Gona, Afar, Ethiopia. *Journal of Human Evolution* 45:169–77.

Semendeferi, K. & Damasio, H. 2000. The brain and its main anatomical subdivisions in living hominoids using magnetic resonance imaging. *Journal of Human Evolution* 38:317–32.

Semendeferi, K., Damasio, H., & Frank, R. 1997. The evolution of the frontal lobes: a volumetric analysis based on three-dimensional reconstructions of magnetic resonance scans of human and ape brains. *Journal of Human Evolution* 32:375–88.

Semmann, D., Krambeck, H. J. R., & Milinski, M. 2003. Volunteering leads to rock-paper-scissors dynamics in a public goods game. *Nature* 425:390–3.

Senut, B., Pickford, M., Gommery, D., Mein, P., Cheboi, K., & Coppens, Y. 2001. First hominid from the Miocene (Lukeino Formation, Kenya). *Comptes Rendus de l'Académie des Sciences Paris Sciences de la Terre et des planètes* 332:137–44.

Sepkoski, Jr., J. J. 1993. Ten years in the library: new data confirm paleontological patterns. *Paleobiology* 19:43–51.

Sepkoski, Jr., J. J. & Raup, D. M. 1986. Was there 26-Myr periodicity of extinctions? *Nature* 321:533.

Sept, J. M. 1992. Was there no place like home? A new perspective on early hominid archaeological sites from the mapping of chimpanzee nests. *Current Anthropology* 33:187–207.

Seymour, K. 1993. Size change in North American Quaternary jaguars. In *Morphological Change in Quaternary Mammals of North America*, Martin, R. A. & Barnosky, A. D., eds., pp. 343–72. Cambridge: Cambridge University Press.

Shackleton, N. J. 1995. New data on the evolution of Pliocene climatic variability. In *Paleoclimate and Evolution with Emphasis on Human Origins*, Vrba, E. S. *et al.*, eds., pp. 242–8. New Haven, CT: Yale University Press.

Shea, B. T. 1989. Heterochrony in human evolution: the case for human neoteny. *Yearbook of Physical Anthropology* 32:69–101.

Shepherdson, D. J., Mellen, J. D., & Hutchins, M., eds. 1998. *Second Nature. Environmental Enrichment for Captive Animals*. Washington, DC: Smithsonian Institution Press.

Shipley, L. A., Gross, J. E., Spalinger, D. E., Hobbs, N. T., & Wunder, B. A. 1994. The scaling of intake rate in mammalian herbivores. *The American Naturalist* 143:1055–82.

Shipman, P. P. 1981. *Life History of a Fossil: An Introduction to Taphonomy and Paleoecology*. Cambridge, MA: Harvard University Press.

Shipman, P. & Rose, J. 1983. Early hominid hunting, butchering, and carcass-processing behaviors: approaches to the fossil record. *Journal of Archaeological Science* 2:57–98.

Shubin, N., Tabin, C., & Carroll, S. 1997. Fossils, genes and the evolution of animal limbs. *Nature* 388:639–48.

Signor, P. W., III & Lipps, J. H. 1982. Sampling bias, gradual extinction patterns, and catastrophes in the fossil record. In *Geological Implications of Impacts of Large Asteroids and Comets on the Earth*, Silver, L. T. & Schultz, P. H., eds., pp. 291–6. Geological Society of America Special Paper No. 190.

Sikorski, R. 1998. *The Polish House. An Intimate History of Poland*. London: Phoenix.

Silcox, M. T. 2003. New discoveries on the middle ear anatomy of the Paromomyidae (Mammalia, Primates) from ultra high resolution X-ray computed tomography. *American Journal of Physical Anthropology* Supplement 36:191–2.

Silk, J. B. 1993. The evolution of social conflict among female primates. In *Primate Social Conflict*, Mason, W. A. & Mendoza, S. P., eds., pp. 49–83. Albany, NY: SUNY Press.

  2002. Practice random acts of aggression and senseless acts of intimidation: the logic of status contests in social groups. *Evolutionary Anthropology* 11:221–5.

Silk, J. B., Alberts, S. C., & Altmann, J. 2003. Social bonds of female baboons enhance infant survival. *Science* 302:1231–4.

Simberloff, D. S. 1981. Community effects of introduced species. In *Biotic Crises in Ecological and Evolutionary Time*, Nitecki, M. H., ed., pp. 53–81. New York, NY: Academic Press.

Simmons, R. M. T. 1990. Role of the thalamus in the evolution of speech and language in man. In *From Apes to Angels*, Sperber, G. H., ed., pp. 229–47. New York, NY: Wiley-Liss.

Simpson, G. G. 1944. *Tempo and Mode in Evolution*. New York, NY: Columbia University Press.

  1945. The Principles of Classification and a Classification of Mammals. *Bulletin of the American Museum of Natural History* 85:1–350.

  1949. Rates of evolution in animals. In *Genetics, Paleontology, and Evolution*, Jepsen, G. L. *et al.*, eds., pp. 205–28. Princeton, NJ: Princeton University Press.

  1961. *Principles of Animal Taxonomy*. New York, NY: Columbia University Press.

  1971. William King Gregory, 1876–1970. *American Journal of Physical Anthropology* 35:155–74.

Sinclair, A. R. E., Mduma, S., & Brashares, J. S. 2003. Patterns of predation in a diverse predator–prey system. *Nature* 425:288–90.

Singer, R. & Wymer, J. 1982. *The Middle Stone Age at Klasies River Mouth in South Africa*. Chicago, IL: University of Chicago Press.

Skelton, P., ed. 2003. *The Cretaceous World*. Cambridge: Cambridge University Press.

Skinner, M. 1991. Bee brood consumption: An alternative explanation for hypervitaminosis A in KNM-ER 1808 (*Homo erectus*) from Koobi Fora, Kenya. *Journal of Human Evolution* 20:493–503.

Slater, P. J. B. 1985. *An Introduction to Ethology*. Cambridge: Cambridge University Press.

Smith, A. B. 2003. Making the best of a patchy fossil record. *Science* 301:321–2.

Smith, R. J. 1996. Biology and body size in human evolution: statistical inference misapplied. *Current Anthropology* 37:451–81.

Smolker, R., Richards, A., Connor, R., Man, J., & Berggren, P. 1997. Sponge carrying by dolphins (Delphinidae, *Tursiops* sp.): Foraging specialization involving tool use? *Ethology* 103:454–65.

Smuts, B. B. 1985. *Sex and Friendship in Baboons*. New York, NY: Aldine.

Smuts, B. B., Cheney, D. L., Seyfarth, R. M., Wrangham, R. W., & Struhsaker, T. T., eds. 1986. *Primate Societies*. Chicago, IL: University of Chicago Press.

Snider, A. J. 1969. Fossils in Ethiopia. Man's 'new age' exciting find. *Chicago Daily News*, Wednesday April 30, 1969.

Snodgrass, J. J., Leonard, W. R., Sorensen, M. V., & Robertson, M. L. 2001. Variation in field metabolic rates among primates and other mammals: implications for human evolutionary biology. *American Journal of Physical Anthropology* 114 (Supplement 32):140.

Snowdon, C. T. 1982. Linguistic and psycholinguistic approaches to primate communication. In *Primate Communication*, Snowdon, C. T., Brown, C. H., & Petersen, M. R., eds., pp. 212–38. Cambridge: Cambridge University Press.

Snyder, L. H. 1927. Blood grouping and its practical applications. *Archives of Pathology and Laboratory Medicine* 4:215–57.

Soligo, C., Anzenberger, G., & Martin, B. 2002. Into the third millennium: one hundred years of anthropology in Zürich. *Evolutionary Anthropology* 11:1–2.

Sollas, W. J. 1924. *Ancient Hunters and Their Modern Representatives*. 3rd edn. London: Macmillan.

Sparks, C. S. & Jantz, R. L. 2002. A reassessment of human cranial plasticity: Boas revisited. *Proceedings of the National Academy of Sciences U.S.A.* 99:14636–9.

  2003. Changing times, changing faces: Franz Boas's immigrant study in modern perspective. *American Anthropologist* 105:333–7.

Spencer, F. 1995. Pithekos to Pithecanthropus: an abbreviated review of changing scientific views on the relationship of the anthropoid apes to *Homo*. In *Ape, Man, Apeman: Changing Views Since 1600*, Corbey, R. & Theunissen, B., eds., pp. 13–27. Leiden, the Netherlands: Leiden University Press.

Sponheimer, M. & Lee-Thorp, J. A. 1999. Isotopic evidence for the diet of an early hominid, *Australopithecus africanus*. *Science* 283:368–70.

Springer, M. S., Murphy, W. J., Eizirik, E., & O'Brien, S. J. 2003. Placental mammal diversification and the Cretaceous-Tertiary boundary. *Proceedings of the National Academy of Sciences U.S.A.* 100:1056–61.

Stanford, C. B. 1998. *Chimpanzee and Red Colobus: The Ecology of Predator and Prey*. Cambridge, MA: Harvard University Press.

  2002. Brief communication: arboreal bipedalism in Bwindi chimpanzees. *American Journal of Physical Anthropology* 119:87–91.

Stedman, H. H., Kozyak, B. W., Nelson, A. *et al.* 2004. Myosin gene mutation correlates with anatomical changes in the human lineage. *Nature* 428:415–18.

Stephan, H. 1972. Evolution of primate brains: a comparative anatomical investigation. In *The Functional and Evolutionary Biology of Primates*, Tuttle, R. H., ed., pp. 155–74. Chicago, IL: Aldine-Atherton, Inc.

Stephan, H., Frahm, H., & Barton, G. 1981. New and revised data on volumes of brain structures in insectivores and primates. *Folia Primatologica* 35:1–29.

Stern, Jr., J. T. 1975. Before bipedality. *Yearbook of Physical Anthropology* 19:59–68.

2000. Climbing to the top: a personal memoir of *Australopithecus afarensis*. *Evolutionary Anthropology* 9:113–33.

Stern, Jr., J. T. & Susman, R. L. 1983. The locomotor anatomy of *Australopithecus afarensis*. *American Journal of Physical Anthropology* 60:279–317.

Stern, Jr., J. T., Wells, J. P., Vangor, A. K., & Fleagle, J. G. 1977. Electromyography of some muscles of the upper limb in *Ateles* and *Lagothrix*. *Yearbook of Physical Anthropology* 20:498–507.

Steudel, K. L. 1995. Locomotor energetics and hominid evolution. *Evolutionary Anthropology* 5:42–8.

1996. Morphology, bipedal gait, and the energetics of hominid locomotion. *American Journal of Physical Anthropology* 99:345–55.

Steudel, K. & Beattie, J. 1994. Was locomotor efficiency an important adaptive constraint in the evolution of the hominid lower limb? *American Journal of Physical Anthropology* Supplement 18:187.

Steudel-Numbers, K. L. 2001. Role of locomotor economy in the origin of bipedal posture and gait. *American Journal of Physical Anthropology* 116:171–3.

Stevens, G. C. 1989. The latitudinal gradient in geographical range: how so many species coexist in the tropics. *American Naturalist* 133:240–56.

Stigler, S. M. & Wagner, M. J. 1987. A substantial bias in nonparametric tests for periodicity in geophysical data. *Science* 238:940–5.

1988. Reply to Raup and Sepkoski, "Testing for periodicity of extinction." *Science* 241:96–9.

Stillman, J. H. 2003. Acclimation capacity underlies susceptibility to climatic change. *Science* 301:65.

Stiner, M. C. 1994. *Honor among Thieves: A Zooarchaeological Study of Neandertal Ecology*. Princeton, NJ: Princeton University Press.

Stockwell, C. A., Hendry, A. P., & Kinnison, M. T. 2003. Contemporary evolution meets conservation biology. *Trends in Ecology and Evolution* 18:94–101.

Stoczkowski, W. 1995. Le portrait de l'ancêtre en singe. L'hominisation sans évolutionnisme dans la pensée naturaliste du XVIIIe siècle. In *Ape, Man, Apeman: Changing Views Since 1600*, Corbey, R. & Theunissen, B., eds., pp. 141–55. Leiden, the Netherlands: Leiden University Press.

2002. *Explaining Human Origins. Myth, Imagination and Conjecture*. Cambridge: Cambridge University Press.

Stout, D., Toth, N., Schick, K., Stout, J., & Hutchins, G. 2000. Stone tool-making and brain activation: Positron emission tomography (PET) studies. *Journal of Archaeological Science* 27:1215–23.

Stowers, L., Holy, T. E., Meister, M., Dulac, C., & Koentges, G. 2002. Loss of sex discrimination and male–male aggression in mice deficient for TRP2. *Science* 295:1493–500.

Straus, Jr., W. L. 1953. Primates. In *Anthropology Today. An Encyclopedic Inventory*, Kroeber, A. L., ed., pp. 77–92. Chicago, IL: University of Chicago Press.

1962. Primates. In *Anthropology Today. Selections*, Tax, S., ed., pp. 15–30. Chicago, IL: University of Chicago Press.

1963. The classification of *Oreopithecus*. In *Classification and Human Evolution*, Washburn, S. L., ed., pp. 146–177. Chicago, IL: Aldine.

Strier, K. B. 2002. *Primate Behavioral Ecology*, 2nd edn. New York: Allyn & Bacon.

Struhsaker, T. T. & Siex, K. S. 1996. The Zanzibar red colobus monkey: conservation status of an endangered island endemic. *African Primates* 2:54–61.

Struhsaker, T. T., Cooney, D. O., & Siex, K. S. 1997. Charcoal consumption by Zanzibar red colobus monkeys: its function and its ecological and demographic consequences. *International Journal of Primatology* 18:61–72.

Strum, S. C. 1975. Primate predation: interim report on the development of a tradition in a troop of olive baboons. *Science* 187:755–7.

Su, H.-H. 2003. *Within-Group Female-Female Feeding Competition and Inter-Matriline Feeding Competition in Taiwanese Macaques* (Macaca cyclopis) *at Fushan Experimental Forest, Taiwan*. Ph.D. dissertation, Department of Anthropology, Rutgers University, New Brunswick, NJ.

Susman, R. L. 1988. Hand of *Paranthropus robustus* from Member 1, Swartkrans: fossil evidence for tool behavior. *Science* 239:781–4.

1994. Fossil evidence for early hominid tool use. *Science* 265:1570–3.

2004. *Oreopithecus bambolii*: an unlikely case of hominidlike grip capability in a Miocene ape. *Journal of Human Evolution* 46:105–17.

Susman, R. L., de Ruiter, D., & Brain, C. K. 2001. Recently identified postcranial remains of *Paranthropus* and early *Homo* from Swartkrans Cave, South Africa. *Journal of Human Evolution* 41:607–29.

Sutherland, W. J. 2003. Parallel extinction risk and global distribution of languages and species. *Nature* 423:276–9.

Swartz, S. M., Bertram, J. E. A., & Biewener, A. A. 1989. Telemetered in vivo strain analysis of locomotor mechanics of brachiating gibbons. *Nature* 342:270–2.

Swisher, III, C. C., Curtis, G. H., Jacob, T., Getty, A. G., & Suprijo, A. 1994. Age of the earliest known hominids in Java, Indonesia. *Science* 263:1118–21.

Szalay, F. S. 1968. The beginnings of primates. *Evolution* 22:19–36.

Szalay, F. S. & Berzi, A. 1973. Cranial anatomy of *Oreopithecus*. *Science* 180: 183–6.

Szalay, F. S. & Dagosto, M. 1988. Evolution of hallucial grasping in the primates. *Journal of Human Evolution* 17:1–33.

Szalay, F. S., Novacek, M. J., & McKenna, M. C., eds. 1993. *Mammal Phylogeny: Placentals*. New York, NY: Springer-Verlag.

Tanner, N. M. 1981. *On Becoming Human*. Cambridge: Cambridge University Press.

Tanner, N. M. & Zihlman, A. L. 1976. Women in evolution. Part I: Innovation and selection in human origins. *Signs* 1:585–608.

Tappen, N. 1960. Problems of distribution and adaptation of the African monkeys. *Current Anthropology* 1:91–120.

Tardieu, C. 1998. Short adolescence in early hominids: infantile and adolescent growth of the human femur. *American Journal of Physical Anthropology* 107: 163–78.

1999. Ontogeny and phylogeny of femoro-tibial characters in humans and hominid fossils: functional influence and genetic determinism. *American Journal of Physical Anthropology* 110:365–77.

Tardif, S. D. 1994. Relative energetic cost of infant care in small-bodied Neotropical primates and its relation to infant-care patterns. *American Journal of Primatology* 34:133–43.

Tardif, S. D. & Layne, D. G. 2002. Neonatal behavioral scoring of common marmosets (*Callithrix jacchus*): relation to physical condition and survival. *Journal of Medical Primatology* 31:147–51.

Tattersall, I. 1984. The good, the bad, and the synthesis. *American Anthropologist* 86:86–90.

1986. Species recognition in human paleontology. *Journal of Human Evolution* 15:165–75.

Tavaré, S., Marshall, C. R., Will, O., Soligo, C., & Martin, R. D. 2002. Using the fossil record to estimate the age of the last common ancestor of extant primates. *Nature* 416:726–9.

Teaford, M. F., Walker, A. C., & Mugaisi, G. S. 1993. Species discrimination in *Proconsul* from Rusinga and Mfwangano islands, Kenya. In *Species, Species Concepts and Primate Evolution*, Kimbel, W. H. & Martin, L. B., eds., pp. 373–92. New York, NY: Plenum.

Teleki, G. 1973. *The Predatory Behavior of Wild Chimpanzees*. Lewisburg, PA: Bucknell University Press.

Tello, N. S., Huck, M., & Heymann, E. W. 2002. *Boa constrictor* attack and successful group defence in moustached tamarins, *Saguinus mystax*. *Folia Primatologica* 73:146–8.

Templeton, A. R. 1998. Nested clade analyses of phylogeographic data: testing hypotheses about gene flow and population history. *Molecular Ecology* 7:381–97.

2002. Out of Africa again and again. *Nature* 416:45–51.

Terborgh, J. 1983. *Five New World Primates*. Princeton, NJ: Princeton University Press.

1990. Mixed flocks and polyspecific associations: costs and benefits of mixed-groups to birds and monkeys. *American Journal of Primatology* 21:87–100.

The Avian Brain Nomenclature Consortium. 2005. Avian brains and a new understanding of vertebrate brain evolution. *Nature Reviews Neuroscience* 6:151–9.

The International Human Genome Mapping Consortium 2001. A physical map of the human genome. *Nature* 409:934–41.

Thomason, J. J. 1995. To what extent can the mechanical environment of a bone be inferred from its internal architecture? In *Functional Morphology and Vertebrate Paleontology*, Thomason, J. J., ed., pp. 249–63. Cambridge: Cambridge University Press.

Thompson, L. G., Mosley-Thompson, E., Davis, M. E. *et al.* 2002. Kilimanjaro ice core records: evidence of Holocene climate change in tropical Africa. *Science* 298:589–93.

Thompson, P. M., Cannon, T. D., Narr, K. L. *et al.* 2001. Genetic influences on brain structure. *Nature Neuroscience* 4:1253–8.

Theunissen, B. 1989. *Eugène Dubois and the Ape-Man from Java*. Dordrecht, the Netherlands: Kluwer Academic Publishers.

Tiitinen, H., May, P., Reinikainen, K., & Naatanen, R. 1994. Attentive novelty detection in humans is governed by pre-attentive sensory memory. *Nature* 372:90–2.

Tinbergen, N. 1951. *The Study of Instinct*. New York, NY: Oxford University Press.

1963. On aims and methods of ethology. *Zeitschrift für Tierpsychologie* 20:410–33.

Tobias, P. V. 1967. *The Cranium and Maxillary Dentition of* Australopithecus (Zinjanthropus) boisei. Cambridge: Cambridge University Press.

Tobias, P. V. 1971. *The Brain in Hominid Evolution*. New York, NY: Columbia University Press.

1975. Anthropometry among disadvantaged peoples: studies in Southern Africa. In *Biosocial Interrelations in Population Adaptation*, Watts, E. S. *et al.*, eds., pp. 287–305. The Hague, the Netherlands: Mouton Press.

1991. *The Skulls, Endocasts and Teeth of* Homo habilis, 2 vols. Cambridge: Cambridge University Press.

Tomasello, M. & Call, J. 1994. Social cognition of monkeys and apes. *Yearbook of Physical Anthropology* 37:273–305.

1997. *Primate Cognition*. New York, NY: Oxford University Press.

Tomasello, M., Call, J., & Hare, B. 2003. Chimpanzees understand psychological states – the question is which ones and to what extent. *Trends in Cognitive Sciences* 7:153–6.

Tooby, J. & DeVore, I. 1987. The reconstruction of hominid behavioral evolution through strategic modeling. In *The Evolution of Human Behavior: Primate Models*, Kinzey, W. G., ed., pp. 183–237. Albany, NY: SUNY Press.

Torchin, M. E., Lafferty, K. D., Dobson, A. P., McKenzie, V. J., & Kuris, A. M. 2003. Introduced species and their missing parasites. *Nature* 421:628–30.

Toth, N., Schick, K. D., Savage-Rumbaugh, E. S., Sevick, R. A., & Rumbaugh, D. M. 1993. Pan the tool-maker: investigations into the stone tool-making and tool-using capabilities of a bonobo (*Pan paniscus*). *Journal of Archaeological Science* 20:81–91.

Treganza, T. & Butlin, R. K. 1999. Speciation without isolation. *Nature* 400:311–12.

Treves, A. 2000. Theory and method in studies of vigilance and aggregation. *Animal Behavior* 60:711–22.

Treves, A. & Naughton-Treves, L. 1999. Risk and opportunity for humans coexisting with large carnivores. *Journal of Human Evolution* 36:275–82.

Treves, A., Drescher, A., & Ingrisano, N. 2001. Vigilance and aggregation in black howler monkeys (*Alouatta pigra*). *Behavioral Ecology and Sociobiology* 50:90–5.

Treves, A. & Pizzagalli, D. 2002. Vigilance and perception of social stimuli: views from ethology and social neuroscience. In *The Cognitive Animal: Empirical and Theoretical Perspectives on Animal Cognition*, Bekoff, M., Allen, C., & Burghardt, G. M., eds., pp. 463–70. Cambridge, MA: M.I.T. Press.

Trivers, R. L. 1971. The evolution of reciprocal altruism. *Quarterly Review of Biology* 46:35–57.

1972. Parental investment and sexual selection. In *Sexual Selection and the Descent of Man*, Campbell, B., ed. Chicago, IL: Aldine.

Turner, A. & Antón, M. 2004. *Evolving Eden: An Illustrated Guide to the Evolution of the African Large-Mammal Fauna*. New York, NY: Columbia University Press.

Tuttle, R. H. 1967. Knuckle-walking and the evolution of hominoid hands. *American Journal of Physical Anthropology* 26:171–206.

1969. Knuckle-walking and the problem of human origins. *Science* 166:953–61.

1974. Darwin's apes, dental apes, and the descent of man: normal science in evolutionary anthropology. *Current Anthropology* 15:389–426.

1975. Parallelism, brachiation, and hominoid phylogeny. In *Phylogeny of the Primates. A Multidisciplinary Approach*, Luckett, W. P. & Szalay, F. S., eds., pp. 447–80. New York, NY: Plenum Press.

1977. Naturalistic positional behavior of apes and models of hominid evolution, 1929–1976. In *Progress in Ape Research*, Bourne, G., ed., pp. 277–96. New York, NY: Academic Press.

1986. *Apes of the World*. Park Ridge, NJ: Noyes Publications.

1994. Up from electromyography. Primate energetics and the evolution of human bipedalism. In *Integrative Paths to the Past*, Corruccini, R. S. & Ciochon, R. L., eds., pp. 269–84. Englewood Cliffs, NJ: Prentice-Hall.

Tuttle, R. H. & Rogers, C. M. 1966. Genetic and selective factors in reduction of the hallux in *Pongo pygmaeus*. *American Journal of Physical Anthropology* 24:191–8.

Tuttle, R. H., Buxhoeveden, D. P., & Cortright, G. W. 1979. Anthropology on the move: progress in experimental studies of nonhuman primate positional behavior. *Yearbook of Physical Anthropology* 22:187–214.

Tyler, D. E. 1991. The problems of the Pliopithecidae as a hylobatid ancestor. *Human Evolution* 6:73–80.

Vandenberghe, R., Price, C., Wise, R., Josephs, O., & Frackowiak, R. S. J. 1996. Functional anatomy of a common semantic system for words and pictures. *Nature* 383:254–6.

Vander Wall, S. B. 1991. *Food Hoarding in Animals*. Chicago, IL: University of Chicago Press.

van Praag, H., Schinder, A. F., Christie, B. R., Toni, N., Palmer, T. D., & Gage, F. H. 2002. Functional neurogenesis in the adult hippocampus. *Nature* 415:1030–4.

van Schaik, C. P. 1989. The ecology of social relationships amongst female primates. In *Comparative Socioecology*, Standen, V. & Foley, R. A., eds., pp. 195–218. Oxford: Blackwell Scientific.

2004. *Among Orangutans: Red Apes and the Rise of Human Culture*. Cambridge, MA: Harvard University Press.

van Schaik, C. P., Ancrenaz, M., Borgen, G. *et al.* 2003. Orangutan cultures and the evolution of material culture. *Science* 299:102–5.

Van Valen, L. 1965. Treeshrews, primates, and fossils. *Evolution* 19:137–51.

1973a. A new evolutionary law. *Evolutionary Theory* 1:1–30.

1973b. Body size and numbers of plants and animals. *Evolution* 27:27–35.

1976. Energy and evolution. *Evolutionary Theory* 1:179–229.

1978. The beginning of the Age of Mammals. *Evolutionary Theory* 4:45–80.

1994. *The Origin of the Plesiadapid Primates and the Nature of* Purgatorius. *Evolutionary Monographs*, No. 15. Chicago.

Van Valen, L. & Sloan, R. E. 1965. The earliest primates. *Science* 150:743–5.

Van Valkenburgh, B. & Janis, C. M. 1993. Historical diversity patterns in North American large herbivores and carnivores. In *Species Diversity in Ecological Communities: Historical and Geographical Perspectives*, Ricklefs, R. & Schluter, D., eds., pp. 330–40. Chicago, IL: University of Chicago Press.

Vargha-Khadem, F., Gadian, D. G., Copp, A., & Mishkin, M. 2005. *FOXP2* and the neuroanatomy of speech and language. *Nature Reviews Neuroscience* 6:131–8.

Vekua, A., Lordkipanidze, D., Rightmire, P. G. *et al.* 2002. A new skull of early *Homo* from Dmanisi, Georgia. *Science* 297:85–9.

Venter, J. C. 2001. The sequence of the human genome. *Science* 291:1304–51.

Vereecke, E., D'Août, K., & Aerts, P. 2004. The bipedal brachiator: a kinematic analysis of bipedal walking in *Hylobates lar. American Journal of Physical Anthropology* 123 (Supplement 38):200.

Vigilant, L., Hofreiter, M., Siedel, H., & Boesch, C. 2001. Paternity and relatedness in wild chimpanzee communities. *Proceedings of the National Academy of Sciences U.S.A.* 98:12890–5.

Vignaud, P., Duringer, P., Mackaye, H. T. *et al.* 2002. Geology and palaeontology of the Upper Miocene Toros-Menalla hominid locality, Chad. *Nature* 418:152–5.

Visalberghi, E. 1993a. Ape ethnography. Review of *Chimpanzee Material Culture* by W. C. McGrew. *Science* 261:1754.

1993b. Capuchin monkeys: A window into tool use in apes and humans. In *Tools, Language and Cognition in Human Evolution*, Gibson, K. R. & Ingold, T., eds., pp. 138–50. Cambridge: Cambridge University Press.

Vogel, G. 2003. Can great apes be saved from Ebola? *Science* 300:1645.

von Koenigswald, G. H. R. & Weidenreich, F. 1939. The relationship between *Pithecanthropus* and *Sinanthropus. Nature* 144:926–9.

Vrba, E. S. 1992. Mammals as a key to evolutionary theory. *Journal of Mammalogy* 73:1–28.

1995. On the connections between paleoclimate and evolution. In *Paleoclimate and Evolution with Emphasis on Human Origins*, Vrba, E. S., Denton, G. H., Partridge, T. C., & Burckle, L. H., eds., pp. 24–45. New Haven, CT: Yale University Press.

Walker, A. 1984. Extinction in hominid evolution. In *Extinctions*, Nitecki, M. H., ed., pp. 119-152. Chicago, IL: University of Chicago Press.

Walker, A., Zimmerman, M. R., & Leakey, R. E. F. 1982. A possible case of hypervitaminosis A in *Homo erectus. Nature* 296:248–50.

Walker, P. & Murray, P. 1975. An assessment of masticatory efficiency in a series of anthropoid primates with special reference to the Colobinae and Cercopithecinae. In *Primate Functional Morphology and Evolution*, Tuttle, R. H., ed., pp. 135–50. The Hague, the Netherlands: Mouton Press.

Walker, T. D. & Valentine, J. W. 1984. Equilibrium models of evolutionary species diversity and the number of empty niches. *American Naturalist* 124:887–99.

Walsh, P. D., Abernethy, K. A., Bermejo, M. *et al.* 2003. Catastrophic ape decline in western equatorial Africa. *Nature* 422:611–14.

Ward, C. V. 2002. Interpreting the posture and locomotion of *Australopithecus afarensis*: Where do we stand? *Yearbook of Physical Anthropology* 45:185–215.

Ward, C. V., Leakey, M. G., Brown, B., Brown, F., Harris, J., & Walker, A. 1999. South Turkwell: a new Pliocene hominid site in Kenya. *Journal of Human Evolution* 36:69–95.

Warner, M. 1995a. Cannibals and kings. In *Ape, Man, Apeman: Changing Views Since 1600*, Corbey, R. & Theunissen, B., eds., pp. 355–63. Leiden, the Netherlands: Leiden University Press.

1995b. *From the Beast to the Blonde. On Fairy Tales and Their Tellers.* New York, NY: Farrar, Straus and Giroux.

1998. *No Go the Bogeyman. Scaring, Lulling, and Making Mock.* New York, NY: Farrar, Straus and Giroux.

Waser, P. M. 1987. Interactions among primate species. In *Primate Societies*, Smuts, B. B. *et al.*, eds., pp. 210–26. Chicago, IL: University of Chicago Press.

Washburn, S. L. 1951. The new physical anthropology. *Transactions of the New York Academy of Sciences* Series II, 13:298–304.

1953. The strategy of physical anthropology. In *Anthropology Today. An Encyclopedic Inventory*, Kroeber, A. L., ed., pp. 714–27. Chicago, IL: University of Chicago Press.

1959. Speculations on the interrelations of the history of tools and biological evolution. *Human Biology* 31:21–31.

1960. Tools and human evolution. *Scientific American* 239:63–75.

ed. 1961. *The Social Life of Early Man.* New York, NY: Wenner-Gren Foundation.

1968. Speculations on the problem of man's coming to the ground. In *Changing Perspectives on Man*, Rothblatt, B., ed., pp. 193–206. Chicago, IL: University of Chicago Press.

Washburn, S. L. & DeVore, I. 1961a. Social behavior of baboons and early man. In *The Social Life of Early Man*, Washburn, S. L., ed., pp. 91–105. Chicago, IL: Aldine.

1961b. The social life of baboons. *Scientific American* 204(6):62–71.

Washburn, S. L. & Lancaster, C. S. 1968. The evolution of hunting. In *Man the Hunter*, Lee, R. B. & DeVore, I., eds., pp. 293–303. Chicago, IL: Aldine.

Waterston, R. H., Lindblad-Toh, K., & Birney, E. *et al.* 2002. Initial sequencing and comparative analysis of the mouse genome. *Nature* 420:520–62.

Webb, B. 2001. Can robots make good models of biological behaviour? *Behavioral and Brain Sciences* 24(6), online archive: www. bbsonline.org/ (ID Code bbs00000464).

Webb, D. 1994. Why people run and the evolutionary implications of lower limb length. *American Journal of Physical Anthropology* Supplement 18:204–5.

Webb, S. D., Hulbert, Jr., R. C., & Lambert, W. D. 1995. Climatic implications of large-herbivore distributions in the Miocene of North America. In *Paleoclimate and Evolution with Emphasis on Human Origins*, Vrba, E. S., Denton, G. H., Partridge, T. C., & Burckle, l.H., eds., pp. 91–108. New Haven, CT: Yale University Press.

Webb, T. J. & Gaston, K. J. 2003. On the heritability of geographic range sizes. *American Naturalist* 161:553–66.

Weidenreich, F. 1943. *The Skull of Sinanthropus pekinensis; A Comparative Study on a Primitive Hominid Skull. Palaeontologia Sinica* New Series D, no. 10. Lancaster, PA: Lancaster Press, Inc. Published by the Geological Survey of China, Pehpei, Chungking, December, 1943.

1946. *Apes, Giants, and Man.* Chicago, IL: University of Chicago Press.

Weiner, J. S. 1955. *The Piltdown Forgery.* Oxford: Oxford University Press.

Weiner, J. S., Oakley, K. P., & Le Gros Clark, W. E. 1953. The solution to the Piltdown problem. *Bulletin of the British Museum of Natural History (Geology)* 2:141–6.

Weinert, H. 1953. Der fossile Mensch. In *Anthropology Today. An Encyclopedic Inventory*. Kroeber, A. L., ed., pp. 101–19. Chicago, IL: University of Chicago Press.

Weishampel, D. B. 1995. Fossils, function, and phylogeny. In *Functional Morphology in Vertebrate Paleontology*, Thomason, J. J., ed., pp. 34–54. Cambridge: Cambridge University Press.

Weiss, K. M. 1985. "Phenotype amplification" as illustrated by cancer of the gallbladder in New World peoples. In *Etiology of Complex Diseases in Small Populations: Ethnic Differences and Research Approaches*, Chakraborty, P. & Szathmary, E., eds., pp. 179–98. New York, NY: Alan R. Liss.

Weiss, K. M., Ferrel, R. E., & Hanis, C. L. 1984. A New World syndrome of metabolic diseases with a genetic and evolutionary basis. *Yearbook of Physical Anthropology* 27:153–78.

Wellik, D. M. & Capecchi, M. R. 2003. *Hox10* and *Hox11* genes are required to globally pattern the mammalian skeleton. *Science* 301:363–7.

Welt, C. K., Chan, J. L., Bullen, J. *et al.* 2004. Recombinant human leptin in women with hypothalamic amenorrhea. *New England Journal of Medicine* 351:987–97.

West, G. B., Brown, J. H. & Enquist, B. J. 1999. The fourth dimension of life: fractal geometry and allometric scaling of organisms. *Science* 284:1677–9.

West, P. M. & Packer, C. 2002. Sexual selection, temperature, and the lion's mane. *Science* 297:1339–43.

West, S. A., Pen, I., & Griffin, A. S. 2002. Cooperation and competition between relatives. *Science* 296:72–5.

Wetzel, R. M., Dubois, R. E., Martin, R. L., & Myers, F. 1975. *Catagonus*, an "extinct" peccary alive in Paraguay. *Science* 189:379–81.

White, M. J. D. 1978. *Modes of Speciation*. San Francisco, CA: W. H. Freeman.

White, T. 2003. Early hominids – diversity or distortion? *Science* 299:1994–7.

White, T. D., Asfaw, B., De Gusta, D. *et al.* 2003. Pleistocene *Homo sapiens* from Middle Awash, Ethiopia. *Nature* 423:742–7.

Whiten, A. & Byrne, R. W., eds. 1997. *Machiavellian Intelligence II: Extensions and Evaluations*. Cambridge: Cambridge University Press.

Whiten, A., Goodall, J., McGrew, W. C. *et al.* 1999. Cultures in chimpanzees. *Nature* 399:682–5 [and online supplementary information].

Whitlock, M. C. 1996. The red queen beats the jack-of-all-trades: the limitations on the evolution of phenotypic plasticity and niche breadth. *American Naturalist* 148 (supplement): S65–77.

Whitmore, T. C. 1984. *Tropical Rain Forests of the Far East*, 2nd edn. Oxford: Clarendon Press.

    1990. *An Introduction to Tropical Rain Forests*. Oxford: Clarendon Press.

Wiens, J. J. 2003. Incomplete taxa, incomplete characters, and phylogenetic accuracy: is there a missing data problem? *Journal of Vertebrate Paleontology* 23:297–310.

Wikelski, M. & Thom, C. 2000. Marine iguanas shrink to survive El Niño. *Nature* 403:37–8.

Wildman, D. E., Uddin, M., Liu, G. Z., Grossman, L. I., & Goodman, M. 2003. Implications of natural selection in shaping 99.4% nonsynonymous DNA identity between humans and chimpanzees: Enlarging genus *Homo*. *Proceedings of the National Academy of Sciences U. S. A.* 100:7181–8.

Wiley, E. O. 1981. *Phylogenetics: Theory and Practice of Phylogenetic Systematics.* New York, NY: Wiley.

Wilf, P., Johnson, K. R., & Huber, B. T. 2003. Correlated terrestrial and marine evidence for global climate changes before mass extinction at the Cretaceous-Paleogene boundary. *Proceedings of the National Academy of Sciences U.S.A.* 100:599–604.

Williams, G. C. 1966. *Adaptation and Natural Selection. A Critique of Some Current Evolutionary Thought.* Princeton, NJ: Princeton University Press [reprinted in 1996 with a new preface].

1992. *Natural Selection: Domains, Levels, and Challenges.* Oxford: Oxford University Press.

Wilson, E. O. 1971. *The Insect Societies.* Cambridge, MA: Harvard University Press.

1975. *Sociobiology: The New Synthesis.* Cambridge, MA: Harvard University Press.

Windle, C. P., Baker, H. F., Ridley, R. M., Oerke, A. K., & Martin, R. D. 1999. Unrearable litters and prenatal reduction of litter size in the common marmoset (*Callithrix jacchus*). *Journal of Medical Primatology* 28:73–83.

Wing, S. L. *et al.*, eds. 2003. *Causes and Consequences of Globally Warm Climate in the Early Paleogene.* Boulder, CO: Geological Society of America Special Paper No. 369.

Winterhalder, B. 1980. Hominid paleoecology: the competitive exclusion principle and determinants of niche relationships. *Yearbook of Physical Anthropology* 23:43–63.

1981. Hominid paleoecology and competitive exclusion: limits to similarity, niche differentiation, and the effects of cultural behavior. *Yearbook of Physical Anthropology* 24:101–21.

2001. The behavioural ecology of hunter-gatherers. In *Hunter-Gatherers: An Interdisciplinary Perspective*, Panter-Brick, C., Layton, R. H., & Rowley-Conwy, P., eds., pp. 12–38. Cambridge: Cambridge University Press.

Wislocki, G. B. 1939. Observations on twinning in marmosets. *American Journal of Anatomy* 64:445–83.

Witmer, L. M. 1995. The extant phylogenetic bracket and the importance of reconstructing soft tissues in fossils. In *Functional Morphology in Vertebrate Paleontology*, Thomason, J. J., ed., pp. 19–33. Cambridge: Cambridge University Press.

Witmer, L. M. & Rose, K. D. 1991. Biomechanics of the jaw apparatus of the gigantic Eocene bird *Diatryma*: implications for diet and mode of life. *Paleobiology* 17:95–120.

Wolański, N. & Siniarska, A. 2001. Assessing the biological status of human populations. *Current Anthropology* 42:301–8.

Wolfe, L. D. & Gray, J. P. 1982. A cross-cultural investigation into the sexual dimorphism of stature. In *Sexual Dimorphism in* Homo sapiens. *A Question of Size*, Hall, R. L., ed., pp. 197–230. New York, NY: Praeger Publishers.

Wolff, J. O. 1997. Population regulation in mammals: an evolutionary perspective. *Journal of Animal Ecology* 66:1–13.

Wolpoff, M. H. 1999. *Palaeoanthropology*, 2nd edn. New York, NY: McGraw Hill.

Wolpoff, M. H., Thorne, A. G., Jeliňek, J., & Yinyun, Z. 1993. The case for sinking *Homo erectus*: 100 years of *Pithecanthropus* is enough! *Courier Forschungsinstitut Senckenberg* 171:341–61.

Wood, B. A., ed. 1991. Koobi Fora Research Project, Vol. 4. *Hominid Cranial Remains*. Oxford: Clarendon Press.

1992. Origin and evolution of the genus *Homo*. *Nature* 355:783–90.

Wood, B. A. & Collard, M. 1999a. The changing face of genus *Homo*. *Evolutionary Anthropology* 8:195–207.

1999b. The human genus. *Science* 284:65–71.

Wood, E. T. & Jankowski, S. M. 1994. *Karski. How One Man Tried to Stop the Holocaust*. New York, NY: John Wiley & Sons, Inc.

Wood Jones, F. 1926. *Arboreal Man*. New York, NY: Edward Arnold & Company.

Woodroffe, R. & Ginsberg, J. R. 1998. Edge effects and the extinction of populations inside protected areas. *Science* 280:2126–8.

Wrangham, R. & Peterson, D. 1996. *Demonic Males. Apes and the Origins of Human Violence*. Boston, MA: Houghton Mifflin.

Wrangham, R. W., Jones, J. H., Laden, G., Pilbeam, D., & Conklin-Brittain, N. 1999. The raw and the stolen: cooking and the ecology of human origins. *Current Anthropology* 40:567–94.

Wright, S. 1921. Systems of mating. *Genetics* 6:111–78.

1931. Evolution in Mendelian populations. *Genetics* 16:97–159.

Wynn, J. G. 2004. Influence of Plio-Pleistocene aridification on human evolution: evidence from paleosols of the Turkana Basin, Kenya. *American Journal of Physical Anthropology* 123:106–18.

Wynn, T. 1993. Layers of thinking in tool behavior. In *Tools, Language and Cognition in Human Evolution*, Gibson, K. R. & Ingold, T., eds., pp. 389–406. Cambridge: Cambridge University Press.

Wynn-Edwards, V. C. 1962. *Animal Dispersion in Relation to Social Behaviour*. London: Oliver & Boyd.

Xu, X., Zhou, Z. H., Wang, X. L., Kuang, X. W., Zhang, F. C., & Du, X. K. 2003. Four-winged dinosaurs from China. *Nature* 421:335–40.

Yalden, D. W. 2000. Shrinking shrews. *Nature* 403:826.

Yamazaki, N. & Ishida, H. 1984. A biomechanical study of vertical climbing and bipedal walking in gibbons. *Journal of Human Evolution* 13:563–71.

Yamazaki, N., Ishida, H., Kimura, T., & Okada, M. 1979. Biomechanical analysis of primate bipedal walking by computer simulation. *Journal of Human Evolution* 8:337–49.

Yerkes, R. M. 1925. *Almost Human*. New York, NY: Century.

1943. *Chimpanzees. A Laboratory Colony*. New Haven, CT: Yale University Press.

Yerkes, R. M. & Yerkes, A. W. 1929. *The Great Apes. A Study of Anthropoid Life*. New Haven, CT: Yale University Press.

Zihlman, A. L. & Cramer, D. L. 1978. Skeletal differences between pygmy (*Pan paniscus*) and common chimpanzees (*Pan troglodytes*). *Folia Primatologica* 29:86–94.

Zihlman, A. E., Cronin, J. E., Cramer, D. L., & Sarich, V. M. 1978. Pygmy chimpanzee as a possible prototype for the common ancestor of human, chimpanzees and gorillas. *Nature* 275:744–6.

Zihlman, A., Bolter, D., & Boesch, C. 2004. Wild chimpanzee dentition and its implications for assessing life history in immature hominin fossils. *Proceedings of the National Academy of Sciences U.S.A.* 101:10541–3.

Zinner, D., Alberts, S. C., Nunn, C. L., & Altmann, J. 2002. Evolutionary biology (communication arising): significance of primate sexual swellings. *Nature* 420:142–3.

Zito, M., Evans, S., & Weldon, P. J. 2003. Owl monkeys (*Aotus* spp.) self-anoint with plants and millipedes. *Folia Primatologica* 74:159–61.

Zuckerman, S. 1932. *The Social Life of Monkeys and Apes*. London: Kegan Paul.

1950a. South African fossil hominoids. *Nature* 166:158–9.

1950b. South African fossil hominoids. *Nature* 166:953–4.

1978. *From Apes to Warlords*. New York, NY: Harper & Row.

1981. *The Social Life of Monkeys and Apes*, 2nd edn. London: Routledge and Kegan Paul.

Index